The Engineering Guide to LEED—New Construction

McGRAW-HILL'S GREENSOURCE SERIES

Attmann
Green Architecture: Advanced Technologies and Materials

Gevorkian
Solar Power in Building Design: The Engineer's Complete Design Resource
Alternative Energy Systems in Building Design

GreenSource: The Magazine of Sustainable Design
Emerald Architecture: Case Studies in Green Building

Haselbach
The Engineering Guide to LEED—New Construction: Sustainable Construction for Engineers, Second Edition

Luckett
Green Roof Construction and Maintenance

Melaver and Mueller (eds.)
The Green Building Bottom Line: The Real Cost of Sustainable Building

Nichols and Laros
Inside the Civano Project: A Case Study of Large-Scale Sustainable Neighborhood Development

Yudelson
Green Building Through Integrated Design
Greening Existing Buildings

About *GreenSource*
A mainstay in the green building market since 2006, *GreenSource* magazine and GreenSourceMag.com are produced by the editors of McGraw-Hill Construction, in partnership with editors at BuildingGreen, Inc., with support from the United States Green Building Council. *GreenSource* has received numerous awards, including American Business Media's 2008 Neal Award for Best Website and 2007 Neal Award for Best Start-up Publication, and FOLIO magazine's 2007 Ozzie Awards for "Best Design, New Magazine" and "Best Overall Design." Recognized for responding to the needs and demands of the profession, *GreenSource* is a leader in covering noteworthy trends in sustainable design and best practice case studies. Its award-winning content will continue to benefit key specifiers and buyers in the green design and construction industry through the books in the *GreenSource* Series.

About McGraw-Hill Construction
McGraw-Hill Construction, part of The McGraw-Hill Companies (NYSE: MHP), connects people, projects, and products across the design and construction industry. Backed by the power of Dodge, Sweets, *Engineering News-Record* (*ENR*), *Architectural Record*, *GreenSource*, *Constructor*, and regional publications, the company provides information, intelligence, tools, applications, and resources to help customers grow their businesses. McGraw-Hill Construction serves more than 1,000,000 customers within the $4.6 trillion global construction community. For more information, visit www.construction.com.

The Engineering Guide to LEED—New Construction

Sustainable Construction for Engineers

Liv Haselbach

Second Edition

New York Chicago San Francisco
Lisbon London Madrid Mexico City
Milan New Delhi San Juan
Seoul Singapore Sydney Toronto

The McGraw-Hill Companies

Cataloging-in-Publication Data is on file with the Library of Congress

McGraw-Hill books are available at special quantity discounts to use as premiums and sales promotions, or for use in corporate training programs. To contact a representative please e-mail us at bulksales@mcgraw-hill.com.

The Engineering Guide to LEED—New Construction, Second Edition

1 2 3 4 5 6 7 8 9 0 DOC/DOC 1 9 8 7 6 5 4 3 2 1 0

ISBN 978-0-07-174512-3
MHID 0-07-174512-2

The pages within this book were printed on acid-free paper containing 100% postconsumer fiber.

Sponsoring Editor
Joy Bramble

Acquisitions Coordinator
Alexis Richard

Editorial Supervisor
David E. Fogarty

Project Manager
Harleen Chopra, Glyph International

Copy Editor
Alekha Jena

Proofreader
Surendra Nath Shivam, Glyph International

Indexer
Jeff Evans

Production Supervisor
Pamela A. Pelton

Composition
Glyph International

Art Director, Cover
Jeff Weeks

About the Author

Liv Haselbach is currently a faculty member in the Department of Civil & Environmental Engineering at Washington State University. She is a licensed engineer and a U.S. Green Building Council LEED® accredited professional with 30 years of experience in the field. The author of numerous articles for industry journals, Dr. Haselbach was the founding owner of a civil/environmental engineering firm specializing in land development and regulatory compliance.

Contents

Preface

The Engineering Guide to LEED—New Construction is intended as a reference or a textbook to aid in the understanding and application of green building design concepts for the engineering and development community. It focuses on the U.S. Green Building Council (USGBC) Leadership in Energy and Environmental Design® (LEED) rating system as an example format for sustainable vertical construction and has been updated in this second edition to version 3.0 (LEED 2009) through addenda December 2, 2009.

Sustainability has many definitions. The definition that has been generally accepted in the context of human beings building and living in a more "sustainable" world was initially developed at the World Commission on Environment and Development (WCED) in 1987. It is simply this: "Sustainable Development is development that meets the needs of the present without compromising the ability of future generations to meet their own needs."

Sustainable Construction is a subset of sustainability, which focuses more on the built environment, both during the construction phase and during the operational life cycle of the facility. However, both sustainability and sustainable construction are subject to interpretation and are very difficult to define. What may seem sustainable in one culture or per one set of values, may not appear sustainable to another. Likewise, the concepts that may be viewed as more important for sustainability by some people, may not be as important to others.

A very common term used for a major focus area in sustainable construction is *Green Building*. Green Building may not really represent true sustainable construction in some people's opinion, but it is an attempt to approach sustainability in a format that fits readily into our current culture. It is a movement that tries to put some of the concepts of sustainability into the construction or renovation of our buildings and facilities.

In like manner, even though this book has sustainable construction in its title, it by no means represents a fully comprehensive method to construct sustainably. This book is intended to be an introduction to some of the major concepts that are being accepted in methodologies to introduce sustainability into construction practices in the United States. It specifically covers the concepts that are currently being promoted by a rating system developed by the USGBC. The system is entitled LEED and has grown out of energy-saving efforts in the United States.

This book focuses more on the "Environmental" aspects of LEED. The "Energy" aspects have been fairly well developed and might require another volume or two to adequately explain in greater detail. In addition, this book expands on some of the environmental issues that are focused on in LEED and gives some direction into means to accomplish the goals, or gives more detailed background information on the environmental systems and impacts that many of the LEED subcategories are based on. In this way, an engineer or other professional may better grasp the intent of the proposed sustainability methods. A better understanding may result in better and more comprehensive, or perhaps alternative designs to obtain the goals. This book is meant to be a guide for all professionals working on sustainable construction in the United States.

The book is divided into three different types of sections. The first section, which consists of Chap. 1, gives an introduction to both sustainable construction and the USGBC LEED–New Construction rating system. The second section, Chaps. 2 through 7, goes into detail on many of the prerequisites and credits used by the USGBC for certification through its version 3 (LEED 2009) rating system for new construction and major renovations. The last section, Chaps. 8, 9, and 10, consists of several distinct parts broadening out from the specific rating system items with emphasis on application to various sectors of particular interest to many groups, management, the military, and low-impact development (LID). Chapter 8 includes the Minimum Program Requirements as specified for a project to be eligible to register for LEED 2009, and gives some overviews of construction management and organization for helping a *green* professional make a sustainable construction project successful particularly if based on the LEED rating system. Chapter 9 gives a view of how the Federal Government and one of its largest departments, the Department of Defense (DoD), have evolved as major contributors to the sustainable development movement in the United States. This chapter starts with a broad overview of how the DoD is involved in the movement and gradually details it down to the example of how the DoD is starting to address the improving of indoor air quality. Chapter 10 gives a broad overview of how this rating system may effectively intersect with another growing movement in the sustainable construction arena in the United States, Low Impact Development (LID), which focuses on the outdoor impacts of construction, particularly with respect to stormwater.

The Engineering Guide to LEED—New Construction can be used either as a textbook or a reference book. The exercises in the chapters help to develop sustainable skills and understanding for the students, while also allowing them to research the new principles and guidances that may evolve after publication of the book. I hope that others find as great a satisfaction and enjoyment in becoming more *green* as I have over the past few years.

Liv Haselbach

Acknowledgments

I would like to thank the people who have helped educate me on various aspects of green building, politely listened to me talk excitedly about sustainability, and have helped me research or review some of the specific items in the text over the past few years. Of particular note in this second edition is my youngest daughter Heidi Marie Brakewood who contributed many of the photos, including the cover photo which is of the Shell Woodcreek Office and Multipurpose Building Complex in Houston, TX. Then there is my husband Mike Navarro, who makes my work a pleasure by constantly encouraging me to continue in my efforts to teach and research "sustainability," even incorporating many of the green concepts in his daily work.

CHAPTER 1
Introduction

1.1 An Introduction to Sustainable Development

There are many definitions of *sustainability* and *sustainable development*. What is sustainable for one group may not be as sustainable for another. The most accepted worldwide definition of *sustainable development* is "development that meets the needs of the present without compromising the ability of future generations to meet their own needs" (Brundtland Commission, 1987). However, in an effort to address some of the sustainability ideals and goals in the United States, a subset of sustainability referred to as *sustainable construction* and its subcomponent called *green building* are emerging. They offer a look at sustainability in terms of a smaller scope, such as typical building or construction in an area of the developed world, and in the shorter term, perhaps the life of a building.

Sustainable construction is any construction, while green building focuses on vertical construction. There is also a movement developing to address horizontal construction (transportation and utility corridors). The horizontal construction program related to roads in the United States is usually referred to as *green highways*, although there are also publications that refer to some of the designs as *sustainable streets.* The Green Highway Partnership is an organization which has the U.S. Environmental Protection Agency (EPA) and the Federal Highway Administration (FHWA) among its partners. This partnership began in the Washington, D.C., area and is expanding nationally. The American Society of Civil Engineers (ASCE) became a leader in this field in 2009 and 2010 with the establishment of committees in both its Transportation and Development Institute (T&DI), and Environmental and Water Resources Institute (EWRI) under the EWRI Low Impact Development (LID) umbrella. The first T&DI/ASCE Green Streets and Highways Conference was scheduled for November 14–17, 2010, in Denver, CO. A Greenroads sustainability performance metric rating system is also being developed at the University of Washington. Another initiative relating mainly to the outdoors and horizontal construction is the Sustainable Sites Initiative (SSI), which is a collaboration between the American Society of Landscape Architects (ASLA), the Lady Bird Johnson Wildflower Center (The Wildflower Center), and the United States Botanic Garden (USBG) among others. More information can be found at their website: http://www.sustainablesites.org/.

Sustainable construction is an international concern. Although this book focuses on the process in the United States, great efforts have been made to develop sustainable policies and practices throughout the world. The sustainable practices range from procurement or supply chain practices through wastewater reuse practices in the operational phases. Good solutions and practices for sustainable construction may not be the same in all areas of the world and for all societies. Therefore, the rating systems developed and the techniques used will differ. Energy efficiency is a high priority in many countries, particularly in countries with cold winters. There has been much publication and research into improving energy efficiencies internationally both for new construction and as retrofits in existing buildings.

By looking at the green building movement in the United States, it is obvious that the green rating system as developed by the U.S. Green Building Council (USGBC) has had wide acceptance to date and is becoming more widespread. This system is referred to as LEED®, which stands for Leadership in Energy and Environmental Design.* Since LEED grew out of energy program initiatives, it is already developed in focus areas of both architects and mechanical engineers. Sometimes the construction that focuses on energy efficiency and also water efficiency is referred to as *high-performance* building.

Sustainable construction research and applications are still in their infancy. There is a great need for research, education, and case studies from applications to further develop a more sustainable future in development and construction. Even the definitions of what is or what is not sustainable need to be researched and further evaluated. For instance, certain people may believe that keeping as much as possible of the natural environment pristine is one of the most important goals of sustainability, whereas others may believe that an improved food supply for humans is more important. Thus, not only do the principles behind sustainability differ from group to group, or may be in contrast or seemingly incompatible with one another, but also the ranking or value judgments of the importance of various impact factors of sustainability are very difficult to develop and find consensus on.

To facilitate the further development and implementation of the green building system in the United States, engineers and other professionals must be educated in the rating systems used, and in the parameters and principles that have been established in the use of these rating systems. The intention of the author is to develop a text that educators can use to teach engineers and professionals about sustainable development, particularly the LEED system. The emphasis is on the development of skills to facilitate the use of this system, as well as guidance on potential additional research avenues to further improve the green building movement, with stress on environmental aspects, as future developments and needs arise.

The author likes to explain that a sustainable environmental goal will focus on both how construction impacts the environment and how environmental decisions impact development. It is a two-way road, and things must be looked at from both directions. In addition, other impacts in the entire life cycle of the facilities or practices need to be looked at. What may seem like an environmentally sound practice may in fact not turn out quite that way when input into an anthropogenic world. Two examples which follow are the sustainability of stormwater ponds and the use of certain additives in gasoline.

A common form of stormwater management is the use of a retention or a detention pond for storage and possibly infiltration of additional runoff caused by an increase in paved or roofed areas. Therefore in an approved site plan set, a pond may be called for. However, it has been shown that in many cases the specified ponds are not installed, much less maintained to sustain a more manageable stormwater system. Why?

---

*LEED is a registered trademark of the U.S. Green Building Council.

The reasons are not known, but perhaps the contractor or owner does not understand the environmental impact that the lack of a pond may cause, therefore, installation of one may not seem important.

Then there is the use of MTBE (methyl-tertiary-butyl ether) as an additive in gasoline to improve air quality, which was a common practice in the 1990s. It was considered to be a sound environmental practice with respect to air issues. However, there were some unexpected consequences. Traditionally in the United States, gasoline is made at a refinery, is transported and distributed through a network of petroleum facilities, and ends up at the retail outlet, the gasoline station. Gasoline is usually stored in underground tanks, for many reasons including safety, at this end resale site. Underground tanks are currently fairly well regulated and monitored in the United States, but that was not always the case. In many sites, gasoline has leaked into the ground, and the product sits on or in the groundwater. Most constituents in gasoline are nonpolar organics which do not dissolve readily in polar water, but might prefer to adsorb to organic material in the soil. So even though the groundwater may slowly flow off-site, most of the gasoline contaminants remain closer to the site. However, MTBE is an ether, and ethers are more polar than many gasoline constituents such as benzene. Polar organics tend to be more soluble in water and may have a greater tendency to travel with the groundwater gradient. Therefore, there are many places where some gasoline leakage in the soils at a site did not initially cause an off-site concern until the MTBE was added and moved more rapidly off-site. In this way a practice that is an environmental benefit for one system may not be as sustainable in another.

There are several other green rating systems for vertical construction in the United States, but the LEED system and the alternative Green Globes system are by far the most prominent. Green Globes is distributed by the Green Building Initiative (GBI) in the United States and is based on a system developed in Europe and later also used in Canada. The European version is referred to as the Building Research Establishment's Environmental Assessment Method (BREEAM). Another similar initiative is the California Green Building Standards (CALGreen) Code, which was in draft form in 2010 and scheduled to become effective on January 1, 2011. It has been developed by the California Building Standards Commission. This text is based on the LEED system, but this in no way implies that the other systems are not useful and viable. The LEED rating system was chosen as the focus of this text as it is currently more widespread in the United States. It also does not allow for a self-certification method, which makes it more restrictive and possibly more difficult to complete, but at the same time may also give greater control and consistency to green building.

The text has a heavy focus on some of the civil and environmental aspects of the rating system, since the author is first targeting this engineering community. However, it also addresses in some fashion other issues in the rating system. One of the reasons for its being more all encompassing in the topics reviewed is to educate the civil and environmental engineers about the other credits and criteria so that they can effectively work with other disciplines in a cooperative fashion to better implement sustainable construction practices. Other engineering and nonengineering disciplines have found the text useful, including mechanical engineering, computer science, architecture, and construction management, just to name a few.

There is a need for further involvement from the civil and environmental engineering community in the development of the LEED rating system. The evaluation of best practices and goals for many of these aspects is still in its infancy. It is here that much research and teaching may be needed to further develop the rating system effectively in a timely manner. There has been a substantial amount of green building at the University of South Carolina since 2003, and a few of the categories of sustainable construction in which additional research, the development of best practices, and optimization are needed include the following:

1. Construction and demolition (C&D) debris recycling and reuse. There is a need to optimize construction practices to facilitate C&D debris recycling in an economic fashion and to develop the recycling and reuse infrastructure in many areas of the United States to support these practices. Figure 1.1.1 shows a construction debris waste container yard at a LEED registered project in Columbia, S.C.
2. Design-procurement-construction process integration and optimization.
3. Construction management processes.
4. Stormwater management and low-impact development (LID).

There is a need for research and development of best management practices and integrated management practices with respect to the nonpoint source type of pollutants. One area includes new infrastructure and materials with respect to stormwater management such as pervious pavements. An example of a multipurpose landscape amphitheater and stormwater management feature at a LEED-certified project in Columbia, S.C., is shown in Fig. 1.1.2. Additional information on LID can be found in Chap. 10.

Figure 1.1.1 Construction debris container yard at a LEED-registered project in 2005 at the University of South Carolina in Columbia, S.C. (*Photograph taken by Steve Bruner.*)

FIGURE 1.1.2 Landscaped amphitheater also functions as a stormwater management feature at the LEED-certified West Quad Housing Complex at the University of South Carolina. (*Photograph taken in September 2006.*)

The text is also intended to be used by other engineers and professionals such as mechanical engineers, architects, planners, community leaders, and construction managers, as it does present the overall holistic approach of the LEED rating system and can facilitate the understanding of many of the credits and criteria outside the traditional purview of these disciplines. Again, the intention is to let all the professionals and affected parties learn to "talk the talk" of the other team members and understand the viewpoints and engineering decisions in an interdisciplinary fashion. It is very important that all interested parties, including the community, be involved in the sustainable construction process. Criteria that are important to many participants can be incorporated into the design and implementation of green construction. One example concerns the maintenance of the "historical" feel in a community. Green does not mean the buildings and facilities have to look different. Figure 1.1.3 depicts a structure built in 1939 at the University of South Carolina, and it shows how the new "green" building in Fig. 1.1.2 can still fit in well with the other historic buildings at this university.

There are many other excellent references to consult in the understanding and educating of others about sustainable construction and green building. Several of these are also focused on the USGBC LEED. However, this book is different from many of the other references that cover the LEED rating system. It presents the information and criteria in a fashion that is useful, then doing mathematical and design exercises for further understanding and familiarity with the criteria and parameters in use. Several excellent references on sustainable development, industrial ecology (terminology referring to sustainable development from a more industrial viewpoint), sustainable construction, and green building are listed in the References section of this chapter.

FIGURE 1.1.3 Preston College building at the University of South Carolina, built in 1939. Similar architecture to the new green West Quad Facility opened in 2004 (as shown in Fig. 1.1.2).

There are also many website resources from which sustainable education materials can be accessed. One of note is the ImpEE (Improving Engineering Education) Project website developed by the University of Cambridge in the United Kingdom, and another is the Green Design Institute at Carnegie Mellon.

There are a plethora of current publications in journals based on how universities and other institutions are implementing sustainability into the curricula. For instance, *The International Journal of Engineering Education* focused Part I of its vol. 23, no. 2, 2007 issue on "Educating Students in Sustainable Engineering." Later in 2010, the ASCE *Journal of Professional Issues in Engineering Education and Practice* also put out a call for a special issue on sustainability education.

Two very important references which should be highlighted here are the LEED-NC 2.2 Reference Guide and the LEED Reference Guide for Green Building Design and Construction, 2009 Edition (referred to as the LEED 2009 Reference Guide in this text), as developed by the USGBC. This text in sustainable construction is intended to supplement the subject Reference Guides. The Reference Guides, just like building codes and legislation, present the parameters and suggested practices for implementation of the rating system that other entities can adopt to facilitate building in a more sustainable fashion. The Reference Guides also expand upon the environmental, societal, and economic principles used in the determination of the various prerequisites and credits, and present many excellent examples, case studies, and references which can be used to aid in the implementation of the LEED system. This text takes the basic parameters and principles of the subject Reference Guides, particularly with respect to many of the civil and environmental design issues, and further expands them into equations and teaching formats which may be helpful in a classroom situation. It is with gratitude that the author acknowledges the USGBC in allowing her to use most of the USGBC rating system verbiage for each prerequisite and credit in this text and many of the tables, definitions, and other information from the LEED-NC 2.2 Reference Guide.

LEED-NC 2.2 was the current version of the USGBC rating system for New Construction and Major Renovations through April 2009, and may still be valid for projects which registered prior to that date. The first edition of LEED-NC version 2.2 was released in October 2005. It replaced LEED-NC version 2.1 as the current rating system for newly registered projects at that time. Since that time, there have been several corrections and addenda to version 2.2, and a listing of these can be found as errata available for download to the general public from the USGBC website. The first edition of this text was based on the LEED-NC version 2.2 released in October 2005 and as amended through the second edition dated September 2006 and errata as posted in the spring of 2007. This second edition still not only covers most of the LEED-NC 2.2 version, but also includes the new LEED-NC 3.0 version for New Construction and Major Renovations, commonly referred to as LEED-NC 2009. The modifications are based on the LEED Reference Guide for Green Building Design and Construction, 2009 Edition published by the USGBC. This 2009 Reference Guide covers the 2009 New Construction flagship rating system (LEED-NC 2009) as well as two other USGBC rating systems for new construction, LEED for Schools (K–12) and also LEED for Core and Shell projects (LEED-CS), both also part of the LEED version 3.0 group. There have been many changes to the LEED-NC rating system between versions, and released as errata or addenda during the versions. Practitioners and educators should ensure that these changes are incorporated in future projects and class curricula.

This text focuses on the LEED-NC rating system, which is a premier and heavily used green rating system in the United States. LEED-NC is for new construction and major renovations for mainly commercial, institutional, industrial, and large residential (four stories or more) projects. The USGBC and several other organizations have also developed, or are in the process of developing, rating systems for other types of projects. One other organization of note is the National Association of Home Builders (NAHB), which through the GBI is offering its voluntary Model Green Home Building Guidelines. Some of the other LEED rating systems are listed in Table 1.1.1.

Sustainability is not just an environmental concept. Many are quick to point out that sustainability focuses on a triple bottom line: economic, environmental, and societal. Many of the concepts relating to future economic and societal concerns are dependent on current resource and environmental management and best practices. Hawken, Lovins, and Lovins further expand on this in their book *Natural Capitalism, Creating the Next Industrial Revolution.*

1.2 Introduction to the USGBC LEED-NC Rating System

LEED-NC is applicable to a large portion of the most expensive vertical construction projects in the United States. It basically covers most commercial, institutional, and industrial projects and includes residential construction of facilities of four or more stories.

Guidances

There are many projects which have unique criteria that may not fit well into the credit intents and requirements as listed in LEED-NC 2.2 or 2009. To facilitate LEED certification and sustainable practices for certain categories of projects, the USGBC has developed or is developing guidances specific to many special subsets of project application types.

		Status as of February 2010	Comments (USGBC Associated Guide)
LEED-EB	Existing Building operations and maintenance	Approved 10/22/04 Current version 3.0	For later recertification of buildings or for existing building renovations. (Green Building Operations and Maintenance Reference Guide, 2009 Edition)
LEED-CI	Commercial Interiors	Approved 11/17/04 Current version 3.0	For commercial tenant space in a building which may or may not have the core and shell certified. (Green Interior Design and Construction Reference Guide, 2009 Edition)
LEED-CS	Core and Shell	Approved July 2006 Current version 3.0	For the core and shell of a building which may or may not have the commercial interior tenant space certified. (Green Building Design and Construction Reference Guide, 2009 Edition)
LEED-Homes	Homes	Released January 2008 Errata updated to January 2010	(LEED for Homes Reference Guide v. 2009, and GreenHomeGuide.com)
LEED-ND	Neighborhood Development	In pilot	Integration of smart growth, urbanism, and green building for neighborhood design
LEED for Schools or for Existing Schools	Schools (K–12)	Launched in April 2007 Current version 3.0 (Existing Schools under development)	Based on LEED-NC, but required for K–12 academic buildings. May be used for K–12 nonacademic buildings. (Green Building Design and Construction, 2009 Edition. Existing will be a supplement to the Green Building Operations and Maintenance Reference Guide.)
LEED for Retail or Retail Interiors	Retail construction	In pilot	Two rating systems. One is based on NC and second is based on CI for interiors. (Supplements will be added to Green Building Design and Construction and Green Interior Design and Construction Reference Guides.)
LEED for Healthcare	Healthcare facilities*	Under development	(Supplement will be added to Green Building Design and Construction Reference Guide.)
LEED for Labs	Facilities with laboratories*	Under development	No longer listed in Version 3.0 release

*Can also be viewed as a part of LEED-NC.

Table 1.1.1 Other LEED Rating Systems and Reference Guides

Application	Status as of February 2010
LEED for multiple buildings/campuses	Guide available
LEED for lodging (<4 stories)	Developed by USAF but not adopted by USGBC membership
LEED for healthcare	Under development as its own rating system
LEED for labs	No longer listed on USGBC site

TABLE 1.2.1 Application Guidances

They are referred to as application guides. Table 1.2.1 is a listing of several of the guidances formerly developed or being developed.

These guidances serve many purposes. They serve to aid in modifying or embellishing criteria for specific credits when it may be difficult to apply the credit directly as written for a particular type of project, and they also identify some items that may be used for a particular project type for one of the Innovation & Design credits. Chapter 9 is a special section on the U.S. military and sustainable construction and explores in greater detail how one of the guidances, LEED for Lodging, may relate to LEED-NC, with particular emphasis on indoor air quality. For any project that deals with any of these special applications, the guidances should be used in conjunction with the LEED-NC 2009 rating system.

How LEED-NC Is Set Up

LEED-NC contains minimum program requirements (MPRs), credits, and prerequisites. The minimum program requirements are new to LEED 2009 and cover some minimum conditions for a project to be eligible to register for the LEED-NC certification process. These will be covered in Chap. 8. All eight prerequisites in LEED 2009 (and all seven in LEED 2.2) are mandatory for each project. None of the credits (worth up to 69 points from these credits in LEED 2.2 and up to 110 points in LEED 2009) are mandatory except for a couple of energy efficiency points in LEED 2.2. The credits are worth a certain number of points, and a combination of credit points adds up to a certain level of certification. The levels of certification, the number of points and the number of prerequisites associated with each level are as listed in Table 1.2.2 for both LEED 2.2 and LEED 2009.

LEED-NC is subdivided into five main categories for which there are sometimes prerequisites, subcategories, and credits representing possible points in both versions. Some of the credits have one associated point, but many credits are multipoint. There is an additional category for Innovation in Design (ID) in both the rating systems, and a new category to emphasize credits which are priorities in various regions of the United States (RP) in LEED-NC 2009. The seven categories and their associated numbers of prerequisites, subcategories, and possible points (excluding exemplary performance points associated with that category) in LEED 2009 are as noted (with former values for LEED 2.2 in parentheses):

- **Sustainable Sites** (SS) in LEED 2009
  - Prerequisites: 1 (1 in LEED 2.2)
  - Subcategories: 8 (8 in LEED 2.2)
  - Possible points: 26 (14 in LEED 2.2)

	LEED-NC 2009	LEED-NC 2.2
Prerequisites	8	7
Base Points from Main Credit Subcategories	100	64
Innovation in Design Points	6	5
Regional Priority Points	4	0
Total	110	69
Points for Certified	40–49	26–32
Points for Certified Silver	50–59	33–38
Points for Certified Gold	60–79	39–51
Points for Certified Platinum	80+	52–69

TABLE 1.2.2 Total Prerequisites, Points, and Certification Level Summaries for LEED-NC 2009 and LEED-NC 2.2 New Construction

- **Water Efficiency** (WE) in LEED 2009
  - Prerequisites: 1 (0 in LEED 2.2)
  - Subcategories: 3 (3 in LEED 2.2)
  - Possible points: 10 (5 in LEED 2.2)
- **Energy and Atmosphere** (EA) in LEED 2009
  - Prerequisites: 3 (3 in LEED 2.2)
  - Subcategories: 6 (6 in LEED 2.2)
  - Possible points: 35 (17 in LEED 2.2)
- **Materials and Resources** (MR) in LEED 2009
  - Prerequisites: 1 (1 in LEED 2.2)
  - Subcategories: 7 (7 in LEED 2.2)
  - Possible points: 14 (13 in LEED 2.2)
- **Indoor Environmental Quality** (IEQ) in LEED 2009 (was labeled EQ in LEED 2.2)
  - Prerequisites: 2 (2 in LEED 2.2)
  - Subcategories: 8 (8 in LEED 2.2)
  - Possible points: 15 (15 in LEED 2.2)
- **Innovation and Design Processes** (ID) in LEED 2009
  - Prerequisites: None (none in LEED 2.2)
  - Subcategories: 2 (2 in LEED 2.2)
  - Possible points: 6 (5 in LEED 2.2)
- **Regional Priorities** (RP) in LEED 2009 (not in LEED 2.2)
  - Number of priorities: 6
  - Possible points: 4

The Regional Priority (RP) category is different in that it gives additional points for priority credits from the first five main categories. The four possible points can be chosen from a maximum of six priority credits. More information on the RP category can be found in Chap. 7.

LEED-NC Documents

Information is provided from the USGBC on several different levels. The main level of information provided for LEED-NC 2.2 and LEED-NC 2009 are the governing documents as approved by the USGBC for the rating system. They are entitled *LEED-NC Green Building Rating System for New Construction & Major Renovations Version 2.2 for Public Use and Display* for LEED 2.2, and *LEED 2009 for New Construction and Major Renovations for Public Use and Display* for LEED 2009. They are available free of charge to the general public for download and public display and can be found on the USGBC website (www.usgbc.org). The verbiage from various sections of the LEED-NC 2.2 document is included verbatim throughout this text, with modifications as noted to update to LEED 2009. The exact wording is important to read to adequately interpret the intent and expanse of the requirements listed. Some other items provided to the general public for free from the USGBC website are the errata (LEED 2.2) or addenda (LEED 2009) sheets (which include corrections to both the governing documents and the Reference Guides as described in the following paragraph) and various fact sheets and sample items, including sample templates. Templates were the online format used by the USGBC for credit submittals and calculations in LEED 2.2. LEED 2009 has an updated LEED-Online process for managing the documentation for certifying a project. Submittals include an overall project narrative and then required documentation associated with each prerequisite and credit. Project teams can even opt to have a review after the design phase for information on whether certain credits or prerequisites with a design phase focus are appropriate for certification. LEED-Online also details the project team member required to sign off on certain documents (declarant). There are also some streamlined options where for certain credits a licensed professional such as a licensed professional engineer, licensed architect, or licensed landscape architect may verify compliance without as much supporting documentation being submitted. Similar to the example template, there are "calculators" embedded in LEED-Online for performing many of the calculations. The USGBC also provides a project checklist for overall summary credit and prerequisite tracking.

As mentioned earlier, the USGBC provides reference guides for the LEED-NC 2.2 and 2009 rating systems. These reference guides are supplemental to the governing documents and are available for purchase. They give example calculations, case studies, and additional references, and they expand upon the intent and definitions necessary to better understand the prerequisite and credits for certification. Members of the USGBC and people associated with registered projects can also log on to the system and view the Credit Interpretation Rulings (CIRs), which are replies to Credit Interpretation Requests. The CIRs are further explained in the following sections and allow the members or projects to keep up with future interpretations and enhancements to the rating system.

Credit Formats

The information given in the credit write-ups for the various credits is set in a standard format. The format in version 2009 is slightly different from that of version 2.2, which was slightly modified from version 2.1. The formats used in the governing document (Public Display Portion) and the Reference Guide in versions 2.2 and 2009 are also slightly different and are summarized as follows:

- In Public Display Portion (LEED 2009 and 2.2): Intent
Requirements
(No longer lists submittals in LEED 2.2 and 2009 to allow for greater flexibility)

Potential Technologies and Strategies (Only LEED 2.2)

- In Reference Guide (LEED 2.2):
  - Intent
  - Requirements
  - Summary of Referenced Standards
  - Approach and Implementation
  - Calculations (Yes or No)
  - Exemplary Performance Point (Yes or No)
  - Submittal Documentation
  - Considerations
    - Environmental
    - Economic
    - Community (Sometimes)
  - Resources
  - Definitions
- In Reference Guide (LEED 2009):
  - Intent
  - Requirements
    1. Benefits and Considerations
    2. Related Credits
    3. Summary of Referenced Standards
    4. Implementation
    5. Timeline and Team
    6. Calculations
    7. Documentation Guidance
    8. Examples
    9. Exemplary Performance
    10. Regional Variations
    11. Operations and Maintenance Considerations
    12. Resources
    13. Definitions

Registration, Certification, Membership, and Accreditation

There are two main definitions of various status categories with respect to building projects and the USGBC LEED process: registration and certification. There are currently fees associated with both registration and certification.

Registration with the USGBC was completed at the inception of a project to begin the certification process and was turned over to a different organization, the Green Building Certification Institute (GBCI) in LEED 2009 for third party certification, except for LEED for Homes. The project team then receives access to many of the management features available on the Green Building Certification Institute (GBCI) website to aid in keeping track of the project including LEED-Online. The project is usually registered at the current version of the rating system, and the project is tracked at this level.

Certification at one of the certification levels listed previously is intended for that project after it is complete. Certification is requested through an application process. Certification is received at the version of the rating system for which the project is registered. Figures 1.2.1 and 1.2.2 are photographs of the first private building in South Carolina that received certification under the USGBC LEED-NC rating system.

FIGURE 1.2.1 Cox and Dinkins Engineers and Surveyors in Columbia, S.C.: First private building to receive certification under the USGBC LEED-NC rating system in South Carolina. (*Photograph taken in June 2007.*)

FIGURE 1.2.2 Cox and Dinkins Engineers and Surveyors in Columbia, S.C. USGBC LEED-NC rating system certification plaque. (*Photograph taken in June 2007.*)

Formerly, the application for certification was one step near the end of a project. In LEED-NC 2.2 and 2009, this application process has been optionally divided into two phases. Several of the design phase credits can be applied for prior to completion of the project, with the remaining applied for at the end of the construction phase. Each project is allowed one design phase review. This aids in expediting paperwork and understanding final interpretations of the applicable credits.

There are some additional definitions of various status categories with respect to people and the USGBC LEED process:

Membership to the USGBC is by organization, and then individuals within that organization can become members under the umbrella of their organization (organizations pay dues, individuals do not). A person can also be a **member of a chapter** such as the USGBC-SC (South Carolina) chapter (individuals usually pay dues to a chapter). Students and recent graduates may also become members of the Emerging Green Builders (EGB), a special membership category usually associated with local chapters for students and young professionals.

Accreditation is the mechanism by which individuals could become LEED-Accredited Professionals (LEED-AP) prior to 2009. This process was originally developed by the USGBC and eventually was turned over to a different organization, the GBCI. It is used to determine whether a person can be recognized as a professional with respect to its rating system. It is an exam-based accreditation process. With the adoption of LEED version 3.0 (LEED 2009), the accreditation was divided into three levels, with associated sublevels:

- LEED Green Associate
- LEED Accredited Professional (AP)
  - LEED AP BD&C
  - LEED AP Homes
  - LEED AP ID&C
  - LEED AP ND
  - LEED AP O&M
- LEED Fellow

The associate level is for persons with only a small level of experience in green building, the AP level is for those who have substantial experience in green building, and the fellow level will be for persons who have contributed substantially to the field of green building and sustainability. The acronyms stand for Building Design & Construction (BD&C), Interior Design & Construction (ID&C), Neighborhood Development (ND), and Operations & Maintenance (O&M). All levels are currently available except for LEED-AP ND and LEED Fellow.

Project Checklist, LEED-Online, and Templates

Development of effective project management and green concept integration into construction projects will be a continual challenge as the green movement expands in the United States. Some simplified suggestions for improvement and implementation will be discussed in later chapters. In addition, substantial efforts are under way to research optimization methods for implementation of the green process. There are many tools and suggestions developed by the GBCI to facilitate becoming green, which can be found in example format on the GBCI website (www.gbci.org). Two of the main tools are the project checklist and the templates (2.2) or LEED-Online (2009). The project checklist is used to keep track of overall project success with respect to green construction, and the templates or LEED-Online are used to keep track of individual prerequisites and credits during the project and to submit information to the GBCI during the certification application process.

At the onset of a project after registration, the GBCI supplies the LEED-NC Registered Project Checklist to the project team. Table 1.2.3 shows a blank LEED-NC 2.2 checklist as downloaded from the USGBC website in May 2007. Table 1.2.4 is an example of how a project team might format a checklist for LEED-NC 2009.

LEED for New Construction v2.2 Registered Project Checklist (p.1)					
Project Name and Address:					
Yes	?	No			
				Sustainable Sites	**14 Points**
Y			Prereq 1	**Construction Activity Pollution Prevention**	Required
			Credit 1	**Site Selection**	1
			Credit 2	**Development Density & Community Connectivity**	1
			Credit 3	**Brownfield Redevelopment**	1
			Credit 4.1	**Alternative Transportation**, Public Transportation Access	1
			Credit 4.2	**AlternativeTransportation**, Bicycle Storage & Changing Rooms	1
			Credit 4.3	**Alternative Transportation**, Low-Emitting & Fuel-Efficient Vehicles	1
			Credit 4.4	**Alternative Transportation**, Parking Capacity	1
			Credit 5.1	**Site Development,** Protect of Restore Habitat	1
			Credit 5.2	**Site Development,** Maximize Open Space	1
			Credit 6.1	**Stormwater Design,** Quantity Control	1
			Credit 6.2	**Stormwater Design,** Quality Control	1
			Credit 7.1	**Heat Island Effect,** Non-Roof	1
			Credit 7.2	**Heat Island Effect,** Roof	1
			Credit 8	**Light Pollution Reduction**	1
				Water Efficiency	**5 Points**
			Credit 1.1	**Water Efficient Landscaping**, Reduce by 50%	1
			Credit 1.2	**Water Efficient Landscaping**, No Potable Use or No Irrigation	1
			Credit 2	**Innovative Wastewater Technologies**	1
			Credit 3.1	**Water Use Reduction**, 20% Reduction	1
			Credit 3.2	**Water Use Reduction**, 30% Reduction	1
				Energy & Atmosphere	**17 Points**
Y			Prereq 1	**Fundamental Commissioning of the Building Energy Systems**	Required
Y			Prereq 2	**Minimum Energy Performance**	Required
Y			Prereq 3	**Fundamental Refrigerant Management**	Required
			Credit 1	**Optimize Energy Performance**	1 to 10
				10.5% New Buildings or 3.5% Existing Building Renovations	1
				14% New Buildings or 7% Existing Building Renovations	2
				17.5% New Buildings or 10.5% Existing Building Renovations	3
				21% New Buildings or 14% Existing Building Renovations	4
				24.5% New Buildings or 17.5% Existing Building Renovations	5
				28% New Buildings or 21% Existing Building Renovations	6
				31.5% New Buildings or 24.5% Existing Building Renovations	7
				35% New Buildings or 28% Existing Building Renovations	8
				38.5% New Buildings or 31.5% Existing Building Renovations	9
				42% New Buildings or 35% Existing Building Renovations	10
			Credit 2	**On-Site Renewable Energy**	1 to 3
				2.5% Renewable Energy	1
				7.5% Renewable Energy	2
				12.5% Renewable Energy	3
			Credit 3	**Enhanced Commissioning**	1
			Credit 4	**Enhanced Refrigerant Management**	1
			Credit 5	**Measurement & Verification**	1
			Credit 6	**Green Power**	1

TABLE 1.2.3 Blank LEED-NC 2.2 Project Checklist (Sheet 1)

LEED for New Construction v2.2 Registered Project Checklist (p.2)					
			Materials & Resources		13 Points
Y			Prereq 1	**Storage & Collection of Recyclables**	Required
			Credit 1.1	**Building Reuse,** Maintain 75% of Existing Walls, Floors & Roof	1
			Credit 1.2	**Building Reuse,** Maintain 100% of Existing Walls, Floors & Roof	1
			Credit 1.3	**Building Reuse,** Maintain 50% of Interior Nonstructural Elements	1
			Credit 2.1	**Construction Waste Management,** Divert 50% from Disposal	1
			Credit 2.2	**Construction Waste Management,** Divert 75% from Disposal	1
			Credit 3.1	**Materials Reuse,** 5%	1
			Credit 3.2	**Materials Reuse,** 10%	1
			Credit 4.1	**Recycled Content,** 10% (Postconsumer + ½ Preconsumer)	1
			Credit 4.2	**Recycled Content,** 20% (Postconsumer + ½ Preconsumer)	1
			Credit 5.1	**Regional Materials,** 10% Extracted, Processed & Manufactured Regionally	1
			Credit 5.2	**Regional Materials,** 20% Extracted, Processed & Manufactured Regionally	1
			Credit 6	**Rapidly Renewable Materials**	1
			Credit 7	**Certified Wood**	1
			Indoor Environmental Quality		15 Points
Y			Prereq 1	**Minimum IAQ Performance**	Required
Y			Prereq 2	**Environmental Tobacco Smoke** (ETS) **Control**	Required
			Credit 1	**Outdoor Air Delivery Monitoring**	1
			Credit 2	**Increased Ventilation**	1
			Credit 3.1	**Construction IAQ Management Plan,** During Construction	1
			Credit 3.2	**Construction IAQ Management Plan,** Before Occupancy	1
			Credit 4.1	**Low-Emitting Materials,** Adhesives & Sealants	1
			Credit 4.2	**Low-Emitting Materials,** Paints & Coatings	1
			Credit 4.3	**Low-Emitting Materials,** Carpet Systems	1
			Credit 4.4	**Low-Emitting Materials,** Composite Wood & Agrifiber Products	1
			Credit 5	**Indoor Chemical & Pollutant Source Control**	1
			Credit 6.1	**Controllability of Systems,** Lighting	1
			Credit 6.2	**Controllability of Systems,** Thermal Comfort	1
			Credit 7.1	**Thermal Comfort,** Design	1
			Credit 7.2	**Thermal Comfort,** Verification	1
			Credit 8.1	**Daylight & Views,** Daylight 75% of Spaces	1
			Credit 8.2	**Daylight & Views,** Views for 90% of Spaces	1
			Innovation & Design Process		5 Points
			Credit 1.1	**Innovation in Design**: Provide Specific Title	1
			Credit 1.2	**Innovation in Design**: Provide Specific Title	1
			Credit 1.3	**Innovation in Design**: Provide Specific Title	1
			Credit 1.4	**Innovation in Design**: Provide Specific Title	1
			Credit 2	**LEED® Accredited Professional**	1
			Project Totals (pre-certification estimates)		69 Points
Certified: 26–32 points, **Silver:** 33–38 points, **Gold:** 39–51 points, **Platinum:** 52–69 points					

TABLE 1.2.3 Blank LEED-NC 2.2 Project Checklist (Sheet 2)

Templates, or letter templates as they were frequently referred to in LEED-NC 2.1, were provided for use by the project team after project registration as is the full LEED-Online tool for LEED 2009. They provide an interactive format for many of the calculations needed for credit verification. They are used for summary calculations and as a tool to track the necessary documentation for submittals for each prerequisite and credit applied for. Figure 1.2.3 is a summary figure of the format from LEED 2.2 for part of a submittal for a prerequisite that requires drawing submittals and narratives explaining the project adherence to the requirements. It is based on the example template available for download from the USGBC website for the Sustainable Sites prerequisite one (SSp1) for Construction Activity Pollution Prevention. Note that it refers to "credit" requirements, even though it is a prerequisite and does not earn points. Figure 1.2.4 is a

Yes	?	No		Possible Points	Tentative Points
Y			SSp1: Construction Activity Pollution Prevention	Required	–
			SSc1: Site Selection	1	
			SSc2: Development Density & Community Connectivity	5	
			SSc3: Brownfield Redevelopment	1	
			SSc4.1: Alternative Transportation—Public Transportation Access	6	
			SSc4.2: Alternative Transportation—Bicycle Storage and Changing Rooms	1	
			SSc4.3: Alternative Transportation—Low-Emitting & Fuel-Efficient Vehicles	3	
			SSc4.4: Alternative Transportation—Parking Capacity	2	
			SSc5.1: Site Development—Protect or Restore Habitat	1	
			SSc5.2: Site Development—Maximize Open Space	1	
			SSc6.1: Stormwater Design—Quantity Control	1	
			SSc6.2: Stormwater Design—Quality Control	1	
			SSc7.1: Heat Island Effect—Non-roof	1	
			SSc7.2: Heat Island Effect—roof	1	
			SSc8: Light Pollution Reduction	1	
Y			WEp1: Water Use Reduction Required	Required	–
			WEc1: Water Efficient Landscaping [2 pts-50%; 4 pts-100%]	2–4	
			WEc2: Innovative Wastewater Technologies	2	
			WEc3: Water Use Reduction [2 pts-30%; 3 pts-35%; 4 pts-40%]	2–4	
Y			EAp1: Fundamental Commissioning of Building Energy Systems	Required	–
Y			EAp2: Minimum Energy Performance	Required	–
Y			EAp3: Fundamental Refrigerant Management	Required	–
			EAc1: Optimize Energy Performance – [New (Existing)] [1 pt-12%(8%); 2 pts-14%(10%); 3 pts-16%(12%); 4 pts-18%(14%); 5 pts-20%(16%); 6 pts-22%(18%); 7 pts-24%(20%); 8 pts-26%(22%); 9 pts-28%(24%); 10 pts-30%(26%); 11 pts-32%(28%); 12 pts-34%(30%); 13 pts-36%(32%); 14 pts-38%(34%); 15 pts-40%(36%); 16 pts-42%(38%); 17 pts-44%(40%); 18 pts-46%(42%); 19 pts-48%(44%)]	1–19	
			EAc2: On-site Renewable Energy [1 pt-1%,2 pts-3%; 3 pts-5%; 4 pts-7%; 5 pts-9%; 6 pts-11%; 7 pts-13%]	1–7	

Table 1.2.4 Blank Example LEED 2009 Checklist for Columbia, SC 29201 (Sheet 1) (*Continued*)

Yes	?	No		Possible Points	Tentative Points
			EAc3: Enhanced Commissioning	2	
			EAc4: Enhanced Refrigerant Management	2	
			EAc5: Measurement and Verification	3	
			EAc6: Green Power (35% - 2 years)	2	
Y			MRp1: Storage & Collection of Recyclables	Required	–
			MRc1.1: Building Reuse-Maintain Existing Walls, Floors & Roof [1 pt-55%; 2 pts-75%; 3 pts-95%]	1–3	
			MRc1.2: Building Reuse-Maintain Existing Interior Nonstructural Elements	1	
			MRc2: Construction Waste Management [1 pt-50%; 2 pts-75%]	1–2	
			MRc3: Materials Reuse [1 pt-5%; 2 pts-10%]	1–2	
			MRc4: Recycled Content [1 pt-10%; 2 pts-20%]	1–2	
			MRc5: Regional Materials [1 pt-10%; 2 pts-20%]	1–2	
			MRc6: Rapidly Renewable Materials	1	
			MRc7: Certified Wood	1	
Y			IEQp1: Minimum Indoor Air Quality Performance	Required	–
Y			IEQp2: Environmental Tobacco Smoke (ETS) Control	Required	–
			IEQc1: Outdoor Air Delivery Monitoring	1	
			IEQc2: Increased Ventilation	1	
			IEQc3.1: Construction IAQ Management Plan, during Construction	1	
			IEQc3.2: Construction IAQ Management Plan, before Occupancy	1	
			IEQc4.1: Low-Emitting Materials, Adhesives and Sealants	1	
			IEQc4.2: Low-Emitting Materials, Paints and Coatings	1	
			IEQc4.3: Low-Emitting Materials, Flooring Systems	1	
			IEQc4.4: Low-Emitting Materials, Composite Wood and Agrifiber Products	1	
			IEQc5: Indoor Chemical and Pollutant Source Control	1	
			IEQc6.1: Controllability of Systems Lighting	1	
			IEQc6.2: Controllability of Systems Thermal Comfort	1	

Table 1.2.4 Blank Example LEED 2009 Checklist for Columbia, SC 29201 (Sheet 2) (*Continued*)

Yes	?	No		Possible Points	Tentative Points
			IEQc7.1: Thermal Comfort—Design	1	
			IEQc7.2: Thermal Comfort Verification	1	
			IEQc8.1: Daylight and Views Daylight	1	
			IEQc8.2: Daylight and Views—Views	1	
			IDc1: Innovation in Design First ID or EP Item________________________	1	
			IDc1: Innovation in Design Second ID or EP Item______________________	1	
			IDc1: Innovation in Design Third ID or EP Item________________________	1	
			IDc1: Innovation in Design Fourth ID or EP Item_______________________	1	
			IDc1: Innovation in Design Fifth ID or EP Item________________________	1	
			IDc2: LEED Accredited Professional	1	
			RP: Regional Priority First RP: Credit (level)_________SSc4.1_________*	1	
			RP: Regional Priority Second RP: Credit (level)_______SSc6.1_________*	1	
			RP: Regional Priority Third RP: Credit (level)_________WEc3 (40%)______*	1	
			RP: Regional Priority Fourth RP: Credit (level)_______EAc1 (28%/24%)___*	1	
			RP: Regional Priority Fifth RP: Credit (level)__________EAc2 (1%)________*	1	
			RP: Regional Priority Sixth RP: Credit (level)_________IEQc7.1_________*	1	
				Total*	

*May only count four of the RP points.

Table 1.2.4 Blank Example LEED 2009 Checklist for Columbia, SC 29201 (Sheet 3) (*Continued*)

LEED-NC 2.2 Submittal Template
SS Prerequisite 1: Construction Activity Pollution Prevention

(Responsible Individual) (Company Name)
I, ________________, from________________________
Verify that the information provided below is accurate to the best of my knowledge.

CREDIT COMPLIANCE
Please select the appropriate compliance path

Option 1: The Erosion and Sedimentation Control Plan (ESC) conforms to the 2003 EPA Construction General Permit, which out lines the provisions necessary to comply with Phase I and Phase II of the National Pollutant Discharge Elimination System (NPDES) program. ________

OR

Option 2: The ESC Plan follows local erosion and sedimentation control standards and codes, which are more stringent than the NPDES program requirements. ________

SUPPORTING DOCUMENTATION
The noted project drawing(s) have been uploaded. The drawing(s) shows the erosion and sedimentation control measures implemented on the site.

Sheet Description Log
Please include sheet name, sheet number and file name for each uploaded, refereneced drawing (e.g. A-101, Site Plan, siteplan.pdf)

_____ I have provided the appropriate supporting documentation in the document upload section of LEED Online. Please refer to the above sheets.

NARRATIVE (Required)
Provide a narrative to describe the Erosion and Sedimentation control measures implemented on the project. If local standard has been followed, please provide specific information to demonstrate that the local standard is equal to or more stringent than the referenced NPDES program.

NARRATIVE (Optional)
Please provide any additional comments or notes regarding special circumstances or considerations regarding the project's credit approach.

________ The project is seeking point(s) for this credit using an alternate compliance approach.The compliance approach including references to any applicable Credit Interpretation Rulings is fully documented in the narrative above.

Project Name:

FIGURE 1.2.3 Portion of example template from the USGBC website for a LEED 2.2 prerequisite which requires drawings and narrative for submittal (SSp1).

summary figure of the format for the first portion of Option 1 for the Energy and Atmosphere credit number 1 (EAc1), Optimize Energy Performance from LEED 2.2. It is depicted to show a template where actual numbers and data are input to the GBCI. The actual online process is interactive and will provide appropriate calculations from some of the data input.

LEED-NC 2.2 Submittal Template
EA Credit 1: Optimize Energy Performance

(Responsible Individual) (Company Name)
I, ________________ , from________________________
Verify that the information provided below is accurate to the best of my knowledge.

CREDIT COMPLIANCE
(Please complete the color coded criteria(s) based on the option path selected)

Please select the appropriate compliance path option
___ Option 1 (Pg2): Performance Rating Method, ASHRAE 90.1-2004 Appendix G or equivalent (upto 10 points possible)
___ Option 2 (Pg14): ASHRAE Advanced Energy Design Guide for Small Office Buildings 2004 (4 points)
___ Option 3 (Pg14): Advanced Buildings Benchmark™ Version 1.1, Basic Criteria & Prescriptive Measures (1 point)

OPTION 1: PERFORMANCE RATING METHOD
___ I Confirm that the energy simulation software used for this project has all capabilities described in EITHER section'G2 Simulation General Requirements' in Appendix G of A SHRAE 90.1-2004 OR the analogous section of the alternative qualifying code used.

___ I Confirm that the baseline building and proposed building in this project's energy simulation runs use the assumptions and modeling methodology described in EITHER Appendix G of ASHRAE 90.1-2004 OR the analogous section of the alternative qualifying energy code used.

Complete the following sections to document compliance using Option1:
Section 1.1 – General Information
Section 1.2 – Space Summary.........

.,...........Section 1.8 – Performance Rating Method Compliance Report

Section 1.1 – General Information
....

Section 1.2 – Space Summary
Provide the space summary for your project
(click "CLEAR" to clear the contents of any row. All numeric entries must be entered as whole numbers without commas):

Table 1.2 – Space Summary

Building Use (Occupancy Type)	Conditioned Area (sf)	Unconditioned Area (sf)	Total Area (sf)	
				CLEAR
				CLEAR
				CLEAR
				CLEAR
				CLEAR
				CLEAR
				CLEAR
Total				

FIGURE 1.2.4 Portion of example template from the USGBC website for a LEED 2.2 credit which requires data and narrative for submittal (EAc1: Option 1). Shaded areas in the table are for user input of data. Blank areas in the table will be filled in automatically.

What to Do if Things Are Not Clear for Your Project

CIRs represent the format that the USGBC has chosen to continually aid projects with interpretations of the rating system for their specific circumstances. Requests for interpretation of credits (Credit Interpretation Requests) are submitted to the USGBC, and the USGBC replies with CIRs. It is a formal process that lists the rulings of the interpretations online for use by other projects to allow for both consistency from project to project and expediency in having the explanation and interpretations readily available without constant revisions to the governing documents. The CIR process is very important to the rating system's success and certification. It is expected that project team members are very familiar with this process. In the LEED-NC 2.1 Reference Guide, the CIR process was summarized in the following four steps:

1. Consult the Reference Guide.
2. Review the Reference Guide and self-evaluate.
3. Review the CIR web page for previously logged CIRs.
4. If still unanswered, submit a CIR with an online form.

The LEED-NC 2.2 Reference Guide summarizes the CIR process in three steps following the project team's being unable to adequately answer a question based on its interpretation of the Reference Guide. The three steps are as follows:

1. Review the CIR web page for previously logged CIRs. Note that some of the CIRs for other rating systems or versions may not be applicable.
2. If no applicable CIR exists, then submit a CIR via the CIR web page. This web page has guidelines for how a CIR should be submitted, with particular information on what should be in the request. The main focus is on the intent of the credit. The CIR will eventually be posted on the CIR web page and is not intended to include a long description of a particular project, but rather a more overall question to be interpreted for application for the project and other similar project circumstances.
3. The USGBC will rule on the CIR.

As mentioned earlier under LEED-NC documents, the USGBC has a process to periodically post errata, or addenda as it is called in LEED 2009, to both the governing document (Public Display Portion) and the Reference Guide, which are also periodically included in new editions of the versions.

1.3 Miscellaneous

Organizations

An important part of specifications and design for construction is the use of established standards. Many organizations are recognized as established sources for many standards and have procedures for accrediting, developing, reviewing, and revising standards. The organization that accredits organizations as a standards developer in the United States is the American National Standards Institute (ANSI). On November 27, 2006, ANSI accredited the U.S. Green Building Council as an official Standards Developing Organization. Some of these organizations that help develop or accredit standards as mentioned in this text include the following:

AIA American Institute of Architects.

ANSI The American National Standards Institute accredits standards for products, services, processes, and systems in the United States; accredits organizations that perform certifications; and coordinates U.S. standards with international standards.

ASHRAE — The American Society of Heating, Refrigerating and Air-Conditioning Engineers, Inc., publishes standards in the areas of HVAC and refrigeration.

ASTM — ASTM International was originally formed in 1898 as the American Society for Testing and Materials. It develops technical standards for materials, products, systems, and services.

IESNA — The Illuminating Engineering Society of North America is recognized as an authority on lighting and illumination standards in the United States.

ISO — The International Organization for Standardization (Organisation Internationale de Normalisation) is the largest international developer of standards. ANSI is the U.S. voting member body for ISO.

Standard 189

In 2006, the USGBC, ASHRAE, IESNA, and AIA joined together to develop Standard 189, a new minimum standard for high-performance green building. The standard is being led by the newly developed ASHRAE Standard Project Committee 189 (SPC 189). The standard was initially proposed in January 2006 and as a standard under development is referred to as SPC 189P. The preliminary title, purpose, and scope as revised through November 8, 2006 read as follows:

> **Standard for the Design of High-Performance, Green Buildings Except Low-Rise Residential Buildings**
>
> **1 Purpose:**
>
> The purpose of this standard is to provide minimum requirements for the design of high-performance, green buildings to:
>
> (a) Balance environmental responsibility, resource efficiency, occupant comfort and well being, and community sensitivity, and
>
> (b) Support the goal of the development that meets the needs of the present without compromising the ability of future generations to meet their own needs.
>
> **2 Scope:**
>
> 2.1 This standard provides minimum criteria that:
>
> (a) Apply to new buildings and major renovation projects (new portions of buildings and their systems): a building or group of buildings, including on-site energy conversion or electric-generating facilities, which utilize a single submittal for a construction permit or which are within the boundary of a contiguous area under single ownership
>
> (b) Address sustainable sites, water use efficiency, energy efficiency, the building's impact on the atmosphere, materials and resources, and indoor environmental quality (IEQ).
>
> 2.2 The provisions of this standard do not apply to:
>
> (a) single-family house, multi-family structures of three stories or fewer above grade, manufactured houses (mobile homes) and manufactured houses (modular).
>
> (b) buildings that do not use either electricity or fossil fuel.
>
> 2.3 This standard shall not be used to circumvent any safety, health or environmental requirements. The ASHRAE Standard Project Committee 189.1 (SPC 189.1) was adopted dated November 2009.

The Carbon Commitment

Global climate change is a major issue of concern in the world. Some details of mechanisms and suspected causes of this phenomenon are treated in subsequent chapters. The USGBC has made a commitment to focus on emphasizing strategies and green building goals which would help lower the emission of carbon dioxide into the atmosphere. Carbon dioxide is recognized as a gas with a global warming potential (GWP) and is thought to be a major contributor to current changes in the atmosphere. Carbon dioxide levels in the atmosphere have increased in recent decades, as has the use of fossil fuels for energy. Fossil fuels are carbon based, and their use is thought to impact the global carbon cycle, with more carbon being released to the atmosphere (in the form of carbon dioxide) than sequestered in the crust, ocean, or flora (usually in dissolved, fossil fuel, mineral rock, or vegetation form). Fuels that cannot be readily reproduced in our human-generation time frames, such as fossil fuels, are referred to as *nonrenewable*, whereas energy sources such as wind or wood which can be regrown are referred to as *renewable*. Even though the burning of wood releases carbon into the air in the form of carbon dioxide, it is considered renewable with respect to the carbon cycle, since new trees recycle carbon from the atmosphere during their growth. Renewable energy sources that are still a part of the carbon cycle such as corn-based ethanol or wood-burning are typically referred to as *carbon-neutral*. Other energy sources such as wind harvesting which do not use carbon-based chemicals as part of the energy transfer mechanism are referred to by the author as *carbon-free*.

In late 2006, the USGBC announced some very progressive goals for reducing carbon emissions. These have resulted in an increased emphasis on energy efficiency and the use of renewable energy in LEED 2009. This is evident in the additional points which are awarded in LEED 2009 for many credits which improve energy efficiency or decrease dependence on fossil fuels. Optimizing energy performance as a part of the carbon commitment will reduce energy use, and as most energy use in the United States is based on fossil fuels, this should result in decreased dependence on fossil fuels and decreased carbon emissions from these nonrenewable fuels.

References

ASCE (2010), http://content.asce.org/conferences/greenstreets-highways2010/, Green Streets-Highways Conference website, American Society of Engineers, accessed February 6, 2010.

ASCE (2010), http://content.ewrinstitute.org, Environmental and Water Resources Institute website, American Society of Engineers, accessed February 6, 2010.

ASCE (2010), http://content.tanddi.org, Transportation and Development Institute website, American Society of Engineers, accessed February 6, 2010.

ASCE (2010), http://pubs.asce.org/journals/professionalissues/default.htm, Journal of Professional Issues in Engineering Education and Practice, website, American Society of Engineers, accessed February 6, 2010.

ASHRAE (2006), http://spc189.ashraepcs.org/, website for Standard Project Committee SPC 189, accessed December 11, 2006.

Bon, R., and K. Hutchinson (2000), "Sustainable Construction: Some Economic Challenges," *Building Research and Information*, 28(5/6): 310–314.

Brundtland Commission (1987), *Our Common Future*, Report by the Brundtland Commission [formally the World Commission on Environment and Development (WCED)], Oxford University Press.

CALGreen (2010), California Green Building Standards (CALGreen) Code, website http://www.bsc.ca.gov/CALGreen/default.htm, accessed March 12, 2010.

du Plessis, C. (2001), "Sustainability and Sustainable Construction: The African Context," *Building Research and Information*, 29(5): 374–380.

Engel-Yan, J., et al. (2005), "Toward Sustainable Neighbourhoods: The Need to Consider Infrastructure Interactions," *Canadian Journal of Civil Engineering*, 32: 45–57.

Fedrizzi, R. (2006), "A Look Back at Greenbuild," *Buildings.com Greener Facilities* Newsletter, November, vol. 4, Issue 11, Stamats Business Media, Cedar Rapids, Iowa.

FMLink Group (2006), "Autodesk and USGBC Partner on Technology to Help Make Building Industry Greener," *Facilities Management News*, November 20, 2006.

Forman, R. T. T., et al. (2003), *Road Ecology: Science and Solutions*, Island Press, Washington, D.C.

GBCI (2010), http://www.gbci.org/, Green Building Certification Institute website, accessed February 6, 2010.

Graedel, T. E., and B. R. Allenby (2002), *Industrial Ecology*, 2d ed., Prentice-Hall, New York.

Green Design Institute (2007), http://www.ce.cmu.edu/GreenDesign/, website at Carnegie Mellon University, accessed May 10, 2007.

Green Highways Partnership (2007), www.greenhighways.org, website accessed May 9, 2007.

Greenroads (2010), Greenroads Sustainability Performance Metric website, http://www.greenroads.us, accessed March 12, 2010.

Haselbach, L. M., and C. M. Fiori (2006), "Construction and the Environment: Research Foci for a Sustainable Future," *Journal of Green Building*, Winter, 1(1).

Haselbach, L. M., S. R. Loew, and M. E. Meadows (2005), "Compliance Rates for Stormwater Detention Facility Installation," *ASCE Journal of Infrastructure Systems*, March, 11(1).

Hawken, P. (1993), *The Ecology of Commerce, A Declaration of Sustainability*, HarperBusiness, New York.

Hawken, P., A. Lovins, and L. H. Lovins (1999), *Natural Capitalism, Creating the Next Industrial Revolution*, Little, Brown, Boston, Ma.

Horman, M. J., et al. (2006), "Delivering Green Buildings: Process Improvements for Sustainable Construction," *Journal of Green Building*, Winter, 1(1).

IESNA (2006), http://www.iesna.org/, Illuminating Engineering Society of North America website, accessed December 12, 2006.

ImpEE (2007), http://www-g.eng.cam.ac.uk/impee/, Improving Engineering Education Project website, accessed May 10, 2007, University of Cambridge, United Kingdom.

ISO (2006), http://www.iso.org/iso/en/ISOOnline.frontpage, International Organization for Standardization website, accessed December 12, 2006.

Jefferson, C., J. Rowe, and C. Brebbia (eds.) (2001), *The Sustainable Street: The Environmental, Human and Economic Aspects of Street Design and Management*, WIT Press, Southampton, United Kingdom.

Jia, H., et al. (2005), "Research on Wastewater Reuse Planning in Beijing Central Region," *Water Science & Technology*, 51: 195–202.

Katz, L., and J. Sutherland (Guest eds.) (2007), "Educating Students in Sustainable Engineering," *The International Journal of Engineering Education*, pt. I, 23(2).

Kibert, C. J. (2005), *Sustainable Construction; Green Building Design and Delivery*, Wiley, New York.

Lomberg, B. (2001), *The Skeptical Environmentalist: Measuring the Real State of the World*, Cambridge University Press, Cambridge, United Kingdom.

Mulder, Karel (ed.) (2006), *Sustainable Development for Engineers, A Handbook and Resource Guide*, Greenleaf Publishing, Sheffield, United Kingdom.

Ofori, G. (2000), "Greening the Construction Supply Chain in Singapore," *European Journal of Purchasing and Supply Management*, 6: 195–206.

Passa, J., and D. Rompf (2007), "Energy Efficient Sustainable Schools in Canada South," *Journal of Green Building*, Spring, 2(2): 14–30.

Pearce, A. R., J. R. DuBose, and S. J. Bosch (2007), "Green Building Policy Options for the Public Sector," *Journal of Green Building*, Winter, 2(1).

Pearce, D. W., G. D. Atkinson, and W. R. Dubourg (1994), "The Economics of Sustainable Development," *Annual Review of Energy and the Environment*, 19: 457–474.

Sahely, H. R., C. A. Kennedy, and B. J. Adams (2005), "Developing Sustainability Criteria for Urban Infrastructure Systems," *Canadian Journal of Civil Engineering*, 32: 72–85.

Shen, L., et al. (2005), "A Computer-Based Scoring Method for Measuring the Environmental Performance of Construction Activities," *Automation in Construction*, 14: 297–309.

Seongwon, S., S. Tucker, and P. Newton (2007), "Automated Material Selection and Environmental Assessment in the Context of 3D Building Modelling," *Journal of Green Building*, Spring, 2(2): 51–61.

Sick, F., and A. Kerschberger (2007), "Innovative Low Energy Renovation of an Office Building: Concept and Simulation," *Journal of Green Building*, Spring, 2(2): 31–41.

SSI (2007), http://www.sustainablesites.org/, Sustainable Sites Initiative website, accessed July 1, 2007.

Straube, J., and C. Schumacher (2007), "Interior Insulation Retrofits of Load-Bearing Masonry Walls in Cold Climates," *Journal of Green Building*, Spring, 2(2): 42–50.

Tam, V. W. Y., and K. N. Le (2007), "Predicting Environmental Performance of Construction Projects by Using Least-Squares Fitting Method and Robust Method," *Journal of Green Building*, Winter, 2(1).

USAF, USGBC, and Paladino and Company, Inc. (2001), *Application Guide for Lodging Using the LEED Green Building Rating System*, U.S. Air Force Center for Environmental Excellence, U.S. Green Building Council, Washington, D.C.

USGBC (2003), *LEED-NC for New Construction, Reference Guide*, Version 2.1, 2d ed., May, U.S. Green Building Council, Washington, D.C.

USGBC (2004–2005), *LEED-CI Green Building Rating System for Commercial Interiors*, Version 2.0, November 2004, updated December 2005, U.S. Green Building Council, Washington, D.C.

USGBC (2004–2005), *LEED-EB Green Building Rating System for Existing Buildings, Upgrades, Operations and Maintenance*, Version 2, October 2004, updated July 2005, U.S. Green Building Council, Washington, D.C.

USGBC (2005), *LEED-NC Application Guide for Multiple Buildings and On-Campus Building Projects, for Use with the LEED-NC Green Building Rating System*, Versions 2.1 and 2.2, October, U.S. Green Building Council, Washington, D.C.

USGBC (2005–2007), *LEED-NC for New Construction, Reference Guide*, Version 2.2, 1st ed., U.S. Green Building Council, Washington, D.C., October 2005 with errata posted through Spring 2007.

USGBC (2006), *LEED Green Building Rating System for Core & Shell Development*, Version 2.0, July, U.S. Green Building Council, Washington, D.C.

USGBC (2006), "USGBC Unveils 8 Climate Actions," News Release at Press Conference, 2 p.m. November 15, 2006, Denver, Colo., US Greenbuild Conference, accessed December 12, 2006, http://www.fypower.org/pdf/USGBC_GHGPlan.pdf.

USGBC (2006), www.usgbc.org, U.S. Green Building Council website, accessed September 2006.

USGBC (2007), *LEED for Homes Program Pilot Rating System*, Version 2.11a, January, U.S. Green Building Council, Washington, D.C.

USGBC (2007), *LEED for Neighborhood Development Rating System*, Pilot, Congress for the New Urbanism, Natural Resources Defense Council, U.S. Green Building Council, Washington, D.C.

USGBC (2007), *LEED for Retail—New Construction and Major Renovations*, Pilot Version 2, April, U.S. Green Building Council, Washington, D.C.

USGBC (2007), *LEED for Schools for New Construction and Major Renovations*, Approved 2007 Version, April, U.S. Green Building Council, Washington, D.C.

USGBC (2009), *LEED for Homes Reference Guide*, 2009 Edition, U.S. Green Building Council, Washington, D.C., April 2009.

USGBC (2009), *LEED Reference Guide for Green Building Design and Construction*, 2009 Edition, U.S. Green Building Council, Washington, D.C., April 2009.

USGBC (2009), *LEED Reference Guide for Green Building Operations and Maintenance*, 2009 Edition, U.S. Green Building Council, Washington, D.C., April 2009.

USGBC (2009), *LEED Reference Guide for Green Interior Design and Construction*, 2009 Edition, U.S. Green Building Council, Washington, D.C., April 2009.

USGBC (2010), http://www.greenhomeguide.com, U.S. Green Building Council website, Washington, D.C., accessed February 6, 2010.

USGBC (2010), http://www.usgbc.org, U.S. Green Building Council website, Washington, D.C., accessed February 6, 2010.

Vanegas, J. A. (ed.) (2004), *Sustainable Engineering Practice: An Introduction*, American Society of Civil Engineers, Reston, Va.

Wikipedia (2006), http://en.wikipedia.org/wiki/ANS I, ANSI information accessed December 12, 2006.

Wikipedia (2006), http://en.wikipedia.org/wiki/ASTM, ASTM information accessed December 12, 2006.

Xiaohua, W., and Z. Feng (2003), "Energy Consumption with Sustainable Development in Developing Country: A Case in Jiangsu, China," *Energy Policy*, 31: 1679–1684.

Yingxin, Z., and L. Borong (2004), "Sustainable Housing and Urban Construction in China," *Energy and Buildings*, 36: 1287–1297.

Exercises

1. What is the current version of LEED-NC and when was it adopted?

2. Download the addenda for LEED-NC 2009. Separate addenda into two categories, addenda to the rating system and addenda to the Reference Guide only. Incorporate these addenda into appropriate sections of this text.

3. Update Table 1.1.1 with the current status of the rating systems listed, and add other or modified rating systems from the USGBC website www.usgbc.org.

4. Update Table 1.2.1 as to the current status of the guidances. Are there additional guidances mentioned on the USGBC website?

5. What are the categories in the BREEAM rating system?

6. What is the current status of the Green Globes rating system? What are the categories in this system?

CHAPTER 2

LEED Sustainable Sites

The Sustainable Sites category of the U.S. Green Building Council (USGBC) rating system for new construction version 2.2 (LEED-NC 2.2) and 2009 (LEED-NC 2009) consist of one prerequisite and eight credit subcategories, some of which are further divided into credit sections which together may earn a possible 14 points in version 2.2 and 26 points in version 2009, not including exemplary performance points. The notation format for the prerequisites and credits is, for example, SSp1 for Sustainable Sites prerequisite 1 and SSc1 for Sustainable Sites credit 1. Also, in this chapter, to facilitate easier cross-referencing between this text and the USGBC rating system, the second digit in a section heading, equation, table, or figure number represents the credit subcategory number for sections that deal directly with a USGBC LEED-NC 2009 credit subcategory.

The Sustainable Sites (SS) portion deals with issues outside of the building, including some of the building exterior, the land that is being developed, and the surrounding community. It is not the only category that includes factors outside of the building, but comprises those exterior issues that do not fit readily into the other categories and those that have a direct impact on the local community or microenvironment. The intent of the Sustainable Sites portion is summarized in LEED-NC 2.2 as follows, and overall has not changed significantly in LEED 2009:

> Project teams undertaking building projects should be cognizant of the inherent impacts of development on the following:
>
> - land consumption
> - ecosystems
> - natural resources
> - energy use
>
> Preference should be given to buildings with high performance attributes in locations that enhance existing neighborhoods, transportation networks and urban infrastructures. During initial project scoping, preference should be given to sites and land use plans that preserve natural ecosystem functions and enhance the health of the surrounding community.

In summary, the Sustainable Sites category encourages revitalizing and using existing infrastructure, and impacting the environment and natural resources as little as possible in the local area (community) near or within the proposed development.

The prerequisite and 14 credit subsections available in the Sustainable Sites category are as follows:

- **SS Prerequisite 1** (SSp1): *Construction Activity Pollution Prevention* (previously referred to as *Erosion and Sedimentation Control* in LEED-NC 2.1)
- **SS Credit 1** (SSc1): *Site Selection*

- **SS Credit 2** (SSc2): *Development Density and Community Connectivity* (previously referred to as *Urban Redevelopment* in LEED-NC 2.1) (EB)
- **SS Credit 3** (SSc3): *Brownfield Redevelopment*
- **SS Credit 4.1** (SSc4.1): *Alternative Transportation: Public Transportation Access* (EB)
- **SS Credit 4.2** (SSc4.2): *Alternative Transportation: Bicycle Storage and Changing Rooms*
- **SS Credit 4.3** (SSc4.3): *Alternative Transportation: Low-Emitting and Fuel-Efficient Vehicles* (previously referred to as *Alternative Transportation, Alternative Fuel Vehicles* in LEED-NC 2.1)
- **SS Credit 4.4** (SSc4.4): *Alternative Transportation: Parking Capacity*
- **SS Credit 5.1** (SSc5.1): *Site Development: Protect or Restore Habitat* (previously referred to as *Reduced Site Disturbance: Protect or Restore Open Space* in LEED-NC 2.1)
- **SS Credit 5.2** (SSc5.2): *Site Development: Maximize Open Space* (previously referred to as *Reduced Site Disturbance: Development Footprint* in LEED-NC 2.1)
- **SS Credit 6.1** (SSc6.1): *Stormwater Management: Quantity Control* (previously referred to as *Stormwater Management: Rate and Quantity* in LEED-NC 2.1)
- **SS Credit 6.2** (SSc6.2): *Stormwater Management: Quality Control* (previously referred to as *Stormwater Management: Treatment* in LEED-NC 2.1)
- **SS Credit 7.1** (SSc7.1): *Heat Island Effect: Non-Roof*
- **SS Credit 7.2** (SSc7.2): *Heat Island Effect: Roof*
- **SS Credit 8** (SSc8): *Light Pollution Reduction*

Table 2.0.0 is a summary of the prerequisites and credits in both versions (2.2 and 2009) with a comparison of the points available and a summary of many of the major differences.

LEED-NC 2.2 and LEED-NC 2009 in the Reference Guides provide tables which summarize some of the prerequisite and credit characteristics, including the main phase of the project for submittal document preparation (Design or Construction).

The prerequisites or credits which are noted as Construction Submittal credits may have major actions going on through the construction phase which, unless documented through that time, may not be able to be verified from preconstruction documents or the built project. In addition, some have activities that are typically contractor-specified, and therefore the construction phase decisions may impact the credit substantially. There are three items in the Sustainable Sites category, which are designated by the USGBC as Construction Submittals: SSp1 (Construction Activity Pollution Prevention), SSc5.1 (Site Development: Protect or Restore Habitat), and SSc7.1 (Heat Island Effect: Non-Roof). These have been so noted in Table 2.0.1. There are many prerequisites or credits which might have variations applicable to the design or implementation of the project related to regional conditions in the United States. All the prerequisites and credits for which regional variations might be applicable are noted in the LEED 2009 Reference Guide and are so listed in Table 2.0.1 for the Sustainable Sites category.

Two of the credits listed previously have the icon EB noted after their title (SSc2 and SSc4.1). This icon is a tool to help those who wish to proceed with the continuing LEED-EB certification (Existing Building) after certification of LEED-NC (New Construction or Major Renovation) is obtained. Those credits noted with this icon are usually significantly more cost-effective to implement during the construction of the building than later during its operation. They are also shown in Table 2.0.1.

Subcategory 2009	Subcategory 2.2	Pts 2009	Pts 2.2	EP Pts 2009	EP Pts 2.2	Major Comments
SSp1: Construction Activity Pollution Prevention	SSp1: Construction Activity Pollution Prevention	na	na	na	na	–
SSc1: Site Selection	SSc1: Site Selection	1	1	–	–	–
SSc2: Development Density & Community Connectivity	SSc2: Development Density & Community Connectivity	5	1	1 (option 1)	1 (option 1)	Point value for similar criteria quintupled. In LEED 2009, for Community Connectivity, services may include 1 provided by the project for mixed-uses, and 2 "anticipated" within the radius if operational within a year of project occupancy.
SSc3: Brownfield Redevelopment	SSc3: Brownfield Redevelopment	1	1			–
SSc4.1: Alternative Transportation—Public Transportation Access	SSc4.1: Alternative Transportation—Public Transportation Access	6	1	1 (1 for overall plan or 1 for SSc4.1)	2 (1 for overall plan, 1 for SSc4.1)	Point value for similar criteria Sextupled.
SSc4.2: Alternative Transportation—Bicycle Storage and Changing Rooms	SSc4.2: Alternative Transportation—Bicycle Storage and Changing Rooms	1	1			In LEED 2009, Commercial and Institutional uses are referred to as Case 1, while Residential uses are referred to as Case 2.
SSc4.3: Alternative Transportation—Low-Emitting and Fuel-Efficient Vehicles	SSc4.3: Alternative Transportation—Low-Emitting and Fuel-Efficient Vehicles	3	1			Point value for similar criteria tripled. LEED 2009 has additional Option 4 to provide LEFE vehicle-sharing (>3% FTE) and other options are renumbered.
SSc4.4: Alternative Transportation—Parking Capacity	SSc4.4: Alternative Transportation—Parking Capacity	2	1			Point value for similar criteria doubled. 2009 addresses mixed-use pkg. requirements and does not provide alternative parking calculations to local code.

Table 2.0.0 SS Summary Table of Major Differences between Versions 2.2 and 2009 and Associated Points (Pts) (*Continued*)

Subcategory 2009	Subcategory 2.2	Pts 2009	Pts 2.2	EP Pts 2009	EP Pts 2.2	Major Comments
SSc5.1: Site Development—Protect or Restore Habitat	SSc5.1: Site Development—Protect or Restore Habitat	1	1	1	1	2009 has additional restrictions to protect or restore >20% of total site area for one point and >30% for EP point. Sites with both Greenfields and previously developed areas should follow requirements for each case.
SSc5.2: Site Development—Maximize Open Space	SSc5.2: Site Development—Maximize Open Space	1	1	1	1	–
SSc6.1: Stormwater Design—Quantity Control	SSc6.1: Stormwater Design—Quantity Control	1	1	1	–	2009 has EP point for plan to capture and treat stormwater beyond requirements.
SSc6.2: Stormwater Design—Quality Control	SSc6.2: Stormwater Design—Quality Control	1	1		–	See SSc6.1 for 2009 EP point. Treatment volume now has time of rain event as 24 hours.
SSc7.1: Heat Island Effect—Non-roof	SSc7.1: Heat Island Effect—Non-roof	1	1	1	1	2009 allows shading by architectural structures with SRI > 29 or solar panels.
SSc7.2: Heat Island Effect—Roof	SSc7.2: Heat Island Effect—Roof	1	1	1	1	–
SSc8: Light Pollution Reduction	SSc8: Light Pollution Reduction	1	1	–	–	2009 has interior lighting restrictions from 1 p.m. to 5 a.m. and allows 2 less restrictive options for doing so, requires exterior LPDs meet ASHRAE 90.1-2007 and addresses driveway exceptions.
	Total	26	14	7	7	

Table 2.0.0 SS Summary Table of Major Differences between Versions 2.2 and 2009 and Associated Points (Pts) (*Continued*)

Credit	Construction Submittal	EB Icon	Regional Variations May Be Applicable in LEED 2009	Location of Site Boundary or Area in Calculations
SSp1: Construction Activity Pollution Prevention	*	NA	Yes	Yes
SSc1: Site Selection			No	Yes
SSc2: Development Density and Community Connectivity		Yes	No	Yes
SSc3: Brownfield Redevelopment			Yes	No
SSc4.1: Alternative Transportation: Public Transportation Access		Yes	No*	Yes
SSc4.2: Alternative Transportation: Bicycle Storage and Changing Rooms			Yes	No
SSc4.3: Alternative Transportation: Low-Emitting and Fuel-Efficient Vehicles			Yes	No
SSc4.4: Alternative Transportation: Parking Capacity			No*	No
SSc5.1: Site Development: Protect or Restore Habitat	*		Yes	Yes
SSc5.2: Site Development: Maximize Open Space			No*	Yes
SSc6.1: Stormwater Management: Quantity Control			Yes	Yes
SSc6.2: Stormwater Management: Quality Control			Yes	Yes
SSc7.1: Heat Island Effect: Non-Roof	*		Yes	No
SSc7.2: Heat Island Effect: Roof			Yes	No
SSc8: Light Pollution Reduction			No	Yes

Table 2.0.1 SS EB Icon, Regional Variation Applicability, and Miscellaneous LEED 2009 (*Although local public transportation availability and code related parking and open space requirements do vary.)

Exemplary performance (EP) points under the Innovation and Design category are available, which relate to several items in the Sustainable Sites category. One is available for the SSc4 Alternative Transportation subcategory as a whole for an overall transportation management plan, and one is available in LEED 2009 for the SSc6 Stormwater Design subcategory as a whole for a stormwater management plan that covers both quantity and quality control beyond the requirements. Six additional point options are also available with one point related directly to six of the individual credits. These are available for exceeding each of the respective credit criteria to a minimum level and meeting other criteria as noted in the credit descriptions for SSc2, SSc4.1, SSc5.1, SSc5.2, SSc7.1, and SSc7.2. However, the EP point for SSc4.1 and the EP point for the overall SSc4 transportation plan cannot both count in LEED 2009. There are no EP points available for the other subcategories or credits in the Sustainable Sites category. The items available for EP points in this category are shown in Table 2.0.0 for both of the versions. Note that only a maximum total of four EP points are available in total for a project in LEED 2.2 and a maximum of three in LEED 2009 and they may be from any of the noted EP options in any of the SS, WE, EA, MR, or IEQ LEED categories (see Chap. 7).

Table 2.0.1 also has a column that notes the importance of the site boundary and site area in the credit calculations or verification. These are variables that should be determined early in a project as they impact many of the credits. Some of the credits are easier to obtain if the site area is less, while others are easier to obtain if the site area is greater. Determination of the site area may sometimes be flexible, particularly for campus locations, but it should be reasonable and must be consistent throughout the LEED process, as well as in agreement with other nearby project submissions. It is mentioned again in Chap. 8 as an important variable to analyze early and set in a project.

Land Area Definitions

Many of the Sustainable Sites credits relate to the site land area and the areas within the sites that are developed or not developed in certain ways (see Fig. 2.0.1). To understand this better, it is helpful to define the various portions of the site. Definitions as given by LEED-NC 2.2 for many of these site areas are given in App. B.

Let the following symbols represent the various areas of a site:

A_T	Total area of the lot (also referred to as the total project site area)
BF	Building footprint—the planar projection of the built structures onto the land
DF	Development footprint—total building and hardscape areas
EQR	Total roof areas that are covered by equipment and/or solar appurtenances and other appurtenances
HS	Hardscape: All nonbuilding manmade hard surface on the lot such as parking areas, walks, drives, and patios. Hardpacked areas such as gravel parking areas are included. Pools, fountains, and similar water-covered areas are included if they are on impervious surfaces which can be exposed when the water is emptied.
LSMAN	Manmade or graded landscaped area totals
LSNAT	Natural landscaped area totals

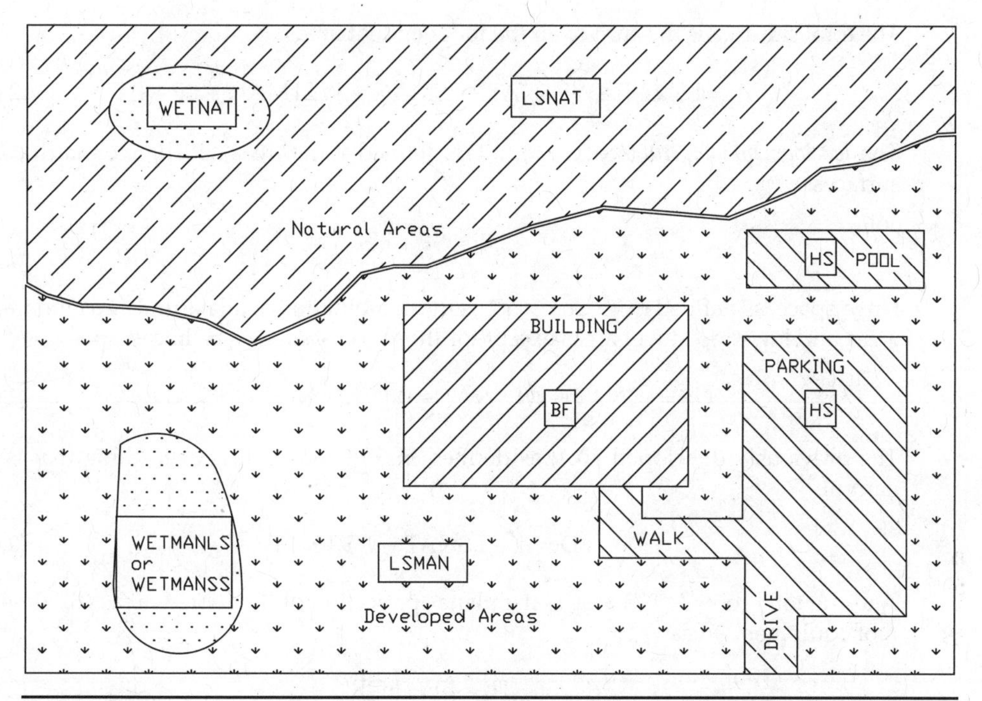

FIGURE 2.0.1 Various basic land areas for the Sustainable Sites calculations.

OS	Open space—all natural areas and landscaped areas, also wet areas with vegetated low slopes
PD	Previously developed—all areas that have been built upon, landscaped, or graded (not including areas that may have been farmed in centuries past and have returned to natural vegetation)
TR	Total roof—area defined as the sum of the horizontal projections of the roofs on the site plan less the sum of the roofed areas that hold equipment, solar energy panels, or other appurtenances (BF – EQR)
WETMAN	Total manmade wetland, ponds, or pool areas
WETMANLS	Total manmade wetland, ponds, or pools with vegetated low slopes (<25 percent) leading to its edges. This does not include pools and fountains with impervious bottoms as these are part of the hardscape.
WETMANSS	Total manmade wetland, ponds, or pools with any steep (>25 percent) or unvegetated slopes leading to its edges. This does not include pools and fountains with impervious bottoms as these are part of the hardscape.
WETNAT	Total natural wetlands, ponds, stream, or pool areas

The various areas relate to each other in the following ways. The total manmade wet areas are the sum of those with low-sloped edges and those which have some steep slopes.

$$\text{WETMAN} = \text{WETMANLS} + \text{WETMANSS} \tag{2.0.1}$$

The total site area is the sum of all the independent areas:

$$A_T = \text{LSNAT} + \text{WETNAT} + \text{LSMAN} + \text{WETMAN} + \text{BF} + \text{HS} \tag{2.0.2}$$

The *development footprint* (see SS Credit 1) is defined as the areas with hard or hardpacked surfaces:

$$\text{DF} = \text{BF} + \text{HS} \tag{2.0.3}$$

Open space, as defined in SS Credit 5.2, is the combination of all natural and landscaped areas and low-sloped wet areas except for those considered to be hardscape:

$$\text{OS} = \text{LSNAT} + \text{WETNAT} + \text{WETMANLS} + \text{LSMAN} \tag{2.0.4}$$

The previously developed area, as defined in SS Credit 5.1, is everything that is not natural:

$$\text{PD} = A_T - \text{LSNAT} - \text{WETNAT} \tag{2.0.5}$$

And the total roofed (TR) area is the planar projection of the roofed areas (BF) less the roof equipment areas:

$$\text{TR} = \text{BF} - \text{EQR} \tag{2.0.6}$$

Areas of natural ledge have not been specifically addressed, but if in the natural areas, these would most likely be included in the open space and excluded from the previously developed areas. If areas of natural ledge are in the developed areas, then they would most likely be included in the hardscape summation. Each case would be subject to interpretation.

Sustainable Sites Prerequisites

SS Prerequisite 1: *Construction Activity Pollution Prevention*

USGBC Rating System

This prerequisite is intended to control erosion and pollutant transport off-site during and after construction to minimize their impact on the surrounding areas and on water and air quality. LEED-NC 2.2 and LEED 2009 list the Intent, Requirements, and Potential Technologies and Strategies (2.2) for this credit as follows:

Intent

Reduce pollution from construction activities by controlling soil erosion, waterway sedimentation and airborne dust generation.

Requirements

Create and implement an Erosion and Sedimentation Control (ESC) Plan for all construction activities associated with the project. The ESC Plan shall conform to the erosion and sedimentation requirements of the EPA Construction General Permit OR local erosion and sedimentation control standards and codes, whichever is more stringent. The Plan shall describe the measures implemented to accomplish the following objectives:

- Prevent loss of soil during construction by stormwater runoff and/or wind erosion, including protecting topsoil by stockpiling for reuse.
- Prevent sedimentation of storm sewer or receiving streams.
- Prevent polluting the air with dust and particulate matter.

The Construction General Permit (CGP) outlines the provisions necessary to comply with Phase I and Phase II of the National Pollutant Discharge Elimination System (NPDES) program. While the CGP only applies to construction sites greater than 1 acre, the requirements are applied to all projects for the purposes of this prerequisite. Information on the EPA CGP is available at http://cfpub.epa.gov/npdes/stormwater/cgp.cfm.

Potential Technologies and Strategies

Create an Erosion and Sedimentation Control Plan during the design phase of the project. Consider employing strategies such as temporary and permanent seeding, mulching, earth dikes, silt fencing, sediment traps and sediment basins.

Calculations and Considerations

This strategy for construction phases of projects was commonly referred to as SEC (soil and erosion control) and now is more commonly known as ESC (erosion and sedimentation control). The current LEED-NC 2009 requirement is based on the USEPA Construction General Permit (CGP), for which information is available at http://cfpub.epa.gov/npdes/stormwater/cgp.cfm. It separates the practices into two overlapping categories, one for the prevention of erosion (stabilization) and the other for the containment or control of eroded sediment to other parts of the site or off-site. Some of the common stabilization methods used include the installation of geotextiles, temporary or permanent seeding or mulching on exposed soils, and installation of a construction vehicle drive. Typical sediment control measures include installation of silt fencing downslope of graded areas, installation of silt barriers around catch basins and other inlets to storm sewers or waterway access points, and installation of sedimentation ponds. The methods used are sometimes referred to as structural for actual items installed or nonstructural for other types of practices, such as scheduling grading activities during a drier season.

The steps for compliance for this credit include having and implementing an ESC Plan. Many of the structural and nonstructural measures used for ESC will be part of the project drawings. These project drawings will usually show the location of the methods on the site plans, list a material schedule for the measures, and include a time schedule for implementation, maintenance, and, if appropriate, removal of the control measure. The ESC Plan and this prerequisite adherence also apply to the demolition phase of the project and to the full extent of the construction, even areas outside of the LEED boundary.

The submittals for the credit include copies of the applicable project drawings and specifications addressing ESC, a statement as to whether the project adheres to the national EPA criteria or a stricter local standard, and a narrative outlining the ESC methods implemented. When a local standard is used, it must be demonstrated in the narrative that it is at least as strict as the NPDES standard. There are no required calculations for this credit other than those that might be necessary to verify that the ESC Plan adheres to the minimum standards required. The submittal is part of the LEED Construction Submittal, as most of the items are a part of the construction phase and adherence to the requirements should be documented.

Many states now have certification programs for ESC professionals, and a listing of these can be found at www.cpesc.net. This is a good resource for a review of common terminology and measures used for ESC in the United States. Additional information on erosion and sediment control can be found in Chap. 10.

2.1 SS Credit 1: *Site Selection*

USGBC Rating System

The intention of this credit is to minimize the impact of building construction on natural resources both on-site and nearby. This credit was worth one point in LEED 2.2 and is worth one point in LEED 2009. LEED-NC 2.2 and 2009 list the Intent, Requirements, and Potential Technologies and Strategies (2.2) for this credit as follows:

Intent

Avoid development of inappropriate sites and reduce the environmental impact from the location of a building on a site.

Requirement

Do not develop buildings, hardscape, roads or parking areas on portions of sites that meet any one of the following criteria:

- Prime farmland as defined by the United States Department of Agriculture in the United States Code of Federal Regulations, Title 7, Volume 6, Parts 400 to 699, Section 657.5 (citation 7CFR657.5).
- Previously undeveloped land whose elevation is lower than 5 feet above the elevation of the 100-year flood as defined by the Federal Emergency Management Agency (FEMA).
- Land which is specifically identified as habitat for any species on Federal or State threatened or endangered lists.
- *Land* within 100 feet of any wetlands as defined by United States Code of Federal Regulations 40 CFR, Parts 230–233 and Part 22, and isolated wetlands or areas of special concern identified by state or local rule, OR within setback distances from wetlands prescribed in state or local regulations, as defined by local or state rule or law, whichever is more stringent.
- Previously undeveloped land that is within 50 feet of a water body, defined as seas, lakes, rivers, streams, tributaries which support or could support fish, recreation or industrial use, consistent with the terminology of the Clean Water Act.
- Land which prior to acquisition for the project was public parkland, unless land of equal or greater value as parkland is accepted in trade by the public landowner (Park Authority projects are exempt).

Potential Technologies and Strategies

During the site selection process, give preference to those sites that do not include sensitive site elements and restrictive land types. Select a suitable building location and design the building with the minimal footprint to minimize site disruption of those environmentally sensitive areas as identified.

Calculations and Considerations

This credit tries to avoid disturbing pristine or special natural resources by selectively locating buildings and the development footprint away from these resources or developing on previously developed areas of the site. To obtain this credit, all six of the criteria must be met. Two important definitions to understand are those of development footprint and previously undeveloped land. *Development footprint* as defined by LEED-NC 2.2 can be found in App. B, as well as the definition of *previously developed sites*, which are assumed to be the opposite of *previously undeveloped land*. The differences between the two are as follows: the development footprint, as defined, does not include landscaped areas, whereas previously developed sites as defined include areas graded or altered by direct human activities which may include landscaped areas.

There are no calculations for this credit other than as appropriate to verify distances, elevations, and setbacks. Confirmation that all six criteria are met must be submitted. The second and fifth criteria are applicable only if the land in question was previously undeveloped, but again this must be shown. There is room for "special circumstances" or "nonstandard compliance paths" in this credit, but these must be described.

Also of note is the first criterion which refers to prime farmland. Not all farmland or potential farmland is prime. The U.S. Department of Agriculture (USDA) definition of *prime agricultural land* as stated in the applicable Code of Federal Regulations (CFR) part is as follows:

> Prime farmland is land that has the best combination of physical and chemical characteristics for producing food, feed, forage, fiber, and oilseed crops, **and is also available for these uses** (the land could be cropland, pastureland, rangeland, forest land, or other land, but not urban built-up land or water). It has the soil quality, growing season, and moisture supply needed to economically produce sustained high yields of crops when treated and managed, including water management, according to acceptable farming methods. In general, prime farmlands have an adequate and dependable water supply from precipitation or irrigation, a favorable temperature and growing season, acceptable acidity or alkalinity, acceptable salt and sodium content, and few or no rocks. They are permeable to water and air. Prime farmlands are not excessively erodible or saturated with water for a long period of time, and they either do not flood frequently or are protected from flooding. Examples of soils that qualify as prime farmland are Palouse silt loam, 0 to 7 percent slopes; Brookston silty clay loam, drained; and Tama silty clay loam, 0 to 5 percent slopes.

The second criterion relates to designated floodplain elevations. The 100-year floodplain is usually as designated on the FEMA Flood Insurance Rate Maps (FIRMs). A 100-year flood is not one that occurs every 100 years, but rather is a flood with a 1 percent probability whose magnitude or greater might occur every year. If the FEMA flood lines are not available, then the Army Corps of Engineers or other recognized authority maps may be substituted. There are some exceptions to building in a floodplain for projects which have some redevelopment in the floodplain area. Another exception is that areas can be "elevated" above the 5-ft minimum if both the 100-year floodplain is not impacted and there are balanced cuts and fills in these areas so as not to impact storage volumes. In LEED-NC 2.1 a credit interpretation ruling (CIR) allowed for the definition of the facilities which needed to be at least 5 ft above the elevation of the 100-year floodplain to be only those included in the building footprint. As the version 2.2 verbiage is more lenient than that of version 2.1 (version 2.1 did not specifically exempt previously developed land), it is reasonable to assume that this definition is also applicable to version 2.2, although it is not specifically mentioned in the LEED 2009 reference guide.

As with the floodplain criterion, already existing facilities that are located within the setbacks as defined for wetlands can remain. Maintenance and minor improvements on these nonconforming existing facilities would normally be appropriate, whereas major renovations or additions would not. Each facility should be examined individually for conformance and intent.

Wetlands are defined by the Code of Federal Regulations 40 CFR Parts 230–233 and Part 22: "Wetlands consist of areas that are inundated or saturated by surface or ground water at a frequency and duration sufficient to support, and that under normal circumstances do support, a prevalence of vegetation typically adapted for life in saturated soil conditions."

The definition of *water body* also needs clarification. Per the LEED 2.2 errata posted by the USGBC in the spring of 2007, small manmade ponds, such as those used in stormwater retention, fire suppression, and recreation, are excluded, whereas manmade wetlands and other water bodies created to restore natural habitat and ecological systems are not exempt.

Glazing on walls that face wetlands and other natural areas may disorient birds. Designers should include special features into such glazed surfaces to avoid these problems.

Special Circumstances and Exemplary Performance

There is no EP point available for this credit.

2.2 SS Credit 2: *Development Density and Community Connectivity*

USGBC Rating System

The intention of this credit is to infill or revitalize urban areas that already have supportive utility, transportation, and service infrastructure, or to increase services near denser residential communities close to other supporting services such as would be found in an urban or dense suburban area. This credit has an EB notation. This credit was worth one point in LEED 2.2 and is worth five points in LEED 2009. LEED-NC 2.2 lists the Intent, Requirements, and Potential Technologies and Strategies for this credit as follows, with noted additions for LEED 2009:

Intent

Channel development to urban areas with existing infrastructure, protect greenfields and preserve habitat and natural resources.

Requirements

OPTION 1—DEVELOPMENT DENSITY

Construct or renovate building on a previously developed site AND in a community with a minimum density of 60,000 square feet per acre net. (Note: density calculation must include the area of the project being built and is based on a typical two-story downtown development.)

OR

OPTION 2—COMMUNITY CONNECTIVITY

Construct or renovate building on a previously developed site AND within ½ mile of a residential zone or neighborhood with an average density of 10 units per acre net AND within ½ mile of at least 10 Basic Services AND with pedestrian access between the building and the services.

Basic Services include, but are not limited to: 1) Bank; 2) Place of Worship; 3) Convenience Store; 4) Day Care; 5) Cleaners; 6) Fire Station; 7) Beauty 'shop'; 8) Hardware; 9) Laundry; 10) Library; 11) Medical/Dental; 12) Senior Care Facility; 13) Park; 14) Pharmacy; 15) Post Office; 16) Restaurant; 17) School; 18) Supermarket; 19) Theater; 20)

Community Center; 21) Fitness Center; 22) Museum. Proximity is determined by drawing a ½ mile radius around the main building entrance on a site map and counting the services within the radius. (Typically the service provided by the proposed project may not be included, except for mixed-use projects, for which one of the project services may count toward the minimum 10 in LEED 2009. In addition, LEED 2009 allows for two "anticipated" services within the radius to count, if there is proof they will be operational within a year of occupancy of the project. Note that any service not specifically as listed in the requirements will be considered only on a case-by-case basis.)

Potential Technologies and Strategies

During the site selection process, give preference to urban sites with pedestrian access to a variety of services.

Calculations and Considerations

Option 1 The submittal requirements for Option 1 of this credit have several calculations associated with them. The minimum development density must be met *both* for the site in question (the development density of the site DD_{site}) and for the surrounding area (the development density of the surrounding area DD_{area}) that is within the density boundary (DB) as defined by the density radius (DR). The following set of symbols is used in the equations, with GFA referring to gross floor area as defined through the errata as posted by the USGBC in the spring of 2007 (see the definitions in App. B):

A_T	Total land area of the project lot or site
A_{Ti}	Total land area of a lot i
DD_{site}	Development density of the site [usually given in square foot per acre (ft²/acre)]
DD_{area}	Development density of the area within the density boundary as determined by the density radius for the density boundary
DDD_{area}	Development density of double the surrounding area as determined by the density radius for double (DRD) the density boundary area (EP point variable)
DR	Density radius for the density boundary area (usually given in feet)
DRD	Density radius for double the density boundary area (usually given in feet) (EP point variable)
GFA_i	Gross floor area of building associated with lot i (square feet)
GFA_P	Gross floor area of the subject project (square feet)

One point can be earned for Option 1 if the inequalities in Eqs. (2.2.1) and (2.2.2) are true, given the definition of the density radius in Eq. (2.2.3):

$$DD_{site} = (GFA_P/A_T) \geq 60{,}000\ \text{ft}^2/\text{acre} \tag{2.2.1}$$

$$DD_{area} = \frac{\sum GFA_i}{\sum A_{T_i}} \geq 60{,}000\ \text{ft}^2/\text{acre} \quad \text{for all applicable buildings and lots within DR including the subject project} \tag{2.2.2}$$

$$DR = 3 \times \sqrt{A_T} \tag{2.2.3}$$

Based on a CIR dated September 22, 2006 and included in LEED 2009, an exemplary performance point can be earned over and above Option 1 for a project which meets one of the two following requirements after all the requirements of Option 1 are met:

- The project itself must have a density of at least double that of the average density within the calculated area [see Eq. (2.2.4)].

or

- The average density within an area twice as large as that for the base credit achievement must be at least 120,000 ft^2/acre. [To double the area, use Eq. (2.2.3), but double the property area first.]

Therefore, to obtain an EP point for SSc2 in addition to the point earned for Option 1, one of the following two options must be valid. Either the inequality in Eq. (2.2.4) must be achieved given the development density of the area as defined in Eq. (2.2.2)

$$DD_{site} = (GFA_P/A_T) \geq 2 = DD_{area} \tag{2.2.4}$$

or both Eqs. (2.2.5) and (2.2.6) must hold. These are based on the two additional variables the development density of double the surrounding area (DDD_{area}) as defined by the density radius for double the area (DRD).

$$DDD_{area} = \frac{\sum GFA_i}{\sum A_{T_i}} \geq 120{,}000 \text{ ft}^2\text{/acre} \quad \text{for all applicable buildings and lots within the DRD, including the subject project} \tag{2.2.5}$$

$$DRD = 3 \times \sqrt{2 \times A_T} \tag{2.2.6}$$

There are several additional definitions of the variables involved that must be explained. The following definitions are per LEED-NC version 2.2 when noted in quotation marks:

Project site area "For projects that are part of a larger property (such as a campus), define the project area as that which is defined in the project's scope. The project area must be defined consistently throughout LEED documentation." This will typically be represented as A_T for most of the credit equations.

Density radius The density radius (DR) is determined by Eq. (2.2.3). A circle of a radius equal to the density radius is then drawn on an area map with its center near the center of the project site.

Density boundary This is the area within the circle as determined by drawing the density radius. "For each property within the density boundary and for those properties that intersect the density boundary, create a table with the building square footage (GFAs) and site area of each property. Include all properties in the density calculations except for undeveloped public areas such as parks and water bodies. Do not include public roads and right-of-way areas." Include the subject project, and include site areas of vacant lots except as previously noted. These GFAs and site areas are used in the summations in the numerator and denominator, respectively, of Eq. (2.2.2). In general the GFA of a building is all the interior space taken from outer wall to outer wall and summed over all the floors, not including unfinished attic/dormer space; and the GFA has also been defined in the errata posted by the USGBC in the spring of 2007 to include up to two stories of parking, but not single-story surface parking [definitions are given in App. B as *site area* and *square footage* of a building (GFA)].

As can be seen in Eq. (2.2.2), the development density of the surrounding area DD_{area} is calculated as the sum of all the GFAs divided by the sum of all the applicable site areas. A common mistake is to take the average of individual development densities for the properties within the density boundary. This type of average would not be appropriately weighted for the sizes of each property and facility and is not correct. The submittals to the USGBC would not allow this error to be made, but it may be made in early estimates not using the templates and is therefore emphasized here.

All the listed equations and tabulations and a copy of the area map with the density boundary drawn must be submitted along with the template for validation of Option 1 of this credit. Figure 2.2.1 shows an example density boundary area map for this credit.

Some exceptions do exist for the numbers in Eqs. (2.2.1) through (2.2.3). For instance, let us say that the circle drawn intersects a very small portion of a large lot

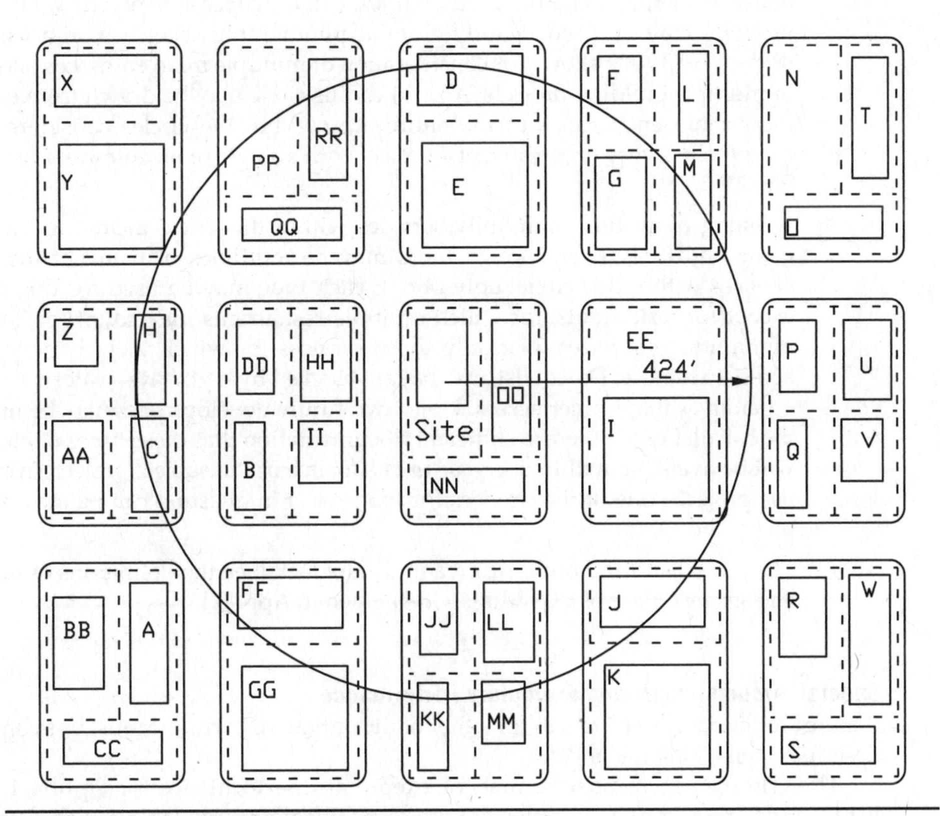

FIGURE 2.2.1 Density boundary area map.

that is not well developed. This small exception at the edge of the density boundary may be the cause for an exception to the summation of area densities, especially if the density is appropriate in other areas outside the circle. This might still meet the intent of the credit. The site itself may sometimes not meet the full density criterion, but if perhaps taken together with other neighboring campus facilities with more than adequate densities, it may still meet the intent. Existing single-family housing and historic sites may not have to be included in the calculations as they are usually considered to be valuable assets to a community. Stadiums may also be exempt if they exist in pedestrian-friendly urban areas. In addition, future development may be used in the calculations if there is sufficient documentation that these types of facilities will be built, but this has not been specifically addressed in the LEED 2009 Reference Guide.

Option 2 The submittal requirements for Option 2 of credit SSc2 include the following:

- A site vicinity map showing the parcel and building in question (project site) with a ½-mi-radius circle drawn from the main entrance, or the drawing scale noted on the map. The location of all services within this circle shall be shown. The location of the residential neighborhoods shall also be shown with the dense residential neighborhood of at least 10 units per acre noted. Many types of maps may be used including aerial photographs, sketches, and assessor maps. LEED 2009 allows for consideration of multiple main entrances on single or multiple building projects. A ½-mi-radius circle may be drawn from each of these main entrances and the entire area within the circles considered. An example in a very urban area of a project with a single main entrance is given in Fig. 2.2.2.
- A listing of all the community services within this circle including business name and type. Even though there may be multiples of many of the basic services within this circle, only one of each type may be used for this credit, except for restaurants, for which multiple restaurants may count for 2 of the minimum 10 services. List only those services for which there is pedestrian access to the site. Do not list any that are blocked by highways, waterways, and other areas that hinder access. Up to two future developments may be used in the calculations if there is sufficient documentation that these types of facilities will be available within one year of occupance of the subject project. In mixed use projects, one of the services for the project in question can be included as one of the 10.
- The project site and building area in square feet. [See the definitions of *site area* and *square footage* of a building (GFA) given in App. B.]

Special Circumstances and Exemplary Performance

If there are special circumstances for either of the options, a narrative must be submitted describing these circumstances.

The criteria which must be met for exemplary performance for Option 1 were listed under the Option 1 section. There is no EP point available for SS Credit 2 Option 2.

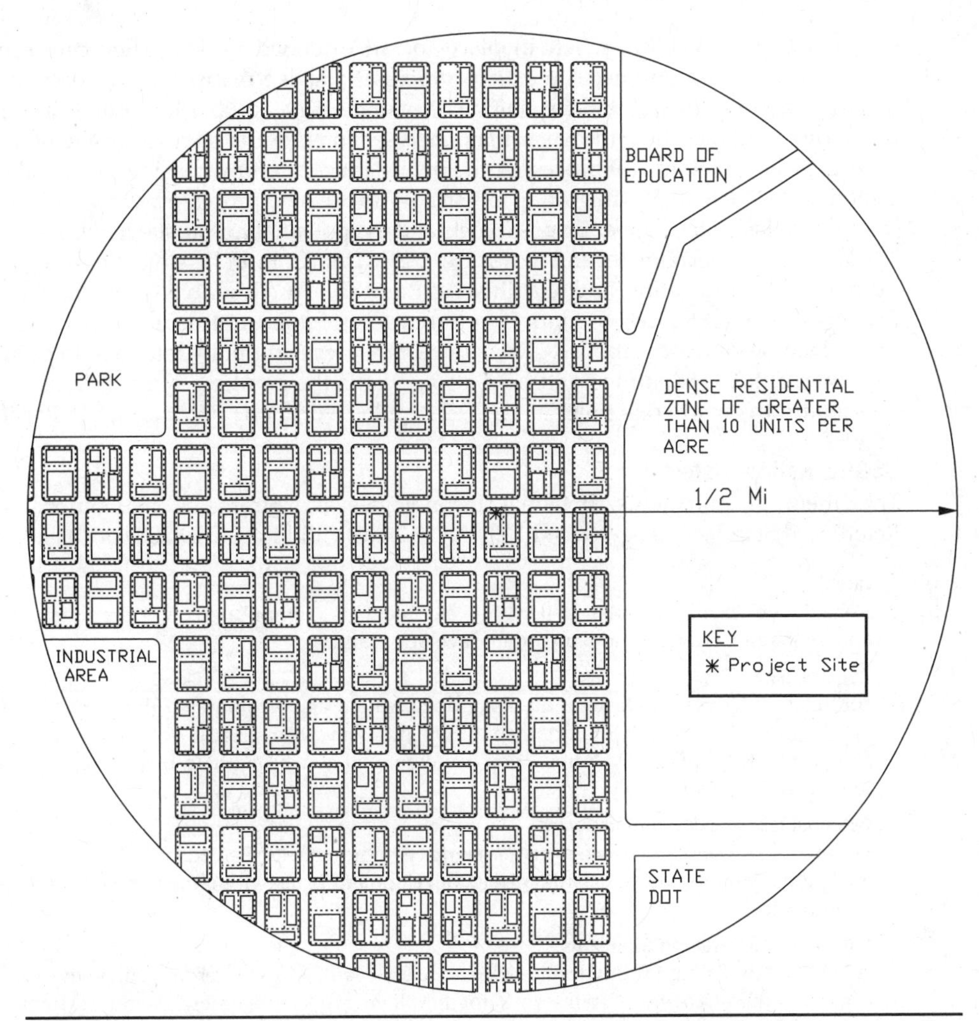

FIGURE 2.2.2 Area map for community connectivity. Locations of all applicable services must also be shown.

2.3 SS Credit 3: *Brownfield Redevelopment*

The intent of this credit is twofold: to reduce developing on undeveloped land and to appropriately redevelop those sites listed as brownfields. The site is first evaluated via a site assessment to determine if there are complications by environmental contamination. Brownfields can be designated for both "real" and "perceived" contamination. A risk assessment is also used to determine if and what remediation may be needed as applicable for proposed development. Remediation means the removal of hazardous materials from the groundwater or the soils. Typical remediation methods include in situ (in place) or ex situ (off-site) technologies. Sometimes the contaminated soils or waters remain on-site and are remediated with facilitated on-site processes such as "pump and treat" or are left in place and allowed to be treated by natural attenuation methods if their presence is not expected to be a liability if left in place for an extended period. They can also be encapsulated if appropriate. The ex situ processes involve many types of options but must be carefully manifested for shipment off-site. It is possible to become liable for pollution on another site if the waste material is sent off-site and not properly remediated or disposed of. With in situ processes, the owner must carefully consider the liability of contaminants potentially migrating off-site. With ex situ processes, the owner must carefully consider the proper disposal costs and the potential for off-site liabilities.

The two applicable federal laws are CERCLA (Comprehensive Environmental Response, Compensation and Liability Act) and RCRA (Resource Conservation and Recovery Act). CERCLA, commonly referred to as Superfund, is a program to help offset the costs of remediation. RCRA is the act that was originally intended to promote recycling, but was expanded and now also governs the handling of hazardous waste. This credit was worth one point in LEED 2.2 and is worth one point in LEED 2009.

USGBC Rating System

The Intent, Requirements, Potential Technologies and Strategies, and Summary of Referenced Standards as given in LEED-NC 2.2 and valid for LEED 2009 are as follows:

Intent

Rehabilitate damaged sites where development is complicated by environmental contamination, reducing pressure on undeveloped land.

Requirements

(LEED 2009 Option 1) Develop on a site documented as contaminated (by means of an ASTM E1903-97 Phase II Environmental Site Assessment or a local Voluntary Cleanup Program) OR (LEED 2009 Option 2) on a site defined as a brownfield by a local, state or federal government agency.

Potential Technologies and Strategies

During the site selection process, give preference to brownfield sites. Identify tax incentives and property cost savings. Coordinate site development plans with remediation activity, as appropriate.

Summary of Referenced Standards

ASTM E1903-97 Phase II Environmental Site Assessment, ASTM International, www.astm.org. This guide covers a framework for employing good commercial and customary practices in conducting a Phase II environmental site assessment of a parcel of commercial property. It covers the potential presence of a range of contaminants that are within the scope of CERCLA, as well as petroleum products.

EPA Brownfields Definition

EPA Sustainable Redevelopment of Brownfields Program, www.epa.gov/brownfields. With certain legal exclusions and additions, the term "brownfield site" means real property, the expansion, redevelopment, or reuse of which may be complicated by the presence or potential presence of a hazardous substance, pollutant, or contaminant (source: Public Law 107–118, H.R. 2869 – "Small Business Liability Relief and Brownfields Revitalization Act"). See the web site for additional information and resources.

Calculations and Considerations

The required submittals include confirmation that the project site was established as being contaminated either by an ASTM E1903-97 Phase II Environmental Site Assessment or by a local, state, or federal government agency, and a detailed narrative which describes the contamination and the remediation actions being taken. A method or means to monitor the cleanup effort should also be described. There is no need to submit the entire Phase II document; the executive summary is sufficient. No minimum level of contamination is specified. The contamination can be widespread or localized such as surface asbestos contamination or confined leakage from an underground storage tank (UST).

Special Circumstances and Exemplary Performance

There is no EP point available for credit SSc3.

2.4 SS Credit Subcategory 4: *Alternative Transportation*

This credit subcategory has a total of four subsections, each worth one credit point in LEED 2.2, but worth varying points in LEED 2009. The main intention of all four Alternative Transportation credits is to reduce our dependency on single-occupancy vehicles (SOVs), particularly those which use gasoline and petroleum diesel fuels. The goals are to reduce the need for building more road infrastructure, reduce motor vehicle pollution, decrease the dependency on foreign oil, and reduce the vast amount of paved surfaces found in urban areas that may contribute to nonpoint source pollution from stormwater runoff and the urban heat island effect. The four subsections (credits) are as follows:

- **SS Credit 4.1** (SSc4.1): *Alternative Transportation: Public Transportation Access* (EB)
- **SS Credit 4.2** (SSc4.2): *Alternative Transportation: Bicycle Storage and Changing Rooms*
- **SS Credit 4.3** (SSc4.3): *Alternative Transportation: Low-Emitting and Fuel-Efficient Vehicles* (previously referred to as *Alternative Transportation, Alternative Fuel Vehicles* in LEED NC 2.1)
- **SS Credit 4.4** (SSc4.4): *Alternative Transportation: Parking Capacity*

There are two possible exemplary performance points available under Alternative Transportation for both LEED 2.2 and LEED 2009, but only one of them may be counted toward certification in LEED 2009. The EP point available in the Innovation and Design category for Alternative Transportation as per the LEED-NC 2.2 first edition may be awarded by "instituting a comprehensive transportation management plan that demonstrates a quantifiable reduction in personal automobile use through the implementation of multiple alternative options." It is related to the overall subcategory, and not to each individual credit. In addition, another EP point can be earned according to the following language from a USGBC CIR dated September 11, 2006, which was then also included in LEED 2009:

> Based on evidence that locations with higher transit density can achieve substantially and a quantifiably higher environmental benefit, meeting the following threshold qualifies a project for exemplary performance Innovation Credit. This follows the Center for Clean Air Policy's finding that average transit ridership increases by 0.5% for every 1.0% increase in growth of transit service levels, which leads to the conclusion that quadrupling transit service generally doubles transit ridership.
>
> To accomplish this quadrupling of service and doubling of ridership, at a minimum:
>
> - Locate the project within ½ mile of at least two existing commuter rail, light rail, or subway lines, OR locate project within ¼ mile of at least two or more stops for four or more public or campus bus lines usable by building occupants;

AND

- Frequency of service must be such that at least 200 transit rides per day are available in total at these stops. A combination of rail and bus is allowable. This strategy is based on the assumption that the threshold of the base credit would provide, in most cases, at least 50 transit rides per day (half-hourly service 24 hours per day or more frequent service for less than 24 hours per day). If, on average, transit ridership increases by 0.5 percent for every 1.0% increase in transit service, then quadrupling the number of rides available would, on average, double the transit ridership (4 × 50 rides = 200 rides). Include a transit schedule and map within your LEED certification submittal.

A summary of the credit criteria and exemplary performance criteria for the entire Alternative Transportation subcategory is given in Table 2.4.1.

SS Credit 4.1: *Alternative Transportation—Public Transportation Access*

USGBC Rating System

The first Alternative Transportation credit, 4.1, provides for access to public transportation as an alternative to SOVs. This credit is noted with the icon EB as it may be very costly to provide these services at a site where they are not already available. The applicable forms of public transport are commuter-type rail service and bus lines. The credit was worth one point in LEED 2.2 and is worth six points in LEED 2009 for essentially the same criteria. LEED-NC 2.2 lists the Intent, Requirements, and Potential Technologies and Strategies for this credit as follows, with minor modifications as noted for LEED 2009:

Intent

Reduce pollution and land development impacts from automobile use.

Requirements

LEED 2009 Option 1: Locate project within ½ mile of an existing—or planned and funded—commuter rail, light rail, or subway station (measured from a main building entrance as per LEED 2009).

OR

LEED 2009 Option 2: Locate project within ¼ mile of one or more stops for two or more public or campus bus lines usable by building occupants (measured from a main building entrance as per LEED 2009).

Potential Technologies and Strategies

Perform a transportation survey of future building occupants to identify transportation needs. Site the building near mass transit.

Calculations and Considerations

To confirm adherence to this credit, a vicinity map should be provided showing the project location and the location of the applicable rail/subway station (Option 1) or the applicable bus stops (Option 2). The ½-mi or ¼-mi distances required, respectively, are usually not a straight line from the property site, but are rather an appropriate pedestrian pathway from the site to the public transportation station or stop. These pathways must be shown on the vicinity map. The submittal also must include a listing of the stations/stops and the pedestrian distances traveled to get there from the site. The rail or subway station must be for use by commuters. The bus lines may or may not be from the same company or agency; the important thing is that they are separate lines and usable by the

Alternative Transportation Credit Section	Case (Uses*)	Option and/or EP	Minimum Criteria
Entire Category	Any	EP	Comprehensive Transportation Management Plan
4.1: *Public Transportation Access*	Any	Opt 1	<½ mi to a commuter rail, light rail, or subway station
4.1: *Public Transportation Access*	Any	Opt 2	<¼ mi to 1 or more stops for 2 public or campus bus lines usable by occupants
4.1: *Public Transportation Access*	Any	Opt 1-EP	<½ mi to 2 commuter rail, light rail, or subway stations and 200 transit rides/day (rail and/or bus)
4.1: *Public Transportation Access*	Any	Opt 2-EP	<¼ mi to 1 or more stops for 4 public or campus bus lines usable by occupants and 200 transit rides/day (rail and/or bus)
4.2: *Bicycle Storage and Changing Rooms*	1: Commercial or institutional	—	Showers = 0.005 × FTE and BR = 0.05 × (Maximum PBU_j) (increase BR for shift overlap)
4.2: *Bicycle Storage and Changing Rooms*	2: Residential uses	—	BSF = 0.15 × (DO)
4.3: *Low-Emitting and Fuel-Efficient Vehicles (LEFEVs) or Alternative Fuel Vehicles (AFVs)*	Any	Opt 1	LEFEVP1 = 0.05 × TP (the number of LEFEV preferred parking spaces provided) or provide minimum 20% discount on parking rates for all LEFEV vehicles
4.3: *Low-Emitting and Fuel-Efficient Vehicles or Alternative Fuel Vehicles*	Any	Opt 2	AFV = 0.03 × TP (the AFV dispensing capability)
4.3: *Low-Emitting and Fuel-Efficient Vehicles or Alternative Fuel Vehicles*	Any	Opt 3	LEFEVP3 = 0.03 × FTE (both the number of LEFEV vehicles provided and the number of LEFEV preferred parking spaces provided)
4.3: *Low-Emitting and Fuel-Efficient Vehicles or Alternative Fuel Vehicles*	Any	Opt 4	LEFEVP4 = FTE/267 (both the number of shared LEFEV vehicles and the preferred parking spaces provided)
4.4: *Parking Capacity*†	1: Nonresidential	Opt 1	Do not exceed minimum parking and CVPP1 = 0.05 × TP (Van/carpool preferred parking provided)

Table 2.4.1 Summary of Alternative Transportation Credits and Exemplary Performance (EP) Points (*Continued*)

Alternative Transportation Credit Section	Case (Uses*)	Option and/or EP	Minimum Criteria
4.4: *Parking Capacity*†	1: Nonresidential	Opt 2	If TP < 0.05 × FTE, CVPP2 = 0.05 × TP (Van/carpool preferred parking provided)
4.4: *Parking Capacity*	1: Nonresidential	Opt 3	No new parking
4.4: *Parking Capacity*†	2: Residential	Opt 1	Do not exceed minimum parking and have ride-sharing programs/ infrastructure
4.4: *Parking Capacity*	2: Residential	Opt 2	No new parking

*For uses other than K–12 schools.
†For mixed uses adhere to the requirements for each use except for mixed residential buildings with less than 10 percent of the floor area commercial/retail, which will then be considered residential.

Table 2.4.1 Summary of Alternative Transportation Credits and Exemplary Per formance (EP) Points (*Continued*)

building occupants. Figure 2.4.1 shows an example area map which fulfills both options, although only one is needed for the credit. Note that if a building has multiple main entrances, the requirement may be met by meeting the minimum pedestrian distance from any of the main entrances for either of the options as per LEED 2009.

In the case of Option 1, note that if a train station is not available, but is *both* planned and funded, then this will count. If something is funded, then there is a high likelihood that it will actually be built in the near future.

In either case, a permanent, private shuttle may also be established to the bus or train stops to potentially qualify if the minimum distance requirements cannot be met. Details, accessibility, and availability should be described. LEED 2009 requires that this private shuttle operate during the most frequent commute hours and be available to the building occupants.

Special Circumstances and Exemplary Performance

As noted previously, an exemplary performance (EP) point based on SSc4.1 was approved in a Credit Ruling dated September 11, 2006 and is also applicable to LEED 2009. To obtain this additional point, there must be double the number of either rail or bus lines within the requisite distances and there must be a combined frequency of service such that at least 200 transit rides per day are available at these stops; a combination of rail and bus is allowable. Transit rides available per day are not the seats, but the number of buses or rails that stop within the requisite areas daily. For instance, for at least 50 transit rides per day, it would be just less than half-hourly service 24 h/day or more frequent service for less than 24 h/day. Transit schedules and maps must be included in the LEED certification submittal. This EP point cannot be counted toward certification in LEED 2009 if the overall transportation plan EP point has been applied.

If there are special circumstances for either of the options, a narrative must be submitted describing these circumstances.

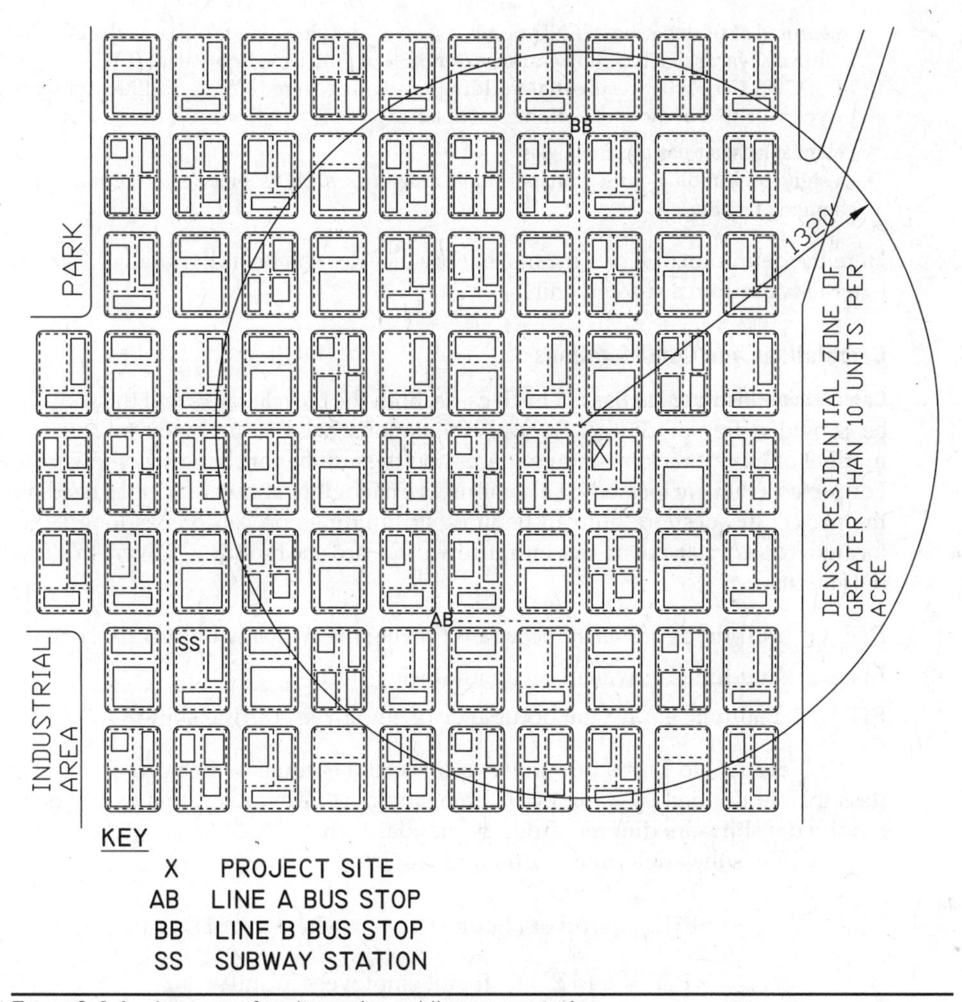

Figure 2.4.1 Area map for alternative public transportation.

SS Credit 4.2: *Alternative Transportation—Bicycle Storage and Changing Rooms*

USGBC Rating System

This credit helps facilitate bicycle transportation as an alternative to SOV transportation. It was worth one point in LEED 2.2 and is worth one point in LEED 2009. LEED-NC 2.2 lists the Intent, Requirements, and Potential Technologies and Strategies for this credit as follows, with minor modifications for LEED 2009:

Intent

Reduce pollution and land development impacts from automobile use.

Requirements

LEED 2009 Case 1: For **commercial or institutional buildings**, provide secure bicycle racks and/or storage (within 200 yards of a building entrance) for 5% or more of all building users

(measured at peak periods), AND provide shower and changing facilities in the building or within 200 yards of a building entrance, for 0.5% of Full-Time Equivalent (FTE) occupants.
LEED 2009 Case 2: For **residential buildings**, provide covered storage facilities for securing bicycles for 15% or more of building occupants.

Potential Technologies and Strategies

Design the building with transportation amenities such as bicycle racks and shower/changing facilities.

Here the definition of *secure* means that the bicycles can be individually locked such as to a rack or within a storage unit.

Calculations and Considerations

Commercial/Institutional Uses In all cases, both the bicycle racks and the showers must be provided free of charge. Several calculations need to be made to determine the number of bicycle rack or storage spaces and the number of changing/shower facilities. For projects that are located on a campus, the changing/shower facilities need not be in the project in question, but can be in other buildings on campus as long as they are located within 200 yd of the entrance to the project building. Given the following definitions:

FTE Full-time equivalent occupant (during the typically busiest part of a day)

FTE_j Full-time equivalent occupant during shift *j*

$FTE_{j,i}$ Full-time equivalent occupancy of employee *i* during shift *j*

$FTE_{j,i}$ is equal to 1 for a full-time employee and is equal to the normal hours worked (less than 8) divided by 8 for a part-time employee. Obviously, this would need to be modified if shifts are different from the standard 8 h.

The FTE is then determined from the following set of equations:

$$FTE_{j,i} = (\text{worker } i \text{ hours})/8\text{ h} \qquad \text{where } 0 < FTE_{j,i} \leq 1 \tag{2.4.1}$$

$$FTE_j = \Sigma\, FTE_{j,i} \qquad \text{for all employees in shift } j \tag{2.4.2}$$

$$FTE = \text{maximum}\,(FTE_j) \tag{2.4.3}$$

This FTE is used to determine the number of changing/shower facilities (Showers) which must be provided for any occupant wishing to bicycle to the location:

$$\text{Showers} = 0.005 \times FTE \tag{2.4.4}$$

The number of showers must always be a whole number, and any fractional results must be rounded up to the next-higher whole number.

The calculations get more complex for the number of secure bicycle racks and/or storage spaces that must be provided (bicycle rack spaces) since many facilities service not only the regular employees but also many others who may frequent the facility, such as clients, customers, or students in an educational facility; and for many industrial applications there are shift overlaps. LEED 2009 specifically mentions that facilities for shift overlap should be considered. Bicycle-riding-related showering facilities would usually not be impacted by shift overlap, but if the showers are used at the end of shifts for other reasons, then obviously this would need to be considered. However, the number of bicycle rack spaces should be increased for the applicable FTE overlap from the peak shift. These additional users are referred to as *transient occupants*. Together the FTE_j and the transient occupants TO_j can be used to determine the peak occupancy. LEED 2009 also allows for some discretion in including transient occupants in certain cases where it would not be reasonable for bicycle use, such as for airline passengers at an airport.

There may be some confusion as to the definition of the peak occupancy for transient occupants. The LEED-NC 2.2 reads as follows: "Estimate the transient occupants such as students, visitors and customers for the **peak period** for the facility." This does not mean to give maximum occupancy, but rather the typical transient occupancy during a busy part of the day. This may be difficult in the cases of facilities with very erratic attendance such as stadiums and concert halls. For instance, a typical baseball game is probably attended by a much smaller crowd than the maximum seating capacity. Using the maximum capacity based on the dozen or so times a year when there are tournaments or special games does not necessarily constitute the intent of this credit, just as popular concerts do not constitute typical occupancies for an auditorium. A reasonable evaluation might be based on the number of games with average attendance and perhaps the number of practice days with typical occupancies during those periods such as team members, trainers, spectators, and other staff. Then an averaging of these values based on frequencies should be attempted to give reasonable numbers for transient occupancies.

If the required number of bicycle rack spaces is defined as BR, the peak transient occupancy during shift or period *j* as TO_j, and the peak building users during shift *j* as PBU_j, then BR can be calculated by using the following two equations:

$$PBU_j = FTE_j + TO_j \tag{2.4.5}$$

$$BR = 0.05 \times (\text{maximum } PBU_j)^* \tag{2.4.6}$$

The number of bicycle rack spaces (BR) must always be a whole number, and any fractional calculations must be rounded up to the next-higher whole number. Figure 2.4.2 depicts an outdoor bicycle rack at a LEED-certified building. Figure 2.4.3 is a photo of a sheltered bicycle parking area.

Submittals for obtaining this credit for commercial/institutional uses (Case 1) must include the calculations plus a map showing that the bicycle and changing/shower facilities are within 200 yd of the entrance to the project building (or within the building).

Residential Uses The secure bicycle spaces for residential projects must also be covered to protect the bicycles from the weather and from theft. They must be provided for a minimum of 15 percent of the building occupants. The building designer should declare the design occupancy (DO) for the buildings. Note that a design occupancy is not the same as a maximum occupancy as may be established by a fire code. LEED 2009 provides a typical residential occupancy as two residents per single bedroom unit, three residents

*Include the FTE*j* of the largest shift before or after the shift used in the maximum PBU*j* if there is a shift overlap.

Figure 2.4.2 Bicycle rack outside of the West Quad Learning Center at the University of South Carolina in September 2006.

per a two bedroom unit, and so on. The submittal should show that there are a minimum of secure *and* covered bicycle storage facility (BSF) spaces for at least 15 percent of this design occupancy as shown in Eq. (2.4.7). Again, the number of bicycle storage spaces must be a whole number. Additional changing/shower facilities are not required as they are already available in the residences.

$$\text{BSF} = 0.15 \times \text{DO} \tag{2.4.7}$$

Figure 2.4.3 Shell Woodcreek office and multipurpose building complex in Houston, TX. Bicycle parking area. (*Photo taken November 11, 2009 by Heidi Brakewood.*)

Special Circumstances and Exemplary Performance

There is not an exemplary performance point tied individually to SSc4.2; however, there is an overall point that can be earned for a transportation plan as previously mentioned.

SS Credit 4.3: *Alternative Transportation—Low-Emitting and Fuel-Efficient Vehicles (LEFEVs) [and also Alternative Fuel Vehicles (AFVs)]*

Air quality and the global climate may be affected by emissions from motor vehicles and the combustion of fossil fuels. This credit seeks to reduce these impacts by promoting motor vehicles that have lower emissions either due to advanced or alternative designs or due to high fuel efficiency. However, full life-cycle analyses should be considered when changing to alternative fuels, as there may be other environment issues relating to the manufacture or use of the fuels. This credit is worth three points in LEED 2009 and was worth one point in LEED 2.2.

USGBC Rating System

LEED-NC 2.2 lists the Intent, Requirements, and Potential Technologies and Strategies for this credit as follows, with modifications to LEED 2009 as noted (the option numbers have changed in the new version and the options have been reordered accordingly):

Intent

Reduce pollution and land development impacts from automobile use.

Requirements

OPTION 1 LEED 2009 (Was OPTION 2 in LEED 2.2)

Provide preferred parking for low-emitting and fuel-efficient vehicles for 5% of the total vehicle parking capacity of the site. (LEED 2009 further clarifies that a discounted parking rate is an acceptable substitute for low-emitting and fuel-efficient preferred parking and requires that it must be discounted at least 20%, available to all customers, not just the calculated 5% of the total vehicle parking capacity, publicly posted at the parking area entrance and available for at least two years.)

OR

OPTION 2 LEED 2009 (Was OPTION 3 in LEED 2.2)

Install alternative-fuel refueling stations for 3% of the total vehicle parking capacity of the site (liquid or gaseous fueling facilities must be separately ventilated or located outdoors).

OR

OPTION 3 LEED 2009 (Was OPTION 1 in LEED 2.2)

Provide low-emitting and fuel-efficient vehicles for 3% of Full-Time Equivalent (FTE) occupants AND provide preferred parking for these vehicles.

OR

OPTION 4 LEED 2009 (New Option in 2009)

This new option requires that a low-emitting or fuel-efficient vehicle sharing program be provided for building occupants. In this program, at least one such vehicle is provided for 3% of the FTE occupants (assuming that the vehicle can carry eight persons this means one vehicle for everyone 267 FTE occupants, or fraction thereof), the contract must be provided for at least two years, customers served must be supported by documentation, a narrative of the program must be submitted and parking for such vehicles must be in preferred spaces.

For the purpose of this credit, low-emitting and fuel-efficient vehicles are defined as vehicles that are either classified as Zero Emission Vehicles (ZEV) by the California Air Resources Board (CARB) or have achieved a minimum green score of 40 on the American Council for an Energy Efficient Economy (ACEEE) annual vehicle rating guide.

"Preferred parking" refers to the parking spots that are closest to the main entrance of the project (exclusive of spaces designated for handicapped) or parking passes provided at a discounted price.

Potential Technologies and Strategies

Provide transportation amenities such as alternative-fuel refueling stations. Consider sharing the costs and benefits of refueling stations with neighbors.

Calculations and Considerations

This credit actually covers two different sets of vehicles: one is the low-emitting and fuel-efficient vehicle (LEFEV) category for which Options 1, 3, and 4 in LEED 2009 apply. The other is the alternative-fuel vehicle (AFV) category for which Option 2 in LEED 2009 applies. There is some overlap between the two sets, and they are further defined in the following sections.

Options 1, 3, and 4 (LEED 2009) For Option 3, determine the FTE by Eqs. (2.4.1) through (2.4.3), and then determine both the minimum number of low-emitting and fuel-efficient vehicles to provide and the minimum number of preferred parking spaces or passes designated for these types of vehicles (LEFEVP3) by Eq. (2.4.8).

$$\text{LEFEVP3} = 0.03 \times \text{FTE} \tag{2.4.8}$$

The number of low-emitting and fuel-efficient vehicles and preferred parking spaces or passes (LEFEVP3) must always be a whole number, and any fractional results must be rounded up to the next-higher whole number.

The submittal for Option 3 should include these LEFEVP3 preferred parking calculations and a site plan showing the number and location of LEFEVP3 preferred parking spaces. The submittal should also include a listing with the number, make, model, manufacturer, and ZEV or ACEEE vehicle score of each low-emitting and fuel-efficient vehicle provided for the FTE occupants.

For the Option 1 submittal, confirm the number of total parking spaces provided for the project (TP), confirm that the minimum number of preferred parking spaces for low-emitting and fuel-efficient vehicles (LEFEVP1) is at least 5 percent of this total, and show the location(s) of the preferred parking for these low-emitting and fuel-efficient vehicles on an area map. The number of LEFEVP1 preferred parking spaces is given by Eq. (2.4.9). It must always be a whole number, and any fractional calculations must be rounded up to the next-higher whole number.

$$\text{LEFEVP1} = 0.05 \times \text{TP} \tag{2.4.9}$$

Alternatively, Option 1 may be obtained by simply providing discounted parking rates to any low-emitting or fuel-efficient vehicle, with at least a 20 percent discount, and available for at least two years.

For the Option 4 submittal, determine the FTE by Eqs. (2.4.1) through (2.4.3), and then determine the number of low-emitting or fuel-efficient shared vehicles to provide and the minimum number of preferred parking spaces designated for these shared vehicles (LEFEVP4) by Eq. (2.4.10).

$$\text{LEFEVP4} = \text{FTE}/267 \tag{2.4.10}$$

The number of low-emitting and fuel-efficient shared vehicles and preferred parking spaces (LEFEVP4) must always be a whole number, and any fractional results must be rounded up to the next-higher whole number.

Options 1, 3, and 4 may be much more viable than it might seem from reading the credit description, as the typical person may assume that the low-emitting and fuel-efficient vehicles are mainly alternative-fuel vehicles, and prior to 2007 not many people in the United States owned such vehicles. However, this category also includes many gasoline-fueled vehicles. Nearly all hybrids fall into this category, but so do many of the smaller, more fuel-efficient cars on the market. The current year's listings and information on how to purchase a book with ratings for prior years are available at www.greenercars.com. For instance, in 2006 in addition to many hybrids, vehicles with ratings above 40 on the ACEEE rating included the following models: Toyota Corolla, Hyundai Accent, Kia Rio/Rio 5, Honda Civic, Pontiac Vibe, Chevrolet Cobalt, and Saturn Ion. Since many people commute in the more energy-efficient vehicles, there might be a great demand for preferred parking spaces, and more importantly, the 5 percent requirement for the preferred parking spaces may be easy to fulfill without leaving parking spaces unoccupied. Similarly, providing fleet vehicles with these vehicles may be more economical than initially thought.

Option 2 LEED 2009 For LEED 2009 SS Credit 4.3 Option 2, the submittal should confirm the number of total parking spaces provided. In addition, according to the version 2.2 and LEED 2009 Reference Guides, there should also be a calculation determining that the number of refueling dispensing locations is at least 3 percent of this number. However, this may not be the proper calculation for all situations, and it may be more dependent on the alternative fuels being dispensed. It is not the intent of the credit to have dozens of refueling stations, but rather to provide alternative fuel for at least 3 percent of the vehicles which could be parked. A dispensing location for some alternative fuels can service many vehicles in a reasonable time and is usually regarded in the industry as the actual nozzle or dispensing device. A standard dispensing machine may service more than one vehicle simultaneously and may service many over a fairly short time, with some exceptions where recharging takes a long time. To take these variations into account, the submittals listed in the LEED-NC 2.2 and 2009 Reference Guides include a plan with the location(s) of the alternative fueling stations, the fuel type being dispensed, the number of dispensing locations, and the fueling capacity for each for an 8-h period. It is the opinion of the author that the submittal should then also include verification that these fueling stations can adequately service the requisite number of vehicles easily within a reasonable time frame that is convenient for the vehicle users. Let AFV be this minimum number of alternative-fuel vehicles which can be adequately serviced and, as previously defined, TP is the total parking provided, then Option 2 will be obtained if Eq. (2.4.11) is valid.

$$\text{AFV} = 0.03 \times \text{TP} \tag{2.4.11}$$

The LEED-NC 2.2 and 2009 Reference Guides provide the following definition: "Alternative Fuel Vehicles are vehicles that use low-polluting, non-gasoline fuels such as electricity, hydrogen, propane or compressed natural gas, liquid natural gas, methanol, and ethanol. Efficient gas-electric hybrid vehicles are included in this group for LEED purposes." However, this is probably an older definition from previous versions of LEED that is not fully intended to be used for Option 2 in this version. It is the opinion of the author that it is not the intent of Option 2 to provide gasoline fueling for hybrids, but to instead provide some alternative fuel. In addition, biodiesel is an alternative fuel that is being accepted for use in many diesel engines. Biodiesel is a blend of vegetable/animal diesel, and petroleum-based diesel can be used in most diesel engines with no modifications up to about 20 percent biodiesel based (B20). Lower percentages of petroleum-based diesel in the mix (up to 0 percent in B100) may need some modified parts in vehicles such as different material hoses. This alternative has not been directly addressed in this credit, but the author believes that it may have the opportunity for acceptance but would need verification based on a CIR.

Special Circumstances and Exemplary Performance

In summary, Options 1, 3, and 4 in LEED 2009 deal with low-emitting and fuel-efficient vehicles, whereas Option 2 deals with alternative-fuel vehicles. It is the opinion of the author that low-emitting and fuel-efficient vehicles should include alternative-fuel vehicles, but that alternative-fuel vehicles do not include many of the low-emitting and fuel-efficient vehicles. Figure 2.4.4 depicts signage for parking spaces reserved for fuel-efficient vehicles.

If there are special circumstances or nonstandard compliance paths taken for either of the options, a narrative must be submitted describing these circumstances. There is not an exemplary performance point tied individually to SSc4.3; however, there is an overall point that can be earned for a transportation plan as previously mentioned.

FIGURE 2.4.4 Shell Woodcreek office and multipurpose building complex in Houston, TX. Reserved parking area for fuel-efficient vehicles. (*Photo taken November 11, 2009 by Heidi Brakewood.*)

SS Credit 4.4: *Alternative Transportation—Parking Capacity*

USGBC Rating System

This credit is worth two points in LEED 2009 and was worth one point in LEED 2.2. The intention of this credit is to limit the sizes of parking facilities and the number of vehicle trip generations to the site. *Trip generation* is a term used in transportation engineering to represent the amount of vehicular traffic generated in a network by an activity at a location. The Institute of Transportation Engineers (ITE) publishes a manual that gives representative values for different site uses and street volumes. This reduction from the estimated trip generation for typical, usually single-occupancy, vehicles is accomplished by encouraging the use of carpooling or other shared-ride methods. LEED-NC 2.2 lists the Intent, Requirements, and Potential Technologies and Strategies for this credit as follows, with modifications for LEED 2009 as noted, including the renumbering of options and segregation by cases:

Intent

Reduce pollution and land development impacts from single occupancy vehicle use.

Requirements

LEED 2009 CASE 1. NON-RESIDENTIAL PROJECTS

OPTION 1—NON-RESIDENTIAL

Size parking capacity to not exceed minimum local zoning requirements AND provide preferred parking for carpools or vanpools for 5% of the total provided parking spaces.

OR

OPTION 2—NON-RESIDENTIAL

For projects that provide parking for less than 5% of FTE building occupants:

Provide preferred parking for carpools or vanpools for 5% of the total provided parking spaces. In both Option 1 and Option 2 of LEED 2009 Case 1, an alternative is to provide a discounted parking rate for vanpools or carpools. This must be at least a 20% discount and be available for all carpool or vanpool vehicles for a minimum of two years and publicly posted at the entrance of the parking area.

OR

OPTION 3—NON-RESIDENTIAL

Provide no new parking.

OR

LEED 2009 CASE 2. RESIDENTIAL PROJECTS

OPTION 1—RESIDENTIAL

Size parking capacity to not exceed minimum local zoning requirements, AND, provide infrastructure and support programs to facilitate shared vehicle usage such as carpool drop-off areas, designated parking for vanpools, or car-share services, ride boards, and shuttle services to mass transit.

OR

OPTION 2—RESIDENTIAL

Provide no new parking.

LEED 2009 CASE 3. MIXED USE (Residential with Commercial/Retail) PROJECTS

OPTION 1—MIXED USE

For mixed-use buildings with less than 10% commercial floor area adhere to the Case 2 Residential Requirements and for all others adhere to the Case 1 and 2 requirements for the respective Non-residential and Residential components.

OPTION 2—MIXED USE

Provide no new parking.

NOTES:

"Preferred parking" refers to the parking spots that are closest to the main entrance of the project (exclusive of spaces designated for handicapped) or parking passes provided at a discounted price.

In LEED 2.2, when parking minimums were not defined by relevant local zoning requirements, or when there were no local zoning requirements, either of the following alternative parking calculation methods (A or B) was required. These are not required in LEED-NC 2009.

A) Meet the requirements of Portland, Oregon, Zoning Code: Title 33, Chapter 33.266 (Parking and Loading)

OR, if this standard is not appropriate for the building type,

B) Install 25% less parking than the building type's average listed in the Institute of Transportation Engineers' Parking Generation study, 3rd Edition.

Potential Technologies and Strategies

Minimize parking lot/garage size. Consider sharing parking facilities with adjacent buildings. Consider alternatives that will limit the use of single occupancy vehicles.

Calculations and Considerations

The intention of this credit is to reduce the use of SOVs, which has other potential environmental benefits in addition to those just listed.

Option 1 in Case 1 and Case 2 in LEED 2009 for SSc4.4 require that the total number of parking spaces (TP) does not exceed the required zoning minimum. Most zoning codes have a methodology for determining the minimum number of parking spaces that should be provided for different uses and different size facilities. The submittals should include the verbiage of the applicable zoning code and the supporting calculations to determine the minimum parking required by local zoning code. The two options for this credit mandate that the maximum total parking provided (TP) be equal to or lower than the minimum as set by the local code, whether the local code has a different maximum parking calculation or not. (Sometimes the parking allowed can be lower than the minimum as set in the zoning code, if the project has been granted a variance by the regulatory authority due to special or exceptional circumstances.) In the past, most zoning codes did not have a calculation for maximum allowed parking, and the parking installed would frequently far exceed the zoning minimum. Recently, many zoning codes have been modified to include both a minimum and a maximum allowable number of parking spaces, mainly to address the environmental concerns arising from the typical impervious surfaces that are used to build parking facilities and their impacts on stormwater runoff and the urban heat island effect. As mentioned in the errata as posted in the spring of 2007 by the USGBC, there are two alternatives to determining the total parking allowed if a minimum is not set by local code or zoning. In these cases the total parking (TP) should not exceed the requirements of the Portland, Oregon, Zoning Code: Title 33, Chapter 33.266 (Parking and Loading) or, if the Portland, Ore., standard is not appropriate for the building type, then the total parking should be at least 25 percent less than the building type's average listed in the document by the Institute of Transportation Engineers (ITE) entitled *Parking Generation, Third Edition,* Washington, D.C., 2004.

For example, in one municipality, a sit-down restaurant use may require a minimum of one-half parking space for each customer seat (most people go to restaurants in groups) plus one for each employee in a shift. The maximum may allow for an increase of 25 percent over this, which is reasonable based on customers who may wish to wait for service. Other zoning codes may have the formula for a similar use based on gross floor area or customer floor area. To obtain zoning approval, the designers must meet this minimum requirement unless they have a valid reason to ask for a variance. If a variance is granted from the minimum by the local municipal planning organization (MPO), then this should be included in the submittal documents. An example of a valid reason for a lower than minimum total parking may be the availability of shared parking off-site.

LEED 2009 Case 1 Option 1 Case 1 Option 1 for credit SSc4.4, is for nonresidential construction and has two requirements. The first is that the total number of parking spaces (TP) does not exceed the required zoning minimum as previously defined. The second requirement in Option 1 for credit SSc4.4 is to designate the minimum number of preferred parking spaces for carpools or vanpools (CVPP1) for 5 percent of the total provided parking spaces as given in Eq. (2.4.12).

$$\text{CVPP1} = 0.05 \times \text{TP} \tag{2.4.12}$$

The minimum number of preferred parking spaces for carpools or vanpools (CVPP1) must always be a whole number, and any fractional calculations must be rounded up to the next-higher whole number. Alternatively, instead of providing these spaces, discounted parking passes may be available for all carpools or vanpools as previously noted.

The submittals for Option 1 of credit SSc4.4 should include the zoning calculations for minimum required parking to obtain the total parking (or the appropriate alternative method based on the Portland, Ore., zoning code or the ITE report in LEED 2.2) and the calculation of the minimum number of preferred parking spaces for carpools or vanpools (CVPP1). A plan should be submitted showing both the total number of parking spaces (TP) and the location of at least the minimum number of preferred parking spaces for carpools or vanpools (CVPP1) as appropriately marked in relation to the other parking spaces and the entrance. Again, instead of providing these spaces, discounted parking passes may be available for all carpools or vanpools as previously noted and provisions for this verified in the submittal.

LEED 2009 Case 1 Option 2 Case 1 Option 2 of credit SSc4.4 for nonresidential uses, is similar to Case 1 Option 1 but is for the special case where very little parking is provided, perhaps representing projects in areas with well-established mass transit or other transportation and parking opportunities. The submittals should first establish that the total parking provided (TP) is less than 5 percent of the FTE building occupants and then provide the calculation for the minimum number of preferred parking spaces for carpools or vanpools (CVPP2) as given in Eq. (2.4.13) and an area map showing their location.

$$\text{CVPP2} = 0.05 \times \text{TP} \tag{2.4.13}$$

The minimum number of preferred parking spaces for carpools or vanpools (CVPP2) must always be a whole number, and any fractional calculations must be rounded up to the next-higher whole number. Again, instead of providing these spaces, discounted parking passes may be available for all carpools or vanpools as previously noted and provisions for this verified in the submittal.

LEED 2009 Case 2 Option 1 Case 2 Option 1 of credit SSc4.4 is for residential uses and mandates not exceeding the local zoning parking requirements and also promoting programs to reduce SOV use such as ride sharing. The submittal should include the verbiage of the applicable zoning code and the supporting calculations to determine the minimum parking required by local zoning code. This credit mandates that the maximum total parking provided (TP) equal the minimum as set by the local code and not greater, whether the local code has a different maximum parking calculation or not. (As mentioned previously, there are two alternatives given for determining the total parking allowed [TP] if parking is not addressed by local zoning for LEED 2.2, but not required in the LEED 2009 calculations.) The submittal should also provide a description of the infrastructure and the programs that are in place or going to be provided to not just promote, but also support ride sharing for the residents.

LEED 2009 Case 3 Case 3 is for mixed uses and follows the applicable cases for either nonresidential or residential uses for the respective portions of the building. If there are less than 10% commercial or retail areas in an otherwise residential facility, then the entire facility adheres to the residential case (2).

All Cases Case 1 Option 3, or Case 2 Option 2, or Case 3 Option 2 of credit SSc4.4 mandate that no new parking be provided. The submittal should include the previously existing site parking plan and the new parking plan indicating the total parking (TP) provided on each.

Special Circumstances and Exemplary Performance

If there are special circumstances or nonstandard compliance paths taken for either of the options, a narrative must be submitted describing these circumstances. There is not an exemplary performance point tied individually to SSc4.4; however, there is an overall point that can be earned for a transportation plan as previously mentioned. If there are special circumstances or nonstandard compliance paths taken for either of the options, a narrative must be submitted describing these circumstances.

2.5 SS Credit Subcategory 5: *Site Development*

SS Credit 5.1: *Site Development—Protect or Restore Habitat*

The intent of this credit is to maintain as much of the native or adapted biological species as possible in a region and to also protect other natural site elements, including water bodies, rock outcroppings, soil conditions, and other ecosystem features. There are many reasons for this. Obviously, native and adapted species would survive well in the local climate without much additional irrigation and maintenance. There is also the desire for biodiversity in this world. Maintaining native vegetation in all regions is important both for plant biodiversity and for the local fauna to survive. It is also desirable to maintain habitat for many biologically diverse organisms, not just for ecological reasons, but also because it has been found that a world with rich biodiversity benefits humans in the form of potential medicines and other products. SSc5.1 is worth one point in LEED 2009 and was worth one point in LEED 2.2.

USGBC Rating System

LEED-NC 2.2 lists the Intent, Requirements, and Potential Technologies and Strategies for this credit as follows, with modifications to LEED 2009 as noted:

Intent

Conserve existing natural areas and restore damaged areas to provide habitat and promote biodiversity.

Requirements

LEED 2009 CASE 1: On **greenfield sites**, limit all site disturbance to 40 feet beyond the building perimeter; 10 feet beyond surface walkways, patios, surface parking, and utilities less than 12 inches in diameter; 15 feet beyond primary roadway curbs and main utility branch trenches; and 25 feet beyond constructed areas with permeable surfaces (such as pervious paving areas, stormwater detention facilities, and playing fields) that require additional staging areas in order to limit compaction in the constructed area.

LEED 2009 CASE 2: **On previously developed or graded sites**, restore or protect a minimum of 50% of the site area (excluding the building footprint), or as additionally required in LEED 2009, a minimum of 20% of the total site area, whichever is the greater with native or adapted vegetation. Native/adapted plants are plants indigenous to a locality or cultivars of native plants that are adapted to the local climate and are not considered invasive species or noxious weeds. Projects earning SS Credit 2 and using vegetated roof surfaces may apply the vegetated roof surfaces to this calculation if the plants meet the definition of native/adapted.

Greenfield sites are those that are not developed or graded and remain in a natural state. Previously Developed Sites are those that previously contained buildings, roadways, parking lots or were graded or altered by direct human activities.

Potential Technologies and Strategies

On greenfield sites, perform a site survey to identify site elements and adopt a master plan for development of the project site. Carefully site the building to minimize disruption to existing ecosystems and design the building to minimize its footprint. Strategies include stacking the building program, tuck-under parking and sharing facilities with neighbors. Establish clearly marked construction boundaries to minimize disturbance of the existing site and restore previously degraded areas to their natural state. For previously developed sites, utilize local and regional governmental agencies, consultants, educational facilities, and native plant societies as resources for the selection of appropriate native or adapted plant materials. Prohibit plant materials listed as invasive or noxious weed species. Native/adapted plants require minimal or no irrigation following establishment, do not require active maintenance such as mowing or chemical inputs such as fertilizers, pesticides, or herbicides, and provide habitat value and promote biodiversity through avoidance of monoculture plantings.

Calculations and Considerations

The LEED definitions for *building footprint, development footprint, native* (or *indigenous*) *plants, adapted* (or *introduced*) *plants, invasive plants,* and *previously developed sites* are given in App. B, and many of the symbols used to represent them are given in the Land Area Definitions section at the beginning of this chapter.

The submittal is part of the LEED Construction Submittal since limiting construction disturbance and protecting native areas are an important part of the construction phase. It should be well documented throughout the construction phase that the limits of construction (disturbance) either do not go beyond the maximum setbacks in the first option or, for previously developed sites, do not disturb existing natural areas. For both cases, greenfield sites or previously developed sites, the submittals should include both an existing conditions map and a project site area map with the site area given and the land areas of the project building footprint, developed portions, and undeveloped and/or restored portions noted. A narrative should also be included that describes the approach taken to obtain this credit with any special circumstances noted.

LEED 2009 CASE 1: Greenfield Sites For the greenfield sites, a grading plan should also be submitted with the designated limits of disturbance highlighted. Note that site disturbance includes any earthwork and clearing of vegetation, even if replanted afterward. Figure 2.5.1 is an example sketch of a greenfield site development and limits of disturbance plan.

LEED 2009 CASE 2: Previously Developed Sites The criteria for sites which were previously developed are based on the building footprint, not on the development footprint. The building footprint is the plan view of the site area covered by the buildings and structures, including overhangs. Shade structures do not have to be included if they are not part of the building. For sites that were previously developed, the total of the areas that have been restored to native/adaptive vegetation should be submitted along with a landscaping plan and schedule for the restored and protected areas. Any areas that were previously graded or landscaped count as previously developed, as do active agricultural uses. Fallow areas are sometimes considered undeveloped. Sports fields do not count as native/adaptive vegetation. (Note that sports fields do count as open space for SSc5.2.) When choosing native or adaptive vegetation, the distinction for appropriate species is very localized. Even the definition of *invasive* is local as some species are considered invasive in some areas of the country and are not considered invasive in others. One example is the ice plant, a ground cover. It is used extensively in California and is considered invasive there, but not in many other states.

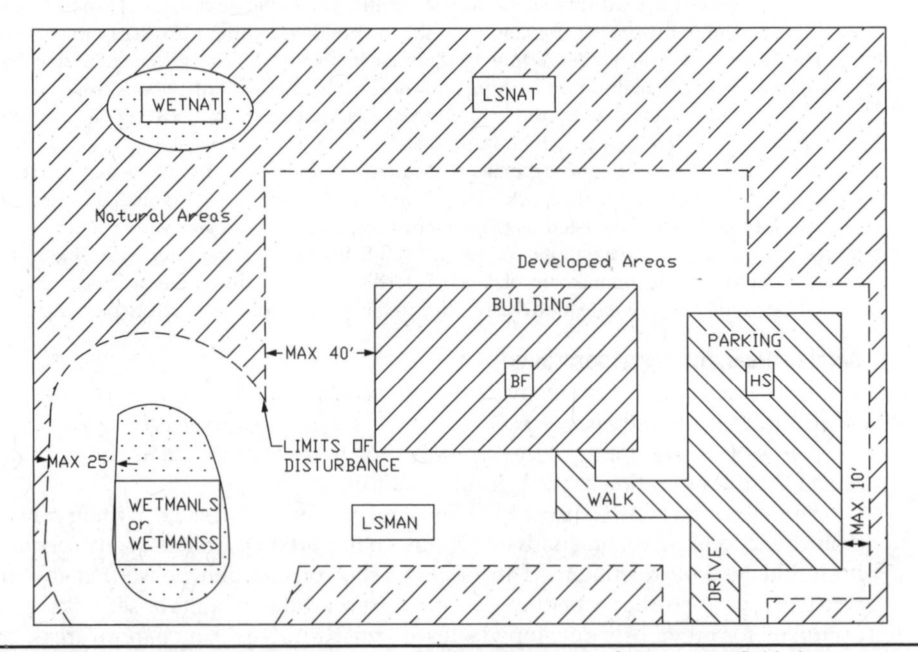

FIGURE 2.5.1 Example maximum setbacks for disturbance at a former greenfield site.

Criteria for	Most Sites	Urban Sites That Also Earn SSc2
Credit point	WETNAT + LSNAT + LSNAP ≥ Maximum of [0.5(A_T – BF) or 0.2 A_T]	WETNAT + LSNAT + LSNAP + VNAPR ≥ Maximum of [0.5(A_T – BF) or 0.2 A_T]
EP point	WETNAT + LSNAT + LSNAP ≥ Maximum of [0.75(A_T – BF) or 0.3 A_T]	WETNAT + LSNAT + LSNAP + VNAPR ≥ Maximum of [0.75(A_T – BF) or 0.3 A_T]

TABLE 2.5.1 Summary of Protected/Restored Habitat Areas for Previously Developed Sites LEED 2009

Note that projects earning SS Credit 2 (Development Density and Community Connectivity) and having vegetated roof surfaces may use these vegetated roof areas in the calculation of the total of the areas restored or protected if these vegetated roof areas meet the native/adapted species criteria. Otherwise, the vegetated roof surfaces are not included. Raised planters were not specifically addressed in LEED-NC 2.2, but it is assumed that they too may count for urban areas, earning SSc2 if appropriately planted, but not for other projects. By raised planters, the author is referring to decorative planters that are substantially raised from the ground surface, not curbed landscape areas. Curbed landscape areas should count in all cases if appropriately planted.

The requirements for previously developed sites can be summarized using the following additional definitions and sets of equations and are summarized in Table 2.5.1.

LSNAP Represents all the previously disturbed areas on the ground that are replanted with native and adapted plants

VNAPR Represents all the roofed areas that are planted with native and adapted plants

One point is earned for most previously developed sites if the following is true:

$$\text{WETNAT} + \text{LSNAT} + \text{LSNAP} \geq \text{Maximum of } [0.5 \times (A_T - \text{BF}) \text{ or } 0.2A_T] \quad (2.5.1)$$

or for the previously developed urban sites where SSc2 is also earned, the following is true:

$$\text{WETNAT} + \text{LSNAT} + \text{LSNAP} + \text{VNAPR} \geq \text{Maximum of } [0.5 \times (A_T - \text{BF}) \text{ or } 0.2A_T] \quad (2.5.2)$$

Special Circumstances and Exemplary Performance

If the project site contains both greenfield areas and previously developed areas, then these areas must each conform to the requirements of Case 1 and Case 2, respectively. The previously developed sites may be awarded one point for exemplary performance if the restored/protected areas total more than 75 percent of the project area, excluding the building footprint, or as further required in LEED 2009, 30 percent of the total project area, whichever is greater.

An additional exemplary performance point is earned for most previously developed sites if

$$\text{WETNAT} + \text{LSNAT} + \text{LSNAP} \geq \text{Maximum of } [0.75 \times (A_T - \text{BF}) \text{ or } 0.3A_T] \quad (2.5.3)$$

or for the previously developed urban sites where SSc2 is also earned, the following is true:

$$\text{WETNAT} + \text{LSNAT} + \text{LSNAP} + \text{VNAPR} \geq \text{Maximum of } [0.75 \times (A_T - \text{BF}) \text{ or } 0.3A_T] \quad (2.5.4)$$

Figure 2.5.2 Landscaped area with native and adapted plants at Cox and Dinkins, Engineers and Surveyors, Columbia, S.C. (*Photograph taken June 2007.*)

Figure 2.5.2 shows some examples of native and adapted plants that were used in a landscape bed in Columbia, S.C.

SS Credit 5.2: *Site Development—Maximize Open Space*

USGBC Rating System

SSc5.2 is worth one point in LEED 2009 and one point in LEED 2.2. LEED-NC 2.2 lists the Intent, Requirements, and Potential Technologies and Strategies for this credit as follows and is similar to LEED 2009 with modifications as noted:

> **Intent**
>
> Provide a high ratio of open space to development footprint to promote biodiversity.
>
> **Requirements**
>
> LEED 2009 CASE 1:
>
> Reduce the development footprint (defined as the total area of the building footprint, hardscape, access roads and parking) and/or provide vegetated open space within the project boundary to exceed the local zoning's open-space requirement for the site by 25%.

LEED 2009 CASE 2:

For areas with no local zoning requirements (e.g., some university campuses and military bases), provide vegetated open-space area adjacent to the building that is equal to the building footprint. (On campus settings this open space need not be necessarily contiguous to the building.)

LEED 2009 CASE 3:

Where a zoning ordinance exists, but there is no requirement for open space (zero), provide vegetated open space equal to 20% of the project's site area.

ALL CASES:

- For projects located in urban areas that earn SS Credit 2, vegetated roof areas can contribute to credit compliance.
- For projects located in urban areas that earn SS Credit 2, pedestrian-oriented hardscape areas can contribute to credit compliance. For such projects, a minimum of 25% of the open space counted must be vegetated.
- Wetlands or naturally designed ponds may count as open space if the side slope gradients average 1:4 (vertical : horizontal) or less and are vegetated.

Potential Technologies and Strategies

Perform a site survey to identify site elements and adopt a master plan for development of the project site. Select a suitable building location and design the building with a minimal footprint to minimize site disruption. Strategies include stacking the building program, tuck-under parking and sharing facilities with neighbors to maximize open space on the site.

Calculations and Considerations

The LEED definitions for *building footprint, development footprint, native* (or *indigenous) plants, adapted* (or *introduced) plants, invasive plants, open space area,* and *previously developed sites* are given in App. B. Note the exceptions for SS Credit 5.2, and also as applicable if SS Credit 2 is obtained, in the definition for *open-space area* as given in LEED-NC 2.2 with the errata as posted in the spring of 2007:

> Open Space Area is as defined by local zoning requirements. If local zoning requirements do not clearly define open space, it is defined for the purposes of LEED calculations as the property area minus the development footprint; and it must be vegetated and pervious, with exceptions only as noted in the credit requirements section. For projects located in urban areas that earn SS Credit 2, open space also includes non-vehicular, pedestrian-oriented hardscape spaces.

The exceptions are with regard to vegetation and for all projects allow for wetlands or naturally designed ponds to be included in open space if they have vegetated sides leading down to them with low slopes (<25 percent). In addition, vegetated playing fields may be included as open space. There is still some question as to whether sports fields within stadiums count as open space. These special exceptions should be reviewed through the USGBC CIR procedure.

For projects located in urban areas that *also* earn SS Credit 2, open space may include vegetated roof areas and nonvehicular, pedestrian-oriented hardscape spaces (with the caveat that at least 25 percent of the open space must be vegetated). Examples of these pedestrian-oriented spaces include items such as pocket parks, accessible roof decks, plazas, and courtyards.

For all the cases, the submittal should include the total site area and the site and landscape plans which highlight the open spaces, both those vegetated and, when allowed, those not vegetated with the associated areas noted. For Case 1, also provide the calculations for the open space as required by local code; and for Case 2, also provide the area of the building footprint. As with SSc5.1, raised planters were not specifically addressed in LEED-NC 2.2, but it is assumed that they too may count for urban areas earning SSc2, but not for other projects. By *raised planters*, the author is referring to decorative planters that are substantially raised from the ground surface, not curbed landscape areas. Curbed landscape areas should count in all cases.

One of the suggested strategies is to stack the building program. This refers to the architectural decision to try to build up, with facilities stacked upon each other, versus building in a more sprawled outward design. Obviously, this would help reduce the building and thus the development footprint.

The requirements to fulfill SSc5.2 for the three cases can be related to equations which are based on the land area definitions as listed in the beginning of this chapter and the three additional definitions as listed here:

OSZONE OSZONE represents the percent of the total area required to be open space by local zoning code, if such a requirement exists.

VR VR is the vegetated roof area (see also SSc7.2).

PED25LS PED25LS represents the special pedestrian areas with a minimum of 25 percent landscaping.

Then each option has the following sets of equations:

LEED 2009 Case 1 One point is earned on most sites if

$$\text{OS} \geq 1.25(\text{OSZONE}/100) \times A_T \tag{2.5.5}$$

Or, on urban sites where SSc2 is also earned,

$$\text{OS} + \text{VR} + \text{PED25LS} \geq 1.25(\text{OSZONE}/100) \times A_T \tag{2.5.6}$$

LEED 2009 Case 2 One point is earned on most sites if

$$\text{OS} \geq \text{BF and is contiguous to it} \tag{2.5.7}$$

Or, on urban sites where SSc2 is also earned,

$$\text{OS} + \text{VR} + \text{PED25LS} \geq \text{BF} \tag{2.5.8}$$

LEED 2009 Case 3 One point is earned on most sites if

$$\text{OS} \geq 0.2A_T \tag{2.5.9}$$

Or, on urban sites where SSc2 is also earned,

$$\text{OS} + \text{VR} + \text{PED25LS} \geq 0.2A_T \tag{2.5.10}$$

Special Circumstances and Exemplary Performance

Many times when open spaces are next to walls with glazing, birds become confused by the reflection of the open spaces and fly into the walls. Care should be taken that glazing next to these open spaces has appropriate features such as etching to lessen these concerns. If there are special circumstances or considerations for any of the cases, a narrative must be submitted describing these circumstances. A point may be given for exemplary performance in the Innovation in Design category if it can be demonstrated that any of the three case requirements for open space are doubled. This means the following for each case:

- For Case 1, exceeding the local zoning's open-space requirement for the site by 50 percent
- For Case 2, providing vegetated open-space areas adjacent to the building that are equal to double the building footprint when the project has no local zoning requirement
- For Case 3, providing vegetated open space equal to 40 percent of the project site area when no open space (zero) is required by local zoning code

For the exemplary performance case, each case has the following sets of equations which are also summarized in Table 2.5.2 along with the regular credit criteria equations.

Case 1 Exemplary Performance One exemplary performance point may be earned on most sites if

$$OS \geq 1.5(OSZONE/100)A_T \tag{2.5.11}$$

or, on urban sites where SSc2 is also earned,

$$OS + VR + PED25LS \geq 1.5(OSZONE/100)A_T \tag{2.5.12}$$

Criteria for	Condition	Most Sites	Urban Sites That Also Earn SSc2
Case 1 credit point	OSZONE in Zoning Code	$OS \geq 1.25(OSZONE/100)A_T$	$OS + VR + PED25LS \geq 1.25(OSZONE/100)A_T$
Case 1 EP point	OSZONE in Zoning Code	$OS \geq 1.5(OSZONE/100)A_T$	$OS + VR + PED25LS \geq 1.5(OSZONE/100)A_T$
Case 2 credit point	No Zoning Code	$OS \geq BF$	$OS + VR + PED25LS \geq BF$
Case 2 EP point	No Zoning Code	$OS \geq 2BF$	$OS + VR + PED25LS \geq 2BF$
Case 3 credit point	OSZONE not in Zoning Code	$OS \geq 0.2A_T$	$OS + VR + PED25LS \geq 0.2A_T$
Case 3 EP point	OSZONE not in Zoning Code	$OS \geq 0.4A_T$	$OS + VR + PED25LS \geq 0.4A_T$

Table 2.5.2 Summary of Open-Space Criteria Equations LEED 2009

Case 2 Exemplary Performance One exemplary performance point may be earned on most sites if

$$\text{OS} \geq 2\text{BF and is contiguous to it} \tag{2.5.13}$$

or, on urban sites where SSc2 is also earned,

$$\text{OS} + \text{VR} + \text{PED25LS} \geq 2\text{BF} \tag{2.5.14}$$

Case 3 Exemplary Performance One exemplary performance point may be earned on most sites if

$$\text{OS} \geq 0.4A_T \tag{2.5.15}$$

or, on urban sites where SSc2 is also earned,

$$\text{OS} + \text{VR} + \text{PED25LS} \geq 0.4A_T \tag{2.5.16}$$

2.6 SS Credit Subcategory 6: *Stormwater Management*

SS Credit 6.1: *Stormwater Management—Quantity Control*

USGBC Rating System

SSc6.1 is worth one point in LEED 2009 and was worth one point in LEED 2.2. LEED-NC 2.2 lists the Intent, Requirements, and Potential Technologies and Strategies for this credit as follows, with modifications as noted for LEED 2009 and redesignated as cases and then options:

Intent

Limit disruption of natural hydrology by reducing impervious cover, increasing on-site infiltration, and managing stormwater runoff.

Requirements

LEED 2009 CASE 1—EXISTING IMPERVIOUSNESS IS LESS THAN OR EQUAL TO 50%

LEED 2009 Option 1: Implement a stormwater management plan that prevents the post-development peak discharge rate and quantity from exceeding the pre-development peak discharge rate and quantity for the one- and two-year, 24-hour design storms.

OR

LEED 2009 Option 2: Implement a stormwater management plan that protects receiving stream channels from excessive erosion by implementing a stream channel protection strategy and quantity control strategies.

LEED 2009 CASE 2—EXISTING IMPERVIOUSNESS IS GREATER THAN 50%

Implement a stormwater management plan that results in a 25% decrease in the volume of stormwater runoff from the two-year, 24-hour design storm.

Potential Technologies and Strategies

Design the project site to maintain natural stormwater flows by promoting infiltration. Specify vegetated roofs, pervious paving, and other measures to minimize impervious surfaces. Reuse stormwater volumes generated for non-potable uses such as landscape irrigation, toilet and urinal flushing and custodial uses.

Calculations and Considerations

To understand this credit and determine potential technologies and strategies, the first thing that needs to be understood is what the term *imperviousness* means. Ordinarily in land development and watershed analyses, the features on the land may be divided into two characteristic types of surfaces: pervious or impervious surfaces. Impervious surfaces are typically roofed and paved areas where the roof or pavement material does not allow any significant infiltration of water into the earth below. Pervious surfaces are everything else. Frequently, it is required to determine the percent pervious or percent impervious areas for a project. In that case, the percent impervious is simply the land areas covered by characteristically impervious surfaces, divided by the total project site area A_T, given in percent. If A_{imp} is the total land areas with impervious surfaces and %Imp is the percent impervious, then

$$\%\text{Imp} = 100A_{\text{imp}}/A_T \tag{2.6.1}$$

However, many surfaces are more or less pervious than others. For instance, hard-packed clay or a soccer field surface may be less pervious than a sandy beach. In addition, the concern here is more with the percent of rainfall that "runs off," not the percent that infiltrates. There are other mechanisms by which rainfall may exit an area other than infiltration and runoff, including evaporation and evapotranspiration. So even if an asphalt shingle roof may be considered an impervious surface, it does not shed all the rainfall as runoff, since some is absorbed or stored in crannies on the surface and eventually evaporated. Therefore, to estimate runoff totals from a development, a simple system was developed to estimate the fraction of rainfall that runs off different types of surfaces. It is referred to as the *rational method*. It has been used for more than half a century to estimate runoff rates from small sites (usually much less than 100 acres).

Rational Method, Percent Imperviousness, and Percent Impervious Many types of surfaces, based on either the material of the surface or use of the surface and other characteristics such as slope, have been given a typical rational method runoff coefficient C_i. This coefficient represents the fractional percentage of a rainfall volume from a 2- to 10-year frequency storm that is estimated to result in runoff from that particular surface. To estimate the overall runoff from these frequency storms for the site in question, the overall runoff coefficient C is calculated as the land-area-weighted average of the individual area coefficients C_i for each individual land area A_i. This overall runoff coefficient represents the estimated fraction of rainfall that will run off a site for a typical 2- to 10-year frequency storm event.

$$C = \frac{\sum C_i A_i}{\sum A_i} \tag{2.6.2}$$

There are many state and local agencies such as the North Carolina Department of Environment and Natural Resources that are also adopting the concept of percent imperviousness along with the variable percent impervious. In version 2.1 of LEED-NC, the definition of imperviousness is given as the rational method coefficient in percent, and the adopting agencies have similarly based definitions. Therefore, percent imperviousness (%Impness) can usually be calculated as

$$\%\text{Impness} = 100C \tag{2.6.3}$$

To determine the percent imperviousness, the following steps should be taken:

- Identify the different surface types on the site.
- Calculate the total area for each of these surface types A_i, using the site drawing.
- Use Table 2.6.1 or other sources of the rational runoff coefficient such as textbooks or manufacturers' information to assign the individual runoff coefficient C_i for each type of surface.
- Finally use Eqs. (2.6.2) and (2.6.3) to determine the percent imperviousness.

The values in Table 2.6.1 have been taken from LEED-NC 2.1 with the pervious pavement grids approximated from values given there, and the pervious concrete value as determined in a laboratory study at the University of South Carolina (Valavala et al., 2006). Any surface that totally captures the rainfall and prevents it from runoff such as a retention basin or area that has the rainfall directed into a drywell or other infiltration system also has a runoff coefficient of zero unless the volume of the structure is exceeded.

Surface Type	Typical Runoff Coefficient
Pavement, asphalt	0.95
Pavement, concrete	0.95
Pavement, brick	0.85
Pervious pavement, plastic grid with grass	0.20
Pervious pavement, concrete grid with grass	0.60
Pervious pavement, concrete paver grid with gravel	0.85
Pervious concrete (0–1% slope)	~0
Turf, flat (0–1% slope)	0.25
Turf, average (1–3% slope)	0.35
Turf, hilly (3–10% slope)	0.40
Turf, steep (>10% slope)	0.45
Roofs, conventional	0.95
Roof, garden roof (<4-in substrate)	0.50
Roof, garden roof (4- to 8-in substrate)	0.30
Roof, garden roof (9- to 20-in substrate)	0.20
Roof, garden roof (>20-in substrate)	0.10
Vegetation, flat (0–1% slope)	0.10
Vegetation, average (1–3% slope)	0.20
Vegetation, hilly (3–10% slope)	0.25
Vegetation, steep (>10% slope)	0.30

TABLE 2.6.1 Typical Runoff Coefficients (Two-Year Storms)

Concept	Symbol	Governing Equation	Subsidiary Equation
Percent impervious	%Imp	$\%\text{Imp} = 100A_{\text{imp}}/A_T$	—
Percent imperviousness	%Impness	$\%\text{Impness} = 100C$	$C = \frac{\sum C_i A_i}{\sum A_i}$

TABLE 2.6.2 Summary of Percent Impervious and Percent Imperviousness Equations

The equations used to determine the two main concepts of percent impervious and percent imperviousness are summarized in Table 2.6.2. Percent impervious is used frequently in other calculations related to green building and sustainable development. Percent imperviousness is used mainly as a concept for this credit, although it is based on the rational method which is used widely in hydrological modeling.

Stormwater Management Plans and Hydrologic Models When the imperviousness of the existing site is determined, then the new design for the proposed site must meet the listed requirement options in the credit based on whether the existing imperviousness is greater than 50 percent. If it is greater than 50 percent, then Case 2 applies and the credit is obtained only if the new design reduces the existing amount of runoff from the 2-year, 24-h design storm, in total volume only, according to the verbiage in the credit. If the existing site has an imperviousness equal to or less than 50 percent, then either of the options in Case 1 may apply. Note that the first requirement of Case 1 applies to both the peak rate *and* the total volume of runoff from both the one-year and two-year storms.

Since the two-year event is greater than the one-year event, it might seem reasonable to assume that calculations on the two-year storms are sufficient for Case 1. However, this is not necessarily true. Statistically, one-year storms are more frequent than two-year storms, and a two-year storm tends to have a higher peak rate and a higher total volume than its equivalent duration one-year storm. However, most current stormwater codes and regulations now have requirements on controlling the runoff from 2-year and 10-year storms, but not usually 1-year storms, and this may result in changes to the outflow from the 1-year storm. One of the most common forms of stormwater management is the stormwater detention pond. The rate is usually controlled by collecting the water from the storm in a pond and releasing it through an outlet structure at a rate less than the peak requirement (two-year usually). Sometimes this is referred to as *release rate rules*. Statistically this flow rate under natural or preexisting conditions would occur only once every two years. However, now with the pond and altered hydrological state on the site, there is the opportunity for this rate to be released much more frequently. The pond can fill to similar levels for many of the smaller storms, and the outlet control is only required to keep the outflow less than the two-year rate. There is a chance, based on pond and outlet structure design, that this equivalent peak two-year rate may now occur for many smaller storms and perhaps occur on a frequency an order of magnitude greater than preexisting. In other words, this two-year outflow rate may occur a dozen or more times in two years. This not only makes the downstream impacts much more frequent, but also gives the system less time to stabilize with vegetation in between major storms. Therefore, this credit requires that the controls also restrict the outflow for smaller one-year storms to the one-year rate. There is still a chance that storms with a frequency of less than a year will now have the larger impacts of the one-year storms based on rate, but at least the impacts of potentially increasing the frequency of the two-year have been drastically reduced.

The two restrictions in Cases 1 and 2 on postdevelopment runoff can be attained with a combination of many strategies. Surfaces can be chosen with lower runoff coefficients. Similarly options to detain and infiltrate or use the rainfall such as retention basins, rainwater harvesting for irrigation or interior uses, or underground infiltration systems can be used to reduce the postdevelopment runoff. The stormwater control methods should be chosen based on environmental, societal, and economic criteria, in addition to local restrictions or preferences on many of the options. These methods and options are frequently referred to as *best management practices* (BMPs). There is also a relatively new and emerging stormwater management concept referred to as low-impact development (LID) which uses a subset of these BMPs, many for multiple purposes. These are sometimes also referred to as *integrated management practices* (IMPs), and the main goal is to mimic the natural hydrologic cycle as much as possible on the site. Usually, the more the site mimics the natural hydrological cycle, the fewer the opportunities for anomalies that can disrupt downstream flows and cause nonpoint source pollution. Chapter 10 deals with other stormwater management options such as LID that may help eradicate some of the potential problems of designing to only to a few specific storm rates and volumes.

The stormwater runoff calculations for Case 2 and the first option of Case 1 should be performed using accepted hydrological models for estimating runoff. The second option of Case 1 requires that the stream protection measures be described in detail and that similar accepted runoff calculations be performed for the predevelopment and postdevelopment runoff rates and quantities. It must be demonstrated that the postdevelopment runoff rates and quantities will be below critical values for the receiving waterways after the measures are in place. No specific frequency storm is given for this requirement. It is assumed that at a minimum, both a one-year and a two-year frequency storm should be used. Again, more details on methods can be found in Chap. 10.

Stream Channel Protection Many stream protection strategies are now using a form of control at the headwaters instead of protection within the channel. There are two main ways to control at the headwaters: either by controlling the hydrology upstream or by modifying the hydraulics at the stream headwaters.

Controlling the hydrology upstream is similar to other on-site stormwater BMPs. Usually retention and infiltration is the preferred strategy. An example is the incorporation of infiltration gardens on projects (bioretention without a subdrain) which may significantly reduce the cumulative runoff based on typical two-year storm-current regulated detention requirements. A good reference for this is the North Carolina Department of Environment and Natural Resources, Division of Water Quality, *Stormwater Best Management Practices Manual*, July 2007.

Controlling the hydraulics at the headwaters is usually by some form of energy dissipation at the culvert outlets. Other than increasing the imperviousness of the watershed, eroding headwater tributaries caused by culvert scour have typically been an observed contributor to the initiation of stream departures from stable stream equilibrium. Energy dissipation techniques are varied and include such items as channels for level spreaders, plunge pools, and rip rap aprons.

Special Considerations and Exemplary Performance

There was no EP point available as related to this credit in LEED 2.2. LEED 2009 allows for an exemplary performance point to be earned if the project team develops a strategy to control both stormwater quantity (SSc6.1) and stormwater quality (SSc6.2) substantially beyond the associated credit criteria.

SS Credit 6.2: *Stormwater Management—Quality Control*

USGBC Rating System

SSc6.2 is worth one point in LEED 2009 and was worth one point in LEED 2.2. LEED-NC 2.2 lists the Intent, Requirements, and Potential Technologies and Strategies for this credit as follows, and is similar for LEED 2009

Intent

Reduce or eliminate water pollution by reducing impervious cover, increasing on-site infiltration, eliminating sources of contaminants, and removing pollutants from stormwater runoff.

Requirements

Implement a stormwater management plan that reduces impervious cover, promotes infiltration, and captures and treats the stormwater runoff from 90% of the average annual rainfall using acceptable best management practices (BMPs).

BMPs used to treat runoff must be capable of removing 80 percent of the average annual postdevelopment total suspended solids (TSS) load based on existing monitoring reports. BMPs are considered to meet these criteria if:

- they are designed in accordance with standards and specifications from a state or local program that has adopted these performance standards,

or

- there exists in-field performance monitoring data demonstrating compliance with the criteria. Data must conform to accepted protocol (e.g., Technology Acceptance Reciprocity Partnership (TARP), Washington State Department of Ecology) for BMP monitoring.

Potential Technologies and Strategies

Use alternative surfaces (e.g., vegetated roofs, pervious pavement or grid pavers) and nonstructural techniques (e.g., rain gardens, vegetated swales, disconnected imperviousness, rainwater recycling) to reduce imperviousness and promote infiltration, thereby reducing pollutant loadings.

Use sustainable design strategies (e.g., Low Impact Development, Environmentally Sensitive Designs) to design integrated natural and mechanical treatment systems such as constructed wetlands, vegetated filters, and open channels to treat stormwater runoff.

Calculations and Considerations

In LEED-NC 2.2 and 2009 the requirements for this credit have been simplified, are easily understood, and are different for different climate zones in the United States. The climate zone criteria for this differentiation are based on the average annual rainfall rates, and the zones have been divided into three watershed categories: humid, semiarid, and arid. By the LEED-NC 2.2 and 2009 definition, *humid* watersheds receive on average at least 40 in of rainfall per year, *semiarid* watersheds receive more than 20 and less than 40 in, and *arid* zones receive less than 20 in of rain annually.

The volume criteria for treating runoff for quality control for this credit are based on runoff from 90 percent of the average annual rainfall. For the purposes of this credit, 90 percent of the average annual rainfall may be estimated in a specific precipitation event as up to 1 in of rainfall for humid areas, 0.75 in of rainfall for semiarid areas, and 0.5 in of rainfall for arid zones. For all events less than the maximum volume criteria as so specified, all the associated runoff would be treated prior to discharge from the site; and for all storm events greater than the maximum specified, the runoff associated with this maximum criterion would need to be treated.

As far as the treatment requirement goes, it may be assumed according to this credit that all rain infiltrated on-site is 100 percent treated and any runoff that is discharged from the site from the specified maximum rainfall amount *must* be treated to attain an *80 percent TSS* removal. The removal method must be one that is either designed by acceptable state or local BMP performance standards or supported by appropriate in-field monitoring data. It is also assumed that any rainwater appropriately reused on-site does not contribute to the TSS load. However, it should be ascertained that the BMPs used adhere to any local and state health and environmental codes. There are many devices, such as rainwater harvesting systems, that have strict health code requirements for items such as length of time in the storage container, minimum treatment required, or differentiation from potable water systems and the "appearance" of potable water systems.

Treating for total phosphorus is no longer a requirement for this credit as it was in earlier versions of LEED. This seems reasonable to the author for a combination of reasons. First, much of the phosphorus is absorbed in, or part of, the suspended solids and will be removed by the TSS process. Second, there are usually two main nutrients that may promote eutrophication or other environmental problems in receiving waters: nitrogen and phosphorus. In general, phosphorus is the limiting nutrient for freshwater systems, and nitrogen is the limiting nutrient for saltwater bodies. Therefore, phosphorus is not always considered an important pollutant in many receiving waters. Many watersheds where phosphorus is the limiting nutrient already have discharge criteria limiting the load of phosphorus and are thus already protected.

A summary of the steps for complying with this credit is as follows:

1. Determine whether your project is in a humid, semiarid, or arid region.
2. Determine the minimum volume of water that the system must be designed to handle with a combination of on-site infiltration, appropriate on-site use, or treatment for 80 percent TSS removal prior to discharge.
   a. If in a humid region, multiply the site area by 1 in to determine this minimum volume of runoff that must be infiltrated, used appropriately on-site, or treated prior to discharge.
   b. If in a semiarid region, multiply the site area by 0.75 in to determine this minimum volume of runoff that must be infiltrated, used appropriately on-site, or treated prior to discharge.
   c. If in an arid region, multiply the site area by 0.5 in to determine this minimum volume of runoff that must be infiltrated, used appropriately on-site, or treated prior to discharge.
3. If WQ_{VH}, WQ_{VS}, and WQ_{VA} are defined as the water quality volumes for humid, semiarid, and arid climates, respectively, then the total rainfall which must be

	Annual Precipitation (in)	90 Percent of Annual Rainfall Estimated per Event (in)	Symbol for Volume to Be "Handled" per Event*	Volume to Be "Handled" per Event*
Humid	>40	1 over total site	WQ_{VH}	$(0.083\ ft)A_T$
Semiarid	20–40	0.75 over total site	WQ_{VS}	$(0.063\ ft)A_T$
Arid	<20	0.5 over total site	WQ_{VA}	$(0.042\ ft)A_T$

*Infiltrated on-site, used on-site, or treated prior to discharge.

TABLE 2.6.3 Regional Precipitation Area Definitions and Water Quality Volumes

infiltrated on-site, used on-site, or treated prior to discharge off-site can be calculated as follows:

$$WQ_{VH} = (0.083\ \text{ft})A_T \tag{2.6.4}$$

$$WQ_{VS} = (0.063\ \text{ft})A_T \tag{2.6.5}$$

$$WQ_{VA} = (0.042\ \text{ft})A_T \tag{2.6.6}$$

These water quality volumes and respective rainfall definitions by area are summarized in Table 2.6.3.

Best management practices (BMPs) are normally separated into two main categories: nonstructural and structural. For the purposes of LEED, nonstructural BMPs include such items as pervious pavements, swales, and disconnection of impervious areas to promote natural infiltration of rainwater into the soils (although others might consider swales and pervious pavements as structural controls). The structural controls include ponds, rainwater harvesting cisterns, constructed wetlands, and treatment devices in the stormwater collection system. Other agencies and organizations segregate the BMPs differently. More information on this topic can be found in Chap. 10.

For any of the methods used that involve infiltration, soil types, and soil infiltration rates must be provided to determine if the soils can handle the infiltration. This credit is based on a set volume and not on a peak rate, so a mass balance under dynamic conditions may not be easily shown. The LEED 2009 reference guide provides a 24-h period for these dynamic calculations. Likewise, for the other structural controls, a period of 24 h is used for any flow rates, discharges, and infiltration rates with respect to the water quality volume.

In all cases, for submittal, a list must be provided of all the BMPs used including a description of each, its pollutant removal capability, and the percent of rainfall that will be handled by the BMP for the site. If there are special circumstances or considerations for this credit, a narrative must be submitted describing these circumstances. Additional information on BMPs and stormwater options can be found in Chap. 10. Figure 2.6.1 shows an example site where constructed wetlands are used for both stormwater quantity and stormwater quality control. Figure 2.6.2 depicts signage by a stormwater pond educating the public on its importance.

FIGURE 2.6.1 Two interconnected constructed wetlands at a West Quad Residential Facility at the University of South Carolina. (*Photograph taken September 2006.*)

FIGURE 2.6.2 Signage educating the public on the importance of a stormwater pond in Bothell, WA. (*Photo taken by David Street on November 30, 2009.*)

Special Circumstances and Exemplary Performance

There was no EP point available as related to this credit in LEED 2.2. LEED 2009 allows for an exemplary performance point to be earned if the project team develops a strategy to control both stormwater quantity (SSc6.1) and stormwater quality (SSc6.2) substantially beyond the associated credit criteria.

2.7 SS Credit Subcategory 7: *Heat Island Effect*

SS Credit 7.1: *Heat Island Effect—Non-Roof*

The urban heat island effect is a phenomenon where daily temperatures are higher in urban areas than in the surrounding suburban and rural areas. It is thought to be caused by many factors some of which are an increase in paved and roofed surfaces that collect warmth from solar radiation, a decrease in foliage which can keep temperatures cooler by evapotranspiration, and an increased energy use in buildings that also warms the ambient air. This effect has many negative impacts including poorer air quality, an increase in energy demand for cooling, an increase in water demand, a reduction in building material durability, and an increase in human health concerns. Therefore, measures and construction practices which reduce the urban heat island effect are encouraged.

USGBC Rating System

SSc7.1 is worth one point in LEED 2009 and was worth one point in LEED 2.2. LEED-NC 2.2 lists the Intent, Requirements, and Potential Technologies and Strategies for this credit as follows, with modifications as noted for LEED 2009:

Intent

Reduce heat islands (thermal gradient differences between developed and undeveloped areas) to minimize impact on microclimate and human and wildlife habitat.

Requirements

OPTION 1

Provide any combination of the following strategies for 50% of the site hardscape (including roads, sidewalks, courtyards and parking lots):

Shade (within 5 years of occupancy for trees, or per LEED 2009 by solar panels, or by architectural shading features with a solar reflectance index [SRI] of at least 29)

Paving materials with a Solar Reflectance Index (SRI) of at least 29

Open grid pavement system (at least 50% pervious)

OR

OPTION 2

Place a minimum of 50% of parking spaces under cover (defined as underground, under deck, under roof, or under a building). Any roof used to shade or cover parking must have an SRI of at least 29, or per LEED 2009 be a green roof or covered by solar panels. LEED 2009 further clarifies that there is no SRI requirement for parking which is underground, under a deck, or under a building.

Potential Technologies and Strategies

Shade constructed surfaces on the site with landscape features and utilize high-reflectance materials for hardscape. Consider replacing constructed surfaces (i.e., roofs, roads, sidewalks, etc.) with vegetated surfaces such as vegetated roofs and open grid paving or specify high-albedo materials to reduce the heat absorption.

Calculations and Considerations

Option 1 Option 1 provides for reduced heat impacts from at least 50 percent of the site non-roof hardscape. This includes all parking areas (spaces, aisles, and other paved areas associated with parking areas), walks, roads, patios, and drives. Even water features are included in the hardscape if they are contained over impervious surfaces which can be exposed when dry. HS is defined as the sum of these non-roof hardscape areas. There are some special cases for which it is difficult to determine if a "hard" surface is a roof or non-roof or considered neither. One example is open stadium seating. These special exceptions should be reviewed through the USGBC CIR procedure.

It is very important to understand the criteria for each of the three pathways to do this. They are shaded, high SRI, or open grid pavement systems.

- *Shading*. Shading can be effected with vegetation, or if tree planting is not available, then architectural shading devices may be substituted, and LEED 2009 requires that these are either covered with solar panels or have an SRI of at least 29. Roofed areas do not count in the hardscape calculation. The vegetation used must be native or adaptive trees, large shrubs, and noninvasive vines (on trellises or other supportive structures as needed). Deciduous trees are many times preferred, especially in northern climates, as they allow for benefits from solar heating of the pavement in the winter months when their leaves are gone, such as reduced icing. Variable S is defined as the effective area of the hardscape that is shaded. It is the arithmetic mean of shade coverage as calculated at 10 a.m., noon, and 3 p.m. on the summer solstice. Although not stated in LEED-NC 2.2 or 2009, the time is probably what is accepted as the local time (daylight or standard) at the location.
- *SRI*. Variable R is defined as the area of the hardscape that has an SRI of at least 29. The definition of the SRI can be found in App. B. It represents a combination of high reflectance and high emissivity. Both emittance and reflectance should be calculated or determined according to the appropriate ASTM (American Society for Testing and Materials) standard as listed in the definition. These values can then be inserted into the LEED submittal template worksheet, and the appropriate SRI will be calculated. Some typical SRI values as given in LEED-NC 2.2 and 2009 are listed in Table 2.7.1. Colored concretes or surface coatings on pavements can be used to increase the values for this credit. Note that the reflectance value given is solar reflectance. There is also a commonly used parameter called *visible*

Material	Emissivity	Reflectance	SRI
Typical new gray concrete	0.9	0.35	35
Typical weathered* gray concrete	0.9	0.20	19
Typical new white concrete	0.9	0.70	86
Typical weathered* white concrete	0.9	0.40	45
New asphalt	0.9	0.05	0
Weathered asphalt	0.9	0.10	6

*Reflectance of surfaces can be maintained with cleaning. Typical pressure washing of cementitious materials can restore reflectance close to the original value. Weathered values are based on no cleaning.

Table 2.7.1 Some Typical SRI Values of Pavements

reflectance. The solar reflectance represents a wider range of wavelengths than the visible reflectance and is often slightly less than the visible reflectance for a specific material. It is important to make sure that the reflectance quoted by a manufacturer is the solar reflectance. If not, then the solar reflectance must be gotten elsewhere. Figures 2.7.1 and 2.7.2 show example photos of weathered

Figure 2.7.1 Example of weathered concrete and asphalt pavement surfaces in Columbia, S.C. The pavements are approximately 15 years old.

Figure 2.7.2 Example of weathered and new concrete pavement surfaces in Columbia, S.C. The weathered surface is approximately 15 years old, and the new surface is less than a month old.

concrete and weathered asphalt pavements, and weathered and new concrete pavement surfaces, respectively.

- *Open grid pavement systems.* Variable O is defined as the area of the hardscape which is covered in open grid pavement systems which are at least 50 percent pervious with vegetation in the open cells.

The three pathways for reducing the heat island effect by keeping at least 50 percent of the non-roof hardscape as shaded, high SRI, or open grid cannot be superimposed on one another. In other words, an area that is both shaded and high SRI cannot be counted twice. There is some discretion with the shading value as it is an arithmetic mean of various areas determined by the angle of the sun, and if there is any question as to whether the appropriate areas are used, then these should be well documented. The easiest way to perform the calculations is to give modified definitions for R and O such that

R_{mod} — R_{mod} is the sum of the areas with high SRI excluding those areas that are shaded.

O_{mod} — O_{mod} is the sum of the areas with open grid pavement systems that are not shaded or do not have high SRI values.

By using these definitions, compliance with Option 1 can be determined if the following inequality is true:

$$HS/2 \leq (S + R_{mod} + O_{mod}) \tag{2.7.1}$$

Option 2 The calculations for Option 2 are simpler. First determine the total number of parking spaces (TP) for the project and then count the number of these total spaces that are under cover (CP) for which the top roofing material above them has an SRI of at least 29 or is covered by solar panels or vegetation if under shading devices. Under cover includes underground, under the building, under a deck, and under shade structures such as canopies. If more than one-half of the parking spaces are under cover as so defined, then the credit is met. Note that this option only refers to the parking spaces themselves and does not mention anything about the parking aisles or other paved areas associated with these spaces. The equation for compliance with Option 2 is

$$CP \geq TP/2 \tag{2.7.2}$$

Special Circumstances and Exemplary Performance

Submittals for both options should include the calculations for each option. In addition, for Option 1, the submittals should include site drawings showing all the non-roof hardscape areas with their heat island–related characteristics (shading, SRI, or open grid) and areas noted. The surface SRIs given may either be the actual for each paving material or as provided for similar mixes and ages on other projects. For Option 2, the submittals should include plans with the number of total parking spaces and parking spaces under cover with the appropriate roofed SRI or solar panel if under canopies. Option 2 is a great example of how design team members need to work together to obtain the LEED credits. The parking layout may sometimes be determined by others than the architectural roofing designer. Only together will this option be met.

Concrete is a common material used for many surface treatments and other structural components of the built facilities. Many concretes are made with a combination

Criteria for	Option 1	Option 2
Credit point	$HS/2 \le (S + R_{mod} + O_{mod})$ (At least one-half the hardscape is specially adapted to reduce the heat island effect.)	$CP \ge TP/2$ (At least one-half of the parking spaces are covered*).
EP point	$HS = (S + R_{mod} + O_{mod})$ (All the hardscape is specially adapted to reduce the heat island effect.)	$CP = TP$ (All the parking spaces are covered*).

*If covered with canopies then these must be covered with solar panels or have an SRI of 29 or greater.

Table 2.7.2 Summary Criteria Equations for Heat Island Effect: Non-Roof LEED 2009

of Portland cement and fly ash (up to 25 percent). This use of fly ash is promoted as an environmental benefit as it recycles waste from another process which might otherwise be landfilled, and it also promotes a reduction in the use of energy and the production of carbon dioxide from the cement-making process. However, concrete with fly ash is usually darker than concrete made with only Portland cement. Therefore, it may be beneficial to use concrete for surface treatments that either have no fly ash or are coated to lighten the surface color and promote reflectivity, whereas it may be good to use concrete with fly ash for other site applications to aid in earning other credits.

If there are special circumstances or considerations for this credit, a narrative must be submitted describing these circumstances.

An EP point may be awarded if it can be demonstrated that either of the options is met with a 100 percent requirement. For Option 1, this means that *all* the non-roof hardscape surfaces have been constructed with high-SRI materials, shaded within 5 years, or open grid paving systems as previously defined. For Option 2, this means that *all* on-site parking spaces are located under cover as required in the credit.

A summary of the equations for credit compliance and awarding of an EP point for either option is presented in Table 2.7.2.

SS Credit 7.2: *Heat Island Effect—Roof*

USGBC Rating System

SSc7.2 is worth one point in LEED 2009 and was worth one point in LEED 2.2. LEED-NC 2.2 lists the Intent, Requirements, and Potential Technologies and Strategies for this credit as follows, with modifications for LEED 2009 as noted:

Intent

Reduce heat islands (thermal gradient differences between developed and undeveloped areas) to minimize impact on microclimate and human and wildlife habitat.

Requirements

OPTION 1

Use roofing materials having a Solar Reflectance Index (SRI) equal to or greater than:

- 78 for low-sloped roofs (slope ≤ 2V:12H)
- 29 for steep-sloped roofs (slope > 2V:12H)

for a minimum of 75% of the roof surface. (LEED 2009 provides a weighted rooftop SRI average for these materials for varying SRI materials. Only the weighted average of the included areas must meet the minimums.)

OR

OPTION 2

Install a vegetated roof for at least 50% of the roof area.

OR

OPTION 3

Install high albedo and vegetated roof surfaces that in combination meet the following criteria:

$$(\text{Area of SRI Roof}/0.75) + (\text{Area of Vegetated Roof}/0.5) \geq \text{Total Roof Area.}$$

Potential Technologies and Strategies

Consider installing high-albedo and vegetated roofs to reduce heat absorption. SRI is calculated according to ASTM E 1980. Reflectance is measured according to ASTM E 903, ASTM E 1918, or ASTM C 1549. Emittance is measured according to ASTM E 408 or ASTM C 1371. Default values will be available in the LEED-NC v2.2 Reference Guide. Product information is available from the Cool Roof Rating Council website, at www.coolroofs.org.

Calculations and Considerations

Note that this credit references solar reflectance which is different from visible reflectance, as solar reflectance covers a wider range of wavelengths than visible reflectance does. Therefore, any reflectance information from manufacturers should be checked to ensure that the proper reflectance is used in the SRI calculation. Table 2.7.3 gives some example SRIs for various roofing materials which have been taken from the referenced table in the LEED-NC 2.2 and 2009 Reference Guides. Roofing materials with the higher SRI values are sometimes referred to as *cool* roofs.

Also notice the big difference in the SRI requirement for low-slope versus steep-slope roofs. Flat roofs have a *much* stricter requirement than sloped roofs. Therefore, alternatives with some sloped roof surfaces may be preferable for obtaining the credit. Also note that cool roofs may not be as energy efficient in the colder months due to the impact of high emittance and low absorption on heating costs; however, on average, the annual overall energy use may still be reduced. In addition, structures in very northern climates tend to be highly insulated to keep warmth in during the summer and may not have any air conditioning for summer months. Thus, darker roofs in the summer would tend to overheat the buildings.

The equation for Option 3 is actually applicable for all three options. It is the only one necessary to use as the limits of this equation for not using any of the appropriately rated SRI roofing material in Option 1 default to the requirement in Option 2, and conversely, not having any vegetated roof area defaults to the requirement in Option 1. Therefore, the following write-up will only address the equation covering Option 3.

The first definition that must be understood is what is meant by *roof area*. As in land areas used in all the other credit calculations, roof areas are assumed to be the horizontal planar surface as would be projected on a site plan, not the actual slanted surface. [This horizontal planar surface is the building footprint (BF) as previously defined.] Therefore, a slanted shingled roof with an area as defined in this credit of 100 ft^2 would actually

Example SRI Values for Generic Material	Solar Reflectance	Infrared Emittance	Temperature Rise (°F)	SRI
Gray EPDM	0.23	0.87	68	21
Gray asphalt shingle	0.22	0.91	67	22
Unpainted cement tile	0.25	0.90	65	25
White granular surface bitumen	0.26	0.92	63	28
Red clay tile	0.33	0.90	58	36
Light gravel on built-up roof	0.34	0.90	57	37
Aluminum coating	0.61	0.25	48	50
White-coated gravel on built-up roof	0.65	0.90	28	79
White coating on metal roof	0.67	0.85	28	82
White EPDM	0.69	0.87	25	84
White cement tile	0.73	0.90	21	90
White coating—1 coat, 8 mils	0.80	0.91	14	100
PVC white	0.83	0.92	11	104
White coating—2 coats, 20 mils	0.85	0.91	9	107

Source: LBNL Cool Roofing Materials Database (http://eetd,lbl.gov/CoolRoofs). These values are for reference only and are not for use as substitutes for actual manufacturer data.

Table 2.7.3 Some Typical SRI Values for Roofing Materials

need well more than 100 squares of shingles to be covered. In addition, any roofed areas that hold equipment, solar energy panels, skylights, or other appurtenances (EQR as previously defined) are not included in the calculations. Using the site plan, the total roof surface area of the project (TR) would equal the sum of the horizontal projections of the roofs on the site plan (BF) less the sum of the roofed areas that hold equipment, solar energy panels, or other appurtenances (EQR). That is,

$$TR = BF - EQR \qquad (2.7.3)$$

In addition, the following variables should be determined from the associated areas on the site plan:

VR Sum of the vegetated roofed areas

RRL Sum of the low-slope roofed areas with an *areal weighted average* SRI greater than 78

RRS Sum of the steep-slope roofed areas with an *areal weighted average* SRI greater than 29

Note that for either type of roofed surfaces, it is not the sum of the areas with the minimum SRI, but rather a sum of areas whose areal weighted average SRI meets the minimum criteria. For example, for two steep-sloped roof areas, each of equal area, one with an SRI of 28 and the other with an SRI of 30, both areas could be counted with an average SRI of 29.

Using these definitions, the project would be eligible for this credit if the following inequality held true.

$$TR \leq \frac{RRL + RRS}{0.75} + \frac{VR}{0.5} \tag{2.7.4}$$

The first half of the right side of Eq. (2.7.3) corresponds to the requirements of Option 1, and the second part of the right side of Eq. (2.7.3) corresponds to the requirements for Option 2.

Special Circumstances and Exemplary Performance

There are some special cases for which it is difficult to determine if a hard surface is a roof or non-roof or considered neither. One example is open stadium seating. These special exceptions should be reviewed through the USGBC CIR procedure. If there are other special circumstances or considerations for this credit, a narrative must be submitted describing these circumstances.

An EP point may be awarded if it can be demonstrated that Option 2 is met with a 100 percent vegetated requirement. This means that *all* the roof areas excluding mechanical equipment, solar panels, skylights, and other similar appurtenances comprise green (vegetated) roof systems.

$$VR = TR \tag{2.7.5}$$

A summary of the criteria equations for SSc7.2 is given in Table 2.7.4.

Many websites are being developed that give alternatives for green roofs. In addition, ASTM International is developing an international standard, WK14283, "Guide for Green Roof Systems," which will aid in understanding concepts and terminology in green roof design. Figure 2.7.3 shows the front of a building which has

Criteria for	Option 1	Option 2	Option 3
Credit point	Given TR = BF – EQR, $TR \leq \frac{RRL + RRS}{0.75}$ (At least 75% of roofed areas meet set SRI criteria.)	Given TR = BF – EQR, $TR \leq \frac{VR}{0.5}$ (At least one-half of the roofed areas are vegetated.)	Given TR = BF – EQR, $TR \leq \frac{RRL + RRS}{0.75} + \frac{VR}{0.5}$ (Combination of Options 1 and 2)
EP point	n/a	Given TR = BF – EQR, VR = TR (All roofs are vegetated.)	n/a

Table 2.7.4 Summary Criteria Equations for Heat Island Effect: Roof LEED 2009

Figure 2.7.3 Front of the Learning Center at the West Quad at the University of South Carolina. The roof is partially vegetated and has walkways and architectural shading structures for pedestrian access and use.

a partially vegetated roof. The roof is used for outdoor receptions as well as a means to control the urban heat island effect.

Typically green roofs are divided into three categories: intensive, semi-intensive, and extensive. Extensive green roofs usually have a small substrate depth (3 to 6 in) for the vegetation to grow in and typically are planted with grasses and sedums which require no irrigation. These roofs do not require high load-bearing structures for support. Intensive green roofs, on the other hand, usually require high load-bearing support structures, can include many types of vegetation such as trees, and have a much deeper substrate (soil) depth, usually more than a foot. Semi-intensive roofs are somewhere in between. Figure 2.7.4 shows an experiment on plants to test for species that might grow best without irrigation on an extensive roof in central South Carolina. Figures 2.7.5, 2.7.6, and 2.7.7 are all photographs of the same new parking garage in Houston, Texas. Figure 2.7.5 shows a northern exposure, whereas Figs. 2.7.6 and 2.7.7 depict exterior and interior views from the southern wall, respectively. The "green wall" on the southern exposure may aid in obtaining many credits related to the sustainable sites category or perhaps an innovative point, even though it is not directly applicable to the SSc7.2 equations.

2.8 SS Credit 8: *Light Pollution Reduction*

USGBC Rating System

SSc8 is worth one point in LEED 2009 and was worth one point in LEED 2.2. LEED-NC 2.2 lists the Intent for this credit as follows:

Figure 2.7.4 Experiment to investigate species that grow well without irrigation on a roof in central South Carolina.

Figure 2.7.5 Shell Woodcreek office and multipurpose building complex in Houston, TX. Northern wall of new parking garage. (*Photo taken November 5, 2009 by Heidi Brakewood.*)

Figure 2.7.6 Shell Woodcreek office and multipurpose building complex in Houston, TX. Exterior view of southern wall of new parking garage showing innovative "green wall." (*Photo taken November 11, 2009 by Heidi Brakewood.*)

Figure 2.7.7 Shell Woodcreek office and multipurpose building complex in Houston, TX. Interior view of southern wall of new parking garage showing innovative "green wall." (*Photo taken November 11, 2009 by Heidi Brakewood.*)

Intent

Minimize light trespass from the building and site, reduce sky-glow to increase night sky access, improve nighttime visibility through glare reduction, and reduce development impact on nocturnal environments.

The requirements for SS credit 8 in LEED 2.2 and LEED 2009 include both Interior Lighting and Exterior Lighting restrictions. The interior lighting requirements are substantially changed from the 2.2 version to the 2009 version and some of the exterior requirements related to lighting power densities are not as stringent in the 2009 version as they were in the 2.2 version. For clarity, the requirements in both versions are presented, first for version 2.2 and then for version 2009.

Requirements

FOR INTERIOR LIGHT

LEED 2.2: The angle of maximum candela from each interior luminaire as located in the building shall intersect opaque building interior surfaces and not exit out through the windows.

OR

LEED 2.2: All non-emergency interior light shall be automatically controlled to turn off during non-business hours. Provide manual override capability for after hours use.

LEED 2009 INTERIOR LIGHTING OPTION 1: This option requires that all non-emergency interior luminaires with a direct line of sight to any openings in the building envelope have automatically reduced input power between 11 p.m. and 5 a.m. and that any manual overrides do not last more than 30 minutes. The openings can be transparent or translucent.

OR

LEED 2009 INTERIOR LIGHTING OPTION 2: This option requires that all non-emergency interior luminaires with a direct line of sight to any openings in the building envelope have automatic shielding that reduces the transmittance by at least 50% between 11 p.m. and 5 a.m. The openings can be transparent or translucent.

AND

FOR EXTERIOR LIGHTING

LEED 2.2 EXTERIOR LIGHTING (first requirement): Only light areas as required for safety and comfort. Do not exceed 80% of the lighting power densities for exterior areas and 50% for building facades and landscape features as defined in ASHRAE/IESNA Standard 90.1-2004, Exterior Lighting Section, without amendments.

LEED 2009 EXTERIOR LIGHTING (first requirement): Do not exceed the lighting power densities in ASHRAE 90.1-2007 (with errata but without addenda, although addenda may be used if consistently used across LEED).

(And)

LEED 2.2 and LEED 2009 EXTERIOR LIGHTING (second requirement): All projects shall be classified under one of the following zones, as defined in IESNA RP-33, and shall follow all of the requirements as listed for that specific zone:

LZ1 - Dark (Park and Rural Settings)

Design exterior lighting so that all site and building mounted luminaires produce a maximum initial illuminance value no greater than 0.01 horizontal and vertical footcandles at the site boundary and beyond. Document that 0% of the total initial designed fixture lumens are emitted at an angle of 90 degrees or higher from nadir (straight down).

LZ2 - Low (Residential Areas)

Design exterior lighting so that all site and building mounted luminaires produce a maximum initial illuminance value no greater than 0.10 horizontal and vertical footcandles at the site boundary and no greater than 0.01 horizontal footcandles 10 feet beyond the site boundary. Document that no more than 2% of the total initial designed fixture lumens are emitted at an angle of 90 degrees or higher from nadir (straight down). For site boundaries that abut public rights-of-way, light trespass requirements may be met relative to the curb line instead of the site boundary.

LZ3 – Medium (Commercial/Industrial, High-Density Residential)
Design exterior light so that all site and building mounted luminaires produce a maximum initial illuminance value no great than 0.20 horizontal and vertical footcandles at the site boundary and no greater than 0.01 horizontal footcandles 15 feet beyond the site. Document that no more than 5% of the total initial designed fixture lumens are emitted at an angle of 90 degrees or higher from nadir (straight down). For site boundaries that abut public rights-of-way, light trespass requirements may be met relative to the curb line instead of the site boundary.

LZ4 – High (Major City Centers, Entertainment Districts)
Design exterior lighting so that all site and building mounted luminaires produce a maximum initial illuminance value no greater than 0.60 horizontal and vertical footcandles at the site boundary and no greater than 0.01 horizontal footcandles 15 feet beyond the site. Document that no more than 10% of the total initial designed site lumens are emitted at an angle of 90 degrees or higher from nadir (straight down). For site boundaries that abut public rights-of-way, light trespass requirements may be met relative to the curb line instead of the site boundary. LEED 2009 also allows for a single luminaire at intersections of private driveways and public roadways in all the zones. At this location, the requisite site boundary is modified so that the maximum allowed footcandles may go beyond the actual site boundary to the centerline of the street for an area with a width of twice the driveway width.

Potential Technologies and Strategies
Adopt site lighting criteria to maintain safe light levels while avoiding off-site lighting and night sky pollution. Minimize site lighting where possible and model the site lighting using a computer model. Technologies to reduce light pollution include full cutoff luminaires, low-reflectance surfaces and low-angle spotlights.

Calculations and Considerations

To obtain a point for this credit in either version, one of the two interior lighting requirement options and *all* the applicable exterior lighting requirements must be met. The interior lighting requirement options are the same regardless of the use of the site or the location of the site. The exterior lighting requirements vary depending on site use, use density, and location. The exterior lighting requirements are divided into two main portions: the first limits the exterior lighting density, and the second limits the light distribution, particularly beyond the site, either over the property lines or upward into the heavens. Note that in all cases, it is important to try to interpret between existing standards and design for the lowest possible light levels while addressing safety, security, access, way finding, identification, and aesthetics.

Interior Lighting Requirement This requirement is very simple and can be met in two different ways. Interior lights should not shine out directly through windows or other openings in the building envelope, be they translucent or transparent. If they do, then all interior lighting other than emergency lighting shall be on automatic controls (with manual overrides) to be turned off during nonbusiness hours for LEED 2.2, or either have automatically reduced input power or shielding at the opening to the building envelope between 11 p.m. and 5 a.m. for LEED 2009.

To verify that the requirement for interior lighting is obtained, it is necessary to get illumination information from the manufacturer on all interior fixtures which may light any windows or other openings in the building. This information should include a lighting intensity photometric table or figure which shows the angle of maximum lighting intensity.

Exterior Lighting Requirements There are two main parts, each with several criteria, to the exterior lighting requirements that *must* be met. The first main part of the exterior lighting requirements is based on the ASHRAE/IESNA Standard 90.1-2004 for LEED 2.2 and 90.1-2007 for LEED 2009, Energy Standard for Buildings Except Low-Rise Residential—Lighting Section 9, Exterior Lighting Section, Table 9.4.5. This table is essentially identical in both versions. However, LEED 2.2 required stricter requirements than in Table 9.4.5 in 90.1, while LEED 2009 has modified the requirement to meet the standards in this table. Note that this standard does not apply to any low-rise residential uses including single-family homes. Therefore, this part is subject to interpretation for low-rise residential facilities (three habitable stories or fewer above grade); nor does this part apply to manufactured houses and buildings that do not use either electricity or fossil fuel. In addition, this does not apply to equipment or portions of building systems that use energy primarily for industrial, manufacturing, or commercial processes. Any project with these items should be examined individually by way of the USGBC CIR process and/or guidances as are being developed.

The second main part of the exterior lighting requirements limits the light distribution, particularly beyond the site, and differs according to the location of the site and the typical zoning uses in the neighborhood.

Exterior Lighting Part 1: For all construction other than low-rise residential uses and as further exempted, the first part of the project exterior lighting requirement for LEED 2009 states that the ASHRAE/IESNA Standard 90.1-2007, Exterior Lighting Section, with errata, but without addenda, Table 9.4.5 lighting powers densities (LPDs) are not exceeded. (LEED 2.2 formerly restricted Table 9.4.5. LPD levels to 50 percent for building facades and landscape features and 80 percent for all other exterior areas.)

The criteria in Table 9.4.5 of the ASHRAE/IESNA Standard 90.1 are given in units of watts per square foot, watts per lineal foot, or watts per unit relating to a specific type of site feature, and are all referred to as *lighting power densities* (LPDs). The criteria are divided into two categories: those that can be traded (flexibility on the areas and exact light locations) and those that cannot be traded. Those that can be traded apply to outside areas in general. The nontradable allowances are in addition to the tradable allowances. For instance, in LEED 2009, if there is an uncovered parking lot next to a building façade, then this parking lot is allowed the parking lot allowance (the value for the Uncovered Parking Lot LPD as given in Table 9.4.5 of the ASHRAE/IESNA Standard 90.1), plus the special building façade light fixtures (totaling, at maximum, the Building Façades LPD as given in Table 9.4.5 of the ASHRAE/IESNA Standard 90.1).

Examples of these LPDs from the ASHRAE/IESNA Standard 90.1 are given in Table 2.8.1. All the wattages for the fixtures in each area should be added up and divided by the appropriate unit: square feet, lineal feet, or each specific item (i.e., drive-up window), and then calculations should confirm that each does not exceed the requirement of the values in Table 2.8.1 for LEED 2009. Again, there are some exemptions listed. Do not

	Application	Lighting Power Densities (LPDs)
Tradable surfaces	Uncovered parking areas	
	- Parking lots and drives	0.15 W/ft^2
	Building grounds	
	- Walkways less than 10 ft wide	1.0 W/lin ft
	- Walkways 10 ft or greater	0.2 W/ft^2
	- Plaza areas	0.2 W/ft^2
	- Special feature areas	0.2 W/ft^2
	- Stairways	1.0 W/ft^2
	Building entrances and exits	
	- Main entries	30 W/lin ft of door width
	- Other doors	20 W/lin ft of door width
	Canopies and overhangs	
	- Canopies (freestanding and attached and overhangs)	1.25 W/ft^2
	Outdoor sales	
	- Open areas (including vehicle sales lots)	0.5 W/ft^2
	- Street frontage for vehicle sales lots in addition to open-area allowance	20 W/lin ft
Nontradable surfaces. These are special allowances that are in addition to the allowances in the tradable section.	Building façades	0.2 W/ft^2 or 5.0 W/lin ft
	ATMs and night depositories	270 W per location plus 90 W per additional ATM
	Guarded gatehouses or entrances	1.25 W/ft^2 of uncovered areas (covered areas are included in canopies)
	Loading areas for police, fire, EMS, or other emergency vehicle	0.5 W/ft^2 of uncovered areas (covered areas are included in canopies)
	Drive-up windows at fast food restaurants	400 W per drive-through
	Parking near 24-h retail entrances	800 W per main entry

Source: ASHRAE/IESNA Std 90.1-2004, Table 9.4.5, Energy Standard for Buildings Except Low-Rise Residential Buildings, Chapter 9: Lighting, ©American Society of Heating, Refrigerating and Air-Conditioning Engineers, Inc.

Table 2.8.1 Lighting Power Densities

include solar power lighting or other lighting that does not use electricity or fossil fuels, and this does not apply to equipment or portions of building systems that use energy primarily for industrial, manufacturing, or commercial processes. Or at least, this is what the author has interpreted from the LEED-NC 2.2 and 2009 Reference Guides.

Exterior Lighting Part 2: This requirement is in addition to one of the interior lighting requirement options and the exterior lighting requirement part 1, previously described. It is intended to limit the lighting distribution beyond the site, both horizontally and vertically. Since different uses have different lighting needs, the project site will have different light distribution requirements according to where it is and how it is used. There are four applicable groupings, and they are referred to as the *outdoor lighting zones* as given in IESNA RP-33 and as defined by the International Dark-Sky Association (IDA) and are given in Table 2.8.2.

The IDA is the International Dark-Sky Association and can be found at www.darksky.org. The Illuminating Engineering Society of North America (IESNA) *Recommended Practice Manual: Lighting for Exterior Environments* (RP-33) is the referenced standard for the definition of the lighting zones. LEED-NC 2.1 gives the 1999 version as the standard, while LEED-NC 2.2 does not reference the updated year. However, the LEED-NC 2.2 and 2009 Reference Guides specifically define the four lighting zones and the definitions as so described from the IDA website are given in Table 2.8.2. Note that Table 2.8.2 also gives examples for very dark lighting zones. These are not really applicable to new construction as they represent undeveloped or specially developed (observatory) areas only. Figure 2.8.1 is a photo of exterior lighting with shielding along the beach in Ft. Lauderdale which reduces light escape beyond the site.

Zone	Ambient Illumination	Typical Examples
LZ0	Very dark	"Critical dark environments, such as especially sensitive wildlife preserves, parks, and major astronomical observatories"
LZ1	Dark	"Developed areas in state and national parks, recreation areas, wetlands and wildlife preserves; developed areas in natural settings; areas near astronomical observatories; sensitive night environments; zoos; areas where residents have expressed the desire to conserve natural illumination levels"
LZ2	Low	"Rural areas, low-density urban neighborhoods and districts, residential historic districts. This zone is intended to be the default condition if a zone has not been established."
LZ3	Medium	"Medium to high-density urban neighborhoods and districts, shopping and commercial districts, industrial parks and districts. This zone is intended to be the default condition for commercial and industrial districts in urban areas."
LZ4	High	"Reserved for very limited applications such as major city centers, urban districts with especially high security requirements, thematic attractions and entertainment districts, regional malls, and major auto sales districts"

TABLE 2.8.2 Outdoor Lighting Zone Definitions from IDA Website (http://www.darksky.org/ordsregs/l-zones.html) as Accessed May 10, 2008

Figure 2.8.1 Shielded exterior lighting along the beach in Fort Lauderdale. Lighting might impact sea turtle habitat. (*Photo taken by David Street on December 28, 2009.*)

Each of the four lighting zones has four specific lighting distribution requirements that must be met. These limit the horizontal lighting intensity at the project property line, the vertical lighting intensity above the property line, the maximum amount of lighting intensity at a specific distance from the property edge, and the maximum amount of exterior lighting that can be directed with any angle which may go upward. These four lighting distribution requirements for each of the lighting zones are summarized in Table 2.8.3. Note that for three of the

Zone	Maximum Initial Horizontal Illuminance Value at Site Boundary (fc)	Maximum Initial Vertical Illuminance Value at Site Boundary (fc)	Number of Feet Beyond Site Boundary That a Maximum Initial Horizontal Illuminance of 0.01 fc Is Allowed	Maximum Allowed Percent of Total Initial Designed Fixture Lumens Emitted at an Angle of 90° or Higher from Nadir (MIDAL)
LZ1-dark	0.01	0.01	0	0
LZ2-low	0.10*	0.10*	10	2
LZ3-medium	0.20*	0.20*	15	5
LZ4-high	0.60*	0.60*	15	10

*For site boundaries that abut public rights-of-way, light trespass requirements may be met relative to the curb line instead of the site boundary. LEED 2009 also allows for additional lighting at private driveways abutting public roadways, with restrictions, in all the zones.

Table 2.8.3 Outdoor Lighting Zone Specific Exterior Lighting Requirements

zones, the site property line can be moved out to the curb line on portions of the lot that abut public rights-of-way. Also, for all zones, there is an allowance for a single luminaire at the intersection of a private driveway and a public roadway in LEED 2009, which can illuminate beyond the property boundary as previously noted.

Submittals for this part of the exterior lighting requirement must include site plans that show both the maximum horizontal and vertical illuminances at the site boundaries as well as values of the maximum horizontal illuminances extending out the stated distance(s) beyond the boundaries. The horizontal lighting intensity plans are typically referred to as *isoluxes* or *isofootcandle* plans and give either contours or spot values of luxes or footcandles. Vertical lighting distributions at all the property lines (or curb lines or as extended at driveways) should also be provided. Isoluxes show the additive effects of all the lighting fixtures in the area, and they cannot be based solely on each individual fixture.

The submittal for the fourth item in this part of the exterior lighting requirements needs to consider *all the exterior lighting fixtures on the site.* It is based on the initial designed fixture lighting capacity (lumens) for each, as most lighting fixtures have a reduction in lighting intensity with age. Determination of whether the requirement has been met, can be performed by following these three steps:

- The total initial designed lumens for all the exterior site fixtures should be added. This will be referred to as the *total initial designed lumens* (TIDL).
- In addition, any of the fixtures that have any light aimed at an angle which breaks above a horizontal plane with the fixture should have the associated lumens that shine in this direction summed over the site. This will be referred to as the *total initial designed angled lumens* (TIDAL).
- If the maximum allowed percent of total initial designed fixture lumens emitted at an angle of 90 degrees or higher from nadir is referred to as the **maximum initial designed angled lumens** (MIDAL), then this requirement can be met if Eq. (2.8.1) is valid. The applicable values for MIDAL for each zone are given in Table 2.8.3. The exception to Eq. (2.8.1) is in cases where no exterior lighting is installed; then the requirement is automatically met.

$$100(\text{TIDAL}/\text{TIDL}) \leq \text{MIDAL} \tag{2.8.1}$$

A strategy for meeting this fourth item in part 2 of the exterior lighting requirements may be to specify and install appropriate shielding and cutoffs on the exterior fixtures.

Special Circumstances and Exemplary Performance

Figure 2.8.2 gives a summary of the interior and exterior requirements necessary for fulfilling this credit in LEED 2009.

If there are special circumstances or considerations for this credit, a narrative must be submitted describing these circumstances. There is no EP point available as related to this credit.

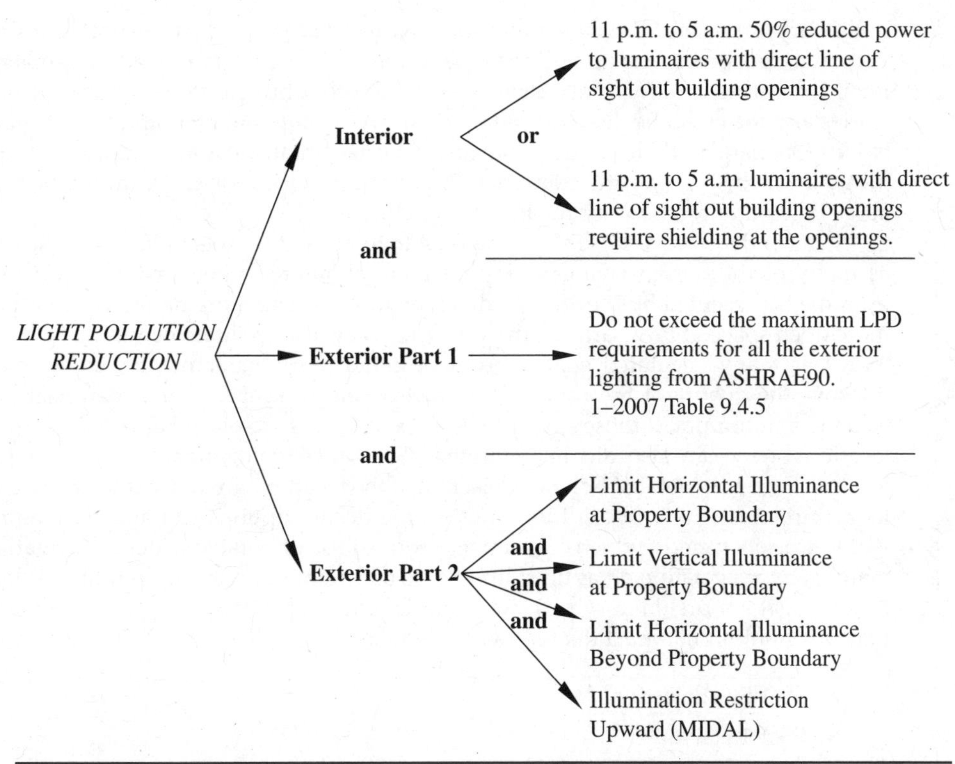

FIGURE 2.8.2 Summary of light pollution reduction credit for LEED 2009.

2.9 Discussion and Overview

Sustainable Site credits may sometimes be mutually exclusive, and some of them may never be applicable to a project or site. For example, if none of the available sites are brownfields, then there is no possibility of obtaining SS Credit 3. However, some seemingly opposing credits may actually be achievable for the same site. For instance, SS Credit 1 encourages spaciousness by keeping certain special areas undeveloped, whereas SS Credit 2 encourages denseness for more efficient use of infrastructure. However, many of the urban areas applicable for SS Credit 2 already meet the six criteria in SS Credit 1 as the area has already been developed and these special resources no longer, or never did, exist. What may be more difficult is to find a situation where both SS Credits 2 and 5 can be met. It may seem very difficult to promote dense urban development and yet create open space, but it is not impossible, especially when the parameters of the project allow for multistory structures. In addition, both SSc5.1 and SSc5.2 have modified requirements in dense urban environments to encourage open space but in a different manner as may be more readily doable in this environment.

It would be very difficult to meet SSc2 in the United States for a traditional retail grocery store, as U.S. customers usually use large shopping carts throughout the structure for shopping for their foodstuffs which are not made for stairwells and escalators. However, in other countries such as Switzerland and Norway, there are multistory food stores in urban areas, some with elevators for the transport of the carts. Although this practice is an unlikely one in the current culture of the United States, a retail food store could be very easily built with other businesses or offices on higher stories and the credit be attained. Why would we want to obtain both of these types of credits (SSc2 and SSc5) in dense urban areas? Dense urban areas and open space are not mutually exclusive, and the combination of both provides for access to infrastructure and a better environment for the human inhabitants.

Some of the credits should be evaluated together with respect to cost for the credit, as many of the design changes may facilitate obtaining other credits. A very good example is presented by the SSc5 credits which encourage more native vegetation and more open space. If these are sought, then they may also facilitate obtaining SSc6.1 and SSc6.2 as many vegetated spaces promote better stormwater management and can improve the quality of the runoff. It may also help in keeping the urban heat island effect to a minimum. If these specially landscaped areas are placed at certain perimeter locations, they can also aid in obtaining SSc8 by reducing lighting needs in these locations and help obtain SSc1, Site Selection, if the locations are where the development footprint is restricted. In addition, these same design options can aid in obtaining a Water Efficiency credit which encourages reduced use of potable water for irrigation as many native vegetation areas need little or no additional irrigation. In summary, taking care to design a preliminary plan with special landscape features at specific locations may at a minimum impact the following credits:

- SSc1 Site Selection
- SSc5.1 Habitat
- SSc5.2 Open Space
- SSc6.1 Stormwater Quantity
- SSc6.2 Stormwater Quality
- SSc7.1 Heat Island (non-roof)
- SSc8 Light Pollution
- WEc1 Water Efficient Landscaping

It was discussed earlier that site area and site boundaries are variables that are important for several credits which should be established early in the design process and used consistently for credit applications. Some other important items to establish early in the design phase are occupancies. These include employees and transient occupancies for commercial uses and design occupancies for residential facilities. These occupancies are not the maximum occupancies that are used for evaluation by fire marshals and other safety purposes. These ones are the expected normal occupancies of the facilities.

References

ACEEE (2007), http://www.aceee.org/, American Council for an Energy Efficient Economy website accessed May 14, 2007.

ACEEE (2007), http://www.greenercars.com, American Council for an Energy Efficient Economy website for green car listings accessed May 14, 2007.

Akan, A. O., and R. J. Houghtalen (2003), *Urban Hydrology, Hydraulics and Stormwater Quality, Engineering Applications and Computer Modeling,* John Wiley & Sons, Hoboken, NJ.

Allan, C. J., and C. J. Estes (2005), "A Morphological and Economic Examination of Plunge Pools as Energy Dissipaters in Urban Stream Channels," *Journal of the American Water Resources Association* (JAWRA), Paper No. 03065.

ASHRAE (2004), ASHRAE 90.1-2004: Energy Standard for Buildings Except Low-Rise Residential Buildings, American Society of Heating, Refrigerating and Air-Conditioning Engineers, Inc., Atlanta, GA.

ASHRAE (2007), 90.1-2007: Energy Standard for Buildings Except Low-Rise Residential Building, American Society of Heating, Refrigerating and Air-Conditioning Engineers, Atlanta, GA.

ASTM (2002), Standard Guide for Environmental Site Assessments: Phase II Environmental Site Assessment Process, E1903-97, accessed May 14, 2007, http://www.astm.org/cgib-in/SoftCart.exe/DATABASE.CART/REDLINE_PAGES/E1903.htm?E+mystore.

ASTM (2007) "Sustainability Subcommittee Launches Development of Proposed Green Roof Guide," Technical News Section of *ASTM Standardization News,* July, 35(7): 14–15.

Atlanta Regional Commission and Georgia Department of Natural Resources (2001), *Georgia Stormwater Management Manual,* vol. 2: *Technical Handbook.*

CARB (2007), http://www.arb.ca.gov/homepage.htm, California Air Resources Board website accessed May 14, 2007.

Carbone, G., and D. Yow (2006), "The Urban Heat Island and Local Temperature Variations in Orlando Florida," *Southeastern Geographer,* 46: 297–321.

CPESC (2007), http://www.cpesc.net/, Certified Professional in Erosion and Sediment Control website accessed May 14, 2007.

Davis, A. P. (2005) "Green Engineering Principles, Promote Low Impact Development," *ES&T,* August 15, pp. 338A–344A.

Davis, A. P., and R. H. McCuen (2005), *Stormwater Management for Smart Growth,* Springer, New York.

Ehrlich, P. R., and E. O. Wilson (1991), "Biodiversity Studies: Science and Policy," *Science,* 253(August 16): 758–761.

EPA (2007), http://cfpub.epa.gov/npdes/index.cfm, EPA website on NPDES accessed May 14, 2007.

EPA (2007), http://cfpub.epa.gov/npdes/stormwater/cgp.cfm, EPA website on NPDES Construction General Permit (CGP) accessed May 14, 2007.

EPA (2007), http://www.epa.gov/air/caa/, EPA website on the Clean Air Act accessed May 14, 2007.

EPA (2007), http://www.epa.gov/brownfields/, EPA website on Brownfields accessed May 14, 2007.

EPA (2007), http://www.epa.gov/heatisland EPA website on Heat Island Effect updated through January 16, 2007, accessed on July 12, 2007.

EPA (2007), http://www.epa.gov/owow/wetlands/regs/, EPA website on Federal Regulations governing wetlands accessed May 14, 2007.

Fay, J. A., and D. S. Golomb (2003), *Energy and the Environment,* Oxford University Press, New York.

FEMA (2007), http://gis1.msc.fema.gov/Website/newstore/viewer.htm, Federal Emergency Management Agency website regarding FIRM (Flood Insurance Rate Maps) accessed May 14, 2007.

Gelt, J. (2006), "Urban Heat Island-Higher Temperatures and Increased Water Use," *Arizona Water Resource,* September–October, 15: 1–12.

Green Roofs (2007), http://www.greenroofs.net/, Green Roofs for Healthy Cities website accessed July 12, 2007.

Gregory, R. L., and C. E. Arnold (1932), "Runoff—Rational Runoff Formulas," *Tranactions of the American Society of Civil Engineers,* 96: 1038.

Haselbach, L. M., S. Valavala, and F. Montes (2006), "Permeability Predictions for Sand Clogged Portland Cement Pervious Concrete Pavement Systems," *Journal of Environmental Management,* October, 81(1): 42–49.

Heat Island Group (2000), http://eetd.lbl.gov/HeatIsland/, Lawrence Berkeley National Laboratory (LBNL) Heat Island Group, August 30, 2000, website accessed July 12, 2007.

ITE (2003), *Trip Generation,* 7th ed., Institute of Transportation Engineers, Washington, D.C.

ITE (2004), *Parking Generation,* 3d ed., Institute of Transportation Engineers, Washington, D.C.

Moran, A. C., W. F. Hunt, and G. D. Jennings (2004), "A North Carolina Field Study to Evaluate Green Roof Quantity, Runoff Quality, and Plant Growth," in *Proceedings of Green Roofs for Healthy Cities Conference,* Portland, Ore.

NCDENR, DLR (2006), *Erosion and Sediment Control Planning and Design Manual,* North Carolina Department of Environment and Natural Resources, Division of Land Resources, Raleigh.

NCDENR, DWQ (2007), *Stormwater Best Management Practices Manual,* July, North Carolina Department of Environment and Natural Resources, Division of Water Quality, Raleigh.

NCHRP (2006), *Evaluation of Best Management Practices for Highway Runoff Control,* National Cooperative Highway Research Program Report 565, Transportation Research Board (TRB), Washington, D.C.

Prince George's County, Department of Environmental Resources (1999), *Low-Impact Development Design Strategies: An Integrated Design Approach,* Prince George's County, Md.

SCDHEC (2003), *South Carolina Stormwater Management and Sediment Control Handbook for Land Disturbance Activities,* August, Accessed July 6, 2007, South Carolina Department of Health and Environmental Control, Bureau of Water, Office of Ocean and Coastal Resource Management. http://www.scdhec.gov/environment/ocrm/pubs/docs/swmanual.pdf

Schueler, T. R. (1987), *Controlling Urban Runoff: A Practical Manual for Planning and Designing Urban BMPs,* Department of Environmental Programs, Metropolitan Washington Council of Governments.

USCFR (2003), *Code of Federal Regulations, Title 7, Volume 6,* revised as of January 1, 2003, http://a257.g.akamaitech.net/7/257/2422/14mar20010800/edocket.access.gpo.gov/cfr_2003/7CFR657.5.htm, Government Printing Office website accessed May 14, 2007.

USGBC (2003), *LEED-NC for New Construction, Reference Guide,* Version 2.1, 2d ed., May, U.S. Green Building Council, Washington, D.C.

USGBC (2005–2007), *LEED-NC for New Construction, Reference Guide,* Version 2.2, 1st ed., U.S. Green Building Council, Washington, D.C., October 2005 with errata posted through Spring 2007.

USGBC (2009), LEED *Reference Guide for Green Building Design and Construction,* 2009 Edition, U.S. Green Building Council, Washington, D.C., April 2009.

Valavala, S., F. Montes, and L. M. Haselbach (2006), "Area Rated Rational Runoff Coefficient Values for Portland Cement Pervious Concrete Pavement," *ASCE Journal of Hydrologic Engineering,* American Society of Civil Engineers, Reston, Va., May–June, 11(3): 257–260.

Zheng, J., H. Nanbakhsh, and M. Scholz (2006), "Case Study: Design and Operation of Sustainable Urban Infiltration Ponds Treating Storm Runoff," *ASCE Journal of Urban Planning and Development,* March, 132(1): 36–41.

Exercises

1. Put together a CPM schedule for items for ESC for a project. (See Chap. 8 for CPM schedules.)

2. Make a site plan with ESC for a project.

3. You are designing a building on 0.4 acre. It will be three stories high, and each story will have about 12,000 ft^2.

A: Determine the development density of the proposed building.
B: Determine the density radius (DR) from this building to be used in the areal development density determination.
C: Using the map in Fig. 2.2.1, please sketch in this density radius.
D: Using the data in Table 2.E.1, calculate the average areal development density around this proposed building. Only use the data that are appropriate.
E: Based on the building's and the average area's development densities, is this project eligible for Credit SSc2 of LEED-NC 2009?

4. You are designing a building on 0.5 acre. It will be four stories high, and each story will have about 11,000 ft^2.

A: Determine the development density of the proposed building.
B: Determine the density radius from this building to be used in the areal development density determination.
C: Using the map in Fig. 2.2.1, sketch in this density radius.
D: Using the data in Table 2.E.1, calculate the average areal development density around this proposed building. Only use the data that are appropriate.
E: Based on the building's and the average area's development densities, is this project eligible for Credit SSc2 of LEED-NC 2009?

5. You are designing a building on 0.3 acre. It will be three stories high, and each story will have about 10,000 ft^2.

A: Determine the development density of the proposed building.
B: Determine the density radius from this building to be used in the areal development density determination.
C: Using the map in Fig. 2.2.1, sketch in this density radius.
D: Using the data in Table 2.E.1, calculate the average areal development density around this proposed building. Only use the data that are appropriate.
E: Based on the building's and the average area's development densities, is this project eligible for Credit SSc2 of LEED-NC 2009?

Building	Bldg. Floor Area (ft²)	Site Area (acres)	Building	Bldg. Floor Area (ft²)	Site Area (acres)
A	33,000	0.4	W	19,000	0.6
B	87,400	1.6	X	6,000	0.3
C	6,000	0.3	Y	5,000	0.3
D	27,900	0.3	Z	4,300	0.2
E	66,000	1.2	AA	33,000	0.4
F	14,000	1.4	BB	87,400	1.6
G	12,900	0.2	CC	6,000	0.3
H	6,000	0.1	DD	27,900	0.3
I	14,000	0.2	EE	Park	0.5
J	29,000	0.4	FF	14,000	1.4
K	17,000	0.3	GG	12,900	0.2
L	9,000	0.3	HH	6,000	0.1
M	24,000	0.6	II	14,000	0.2
N	28,700	0.3	JJ	29,000	0.4
O	6,700	0.2	KK	17,000	0.3
P	39,900	0.4	LL	9,000	0.3
Q	344,000	2.5	MM	24,000	0.6
R	91,300	1.9	NN	33,000	0.4
S	22,000	0.3	OO	66,000	1.2
T	33,000	0.5	PP	Vacant	1.1
U	42,000	0.5	QQ	9,000	0.3
V	14,000	0.8	RR	24,000	0.6

TABLE 2.E.1 Development Density Data for Exercises

6. You are planning to build a textile factory. You plan to have 50 full-time workers and 20 half-time workers on the day shift, 30 full-time workers and 2 half-time workers on the evening shift, and 20 full-time workers on the night shift.

- **A:** Determine the maximum shift FTE.
- **B:** To get SS Credit 4.2, how many bicycles must you provide secure storage for and how many showers must you provide?
- **C:** If the proposed parking by local regulations is 75 spaces, how many of these spaces must be designated for only carpool/vanpool to get SS Credit 4.4?
- **D:** What might be done to obtain SS Credit 4.3?

7. You are planning to build a college hall. There are 55 offices for faculty and support staff, plus a daytime maintenance crew of four employees. There are 10 classrooms. Six of the classrooms typically have a maximum capacity of 30 per fire regulations, and four of the classrooms have a maximum

Surface	Rational Runoff Coefficient	Area
Roofed	0.95	2,000 ft^2
Wooded	0.25	0.25 acre
Landscape Beds	0.4	4,500 ft^2
Asphalt Parking and Drives	0.9	10,500 ft^2
Grass Picnic Area	0.45	900 ft^2
Concrete Drive-Through Lane	0.9	700 ft^2

TABLE 2.E.2 Data for Stormwater Exercise

capacity of 80 people. There are four computer laboratories, each with 20 computers. There are 10 research laboratories with attached offices for two graduate students each. To do the following calculations, first list your assumptions.

A: Determine the maximum shift FTE.

B: To get SS Credit 4.2, how many bicycles must you provide secure storage for and how many showers must you provide?

C: If the proposed parking by local regulations is 75 spaces, how many of these spaces must be designated for only carpool/vanpool to get SS Credit 4.4?

D: What might be done to obtain SS Credit 4.3?

8. You are a consultant for a major fast food restaurant chain which plans to totally rebuild many of its restaurants in the Columbia, S.C., area. Using the data in Table 2.E.2, calculate the LEED percent imperviousness for an existing fast food restaurant site in the northeast section of the town. Which option do you need to follow to go for SS Credit 6.1? Is this also eligible for an RP point? (See Chap. 7.)

9. The fast food chain decides to raze and rebuild another fast food restaurant in the central part of the city which was built similar to the design in Exercise 8, except the concrete drive-through lane is made of pervious concrete and there is an additional 0.10 acre of wooded land on the lot. What is the percent imperviousness of this site?

10. You want to analyze what must be done to obtain LEED 2009 SS Credit 5.1 with the lot in Exercise 8. How much area (if any) would need to be restored to obtain this credit, and what areas would you suggest that you use for it? Draw a quick sketch of the site. (Use your imagination.)

11. You want to analyze what must be done to obtain LEED 2009 SS Credit 5.1 with the lot in Exercise 9. How much area (if any) would need to be restored to obtain this credit, and what areas would you suggest that you use for it? Draw a quick sketch of the site. (Use your imagination.)

12. You want to analyze what must be done to obtain LEED 2009 SS Credit 5.2 with the lot in Exercise 8. The local code requires a minimum of 20 percent open space. Do you meet the LEED open-space requirement with the existing layout? If not, how much additional area would need to be dedicated to open space to obtain this credit, and what areas would you suggest that you use for it? Draw a quick sketch of the site. (Use your imagination.)

13. You want to analyze what must be done to obtain LEED 2009 SS Credit 5.2 with the lot in Exercise 8. The local zoning requirements do not address open space. Do you meet the LEED open-space requirement with the existing layout? If not, how much additional area would need to be dedicated to open space to obtain this credit, and what areas would you suggest that you use for it? Draw a quick sketch of the site. (Use your imagination.)

14. You want to analyze what must be done to obtain LEED 2009 SS Credit 5.2 with the lot in Exercise 8. The local code requires a minimum of 40 percent open space. Do you meet the LEED open-space requirement with the existing layout? If not, how much additional area would need to be dedicated to open space to obtain this credit, and what areas would you suggest that you use for it? Draw a quick sketch of the site. (Use your imagination.)

15. You wish to get credit for the heat island effect by using a light concrete paving material for part of the paved surfaces on a lot. If the parking areas cover 20,000 ft^2 and there is an additional 4000 ft^2 of walks and drives, what minimum area of these paved surfaces must be the lighter paving material?

16. If you instead provide tree shading for 2000 ft^2 of the parking area in Exercise 15, what area in square feet, other than the shaded areas, must have the higher albedo surface to still be eligible for this credit?

17. A site has 20,000 ft^2 of land covered by buildings and hardscape. It also has 21,000 ft^2 of land covered by landscaping. Assume that the average rational runoff coefficients for the development footprint and the landscaped areas are 0.95 and 0.30, respectively. What is the percent imperviousness? What is the percent impervious? Which option is the appropriate one in SSc6.1? Why?

18. Two buildings are being built on a lot. One is a 2000 ft^2 tarred flat roof storage building. The other is an office building complex with a 4:12 pitch (33 percent slope) with a white coating on a metal roof. The actual sloped area of the roof on the office complex is 6400 ft^2. Will you be eligible for SSc7.2?

19. Go outside your classroom building and itemize all the light fixtures that you see. Incorporate these into the categories in Table 2.8.1. Sum the wattages for each category in Table 2.8.1. If wattages are unknown, then use estimated wattages or those suggested by your instructor.

20. Go to your favorite fast food restaurant. Itemize all the exterior light fixtures that you see. Incorporate these into the categories in Table 2.8.1. Sum the wattages for each category in Table 2.8.1. If wattages are unknown, then use estimated wattages or those suggested by your instructor.

21. Go to your bank. Itemize all the exterior light fixtures that you see. Incorporate these into the categories in Table 2.8.1. Sum the wattages for each category in Table 2.8.1. If wattages are unknown, then use estimated wattages or those suggested by your instructor.

22. Which, if any, of the SS credits are available for Regional Priority points in the zip code of your work or school? (See Chap. 7.)

23. Which, if any, of the SS credits are available for Regional Priority points in the following major cities? (See Chap. 7.)

A: New York City
B: Houston, TX
C: Los Angeles, CA
D: Portland, OR
E: Miami, FL
F: Kansas City, MO

CHAPTER 3

LEED Water Efficiency

The Water Efficiency (WE) category of the U.S. Green Building Council (USGBC) rating system for new construction version 2.2 (LEED-NC 2.2) and version 2009 (LEED-NC 2009) consist of three credit subcategories which together may earn a possible five points in version 2.2 and 10 points in version 2009 maximum, not including exemplary performance (EP) points. New to version 2009 is the WE prerequisite 1: Water Use Reduction Required. The notation format for the credits is, for example, WEc1 for Water Efficiency credit 1. Also, in this chapter, to facilitate easier cross-referencing between this text and the USGBC rating system, the second digit in a section heading, equation, table, or figure number represents the credit subcategory number for sections that deal directly with a USGBC LEED-NC 2.2 or 2009 credit subcategory.

The Water Efficiency portion deals with issues that reduce the use of potable water at the site and the discharge of wastewater from the site. This will help limit the amounts of freshwater drawn from our water bodies and aquifers, and then treated for distribution and use, which strain our freshwater supplies and our water infrastructure. It will also serve to reduce the wastewater volumes discharged to these receiving bodies. Although several of the credits in Water Efficiency might include a capital cost investment, all provide a means to potentially reduce the associated utility costs over the life of the building.

The three subcategories are Water Efficient Landscaping (WE credits 1.1 and 1.2 in version 2.2 and WE credit 1 in version 2009), Innovative Wastewater Technologies (WE credit 2 in both versions), and Water Use Reduction (WE credits 3.1 and 3.2 in version 2.2 and WE credit 3 in version 2009). Basically, the first credit subcategory limits the use of potable water for outdoor uses, the second limits the production or discharge of wastewaters (black), and the third limits the use of potable water for indoor uses, as does the new LEED 2009 WEp1. The points for the credits in the Water Efficiency category are not necessarily individually related to a specific credit. Table 3.0.0 is a summary of the prerequisites and credits in both versions (2.2 and 2009) with a comparison of the points available and a summary of many of the major differences.

The prerequisite and credit subcategories available in the Water Efficiency category are as follows:

- **WE Prerequisite 1** (WEp1): Water Use Reduction Required (no previous versions to LEED 2009)
- **WE Credit 1** (WEc1): *Water Efficient Landscaping* (previously referred to as *Water Efficient Landscaping:* Reduce by 50% in LEED-NC 2.2 or *50% Reduction* in LEED-NC 2.1 and also *No Potable Water Use or No Irrigation* in these earlier versions and labeled as WEc1.1 and WEc1.2)
- **WE Credit 2** (WEc2): *Innovative Wastewater Technologies*

Subcategory 2009	Subcategory 2.2	Pts 2009	Pts 2.2	EP Pts 2009	EP Pts 2.2	Major Comments
WEp1: Water Use Reduction Required	None	na	–	na	–	*†Formerly LEED 2.2 WEc1 worth 1 point.
WEc1: Water Efficient Landscaping	WEc1.1: Water Efficient Landscaping. Reduce by 50%	2–4	1	–	–	• In LEED 2009, 2 points are earned for the 50% reduction (formerly WEc1.1). An additional 2 points are earned for No Potable Water use or No Irrigation (LEED 2.2 WEc1.2). • Groundwater seepage pumped away from foundations may be used to replace potable water if stormwater management systems are not affected.
	WEc1.2: Water Efficient Landscaping. No Potable Water Use or No Irrigation		1		–	
WEc2: Innovative Wastewater Technologies	WEc2: Innovative Wastewater Technologies	2	1	1	1	• Point value for similar criteria doubled except for EP. • LEED 2009 allows 3.5 gpf baseline for blowout water closets.
WEc3: Water Use Reduction	WEc3.1: Water Use Reduction. 20% Reduction	2–4	1	1 for 45% reduction. No Process Water reduction EP pt.	1 for 40% reduction and 1 for reducing process water 10%	*†The 20% reduction worth 1 point for LEED 2.2 is now the WEp1 for LEED 2009. Points for LEED 2009 are 2, 3, and 4 for reducing baseline 30, 35, and 40%, respectively.
	WEc3.2: Water Use Reduction. 30% Reduction		1			
	Total	10	5	2	3	

*Calculation variations for 2009 for water use reduction include: Addressing residential showers with multiple showerheads (Can allow additional flow for areas greater than 2500 in^2), baseline faucet flow at 2.2 gpm for private (e.g., hotel room) and residential applications, 0.5 gpm for other commercial lavatory uses with a 0.25 gal baseline for metered faucets and a listing of EPA Watersense options for design including 1.28 gpf WC, 1.5–2 gpm showerheads, and 1.5 gpm private lavatory fixtures. LEED 2009 gives longer residential shower, kitchen sink and lavatory faucet durations (8, 1, and 1 min, respectively) than for commercial.

†FTE calculations use the Default Occupancy Counts in Appendix 1 when occupancy is not known (e.g., Core and shell project prior to acquiring tenants).

Table 3.0.0 WE Summary Table of Major Differences between Version 2.2 and Version 2009 and Associated Points (Pts)

- **WE Credit 3** (WEc3): *Water Use Reduction* (previously referred to as Water Use Reduction: 20% Reduction and additionally Water Use Reduction: 30% Reduction in the earlier versions and labeled as WEc3.1 and WEc3.2)

As in Chap. 2, the submittals for each credit or prerequisite have been sectioned into either Design Submittals and/or Construction Submittals by the USGBC. Even though the entire LEED process encompasses all phases of a project, the prerequisites or credits that are noted as Construction Submittal credits have major actions going on through the construction phase which, unless documented through that time, may not be able to be verified from preconstruction documents or the built project. All the credits in the Water Efficiency category are listed as Design Submittal in the LEED-NC 2.2 Reference Guide and therefore are not noted as Construction Submittals in Table 3.0.1.

None of the credits in the Water Efficiency category have the icon EB noted after their title. This icon is a tool to help those who wish to proceed with the continuing LEED-EB certification (Existing Building) after certification of LEED-NC (New Construction or Major Renovation) is obtained. Those credits noted with this icon are usually significantly more cost-effective to implement during the construction of the building than later during its operation. This is also noted in Table 3.0.1. There are many prerequisites or credits which might have variations applicable to the design or implementation of the project related to regional conditions in the United States. All the prerequisites and credits for which regional variations might be applicable are noted in the LEED 2009 Reference Guide and are so listed in Table 3.0.1 for the Water Efficiency category.

Credit	Construction Submittal	EB Icon	Regional Variations May Be Applicable in LEED 2009	Location of Site Boundary and Site Area Important in Calculations
WEp1: Water Use Reduction Required		No	Yes	Yes
WEc1: Water Efficient Landscaping		No	Yes	Yes
WEc2: Innovative Wastewater Technologies		No	Yes	Perhaps Option 2
WEc3: Water Use Reduction		No	Yes	No

TABLE 3.0.1 WE EB Icon, Regional Variation Applicability, and Miscellaneous LEED 2009

The different credit subcategories in the Water Efficiency category are worth different point values. The first credit subcategory, *Water Efficient Landscaping*, was worth two points in LEED 2.2, one each for reducing potable water use by 50 percent (LEED 2.2 WEc1.1) and one additional for using no potable water or no irrigation (LEED 2.2 WEc1.2). This subcategory is worth from two to four points in LEED 2009 in a combined single subcategory notation (LEED 2009 WEc1), with two points earned for the 50 percent reduction and an additional two points awarded for the cases of no potable water use or no irrigation. The second subcategory, *Innovative Wastewater Technologies*, was worth one point in version 2.2 and two points in version 2009. The third subcategory, *Water Use Reduction*, has also been combined from two sections in LEED 2.2 (LEED 2.2 WEc3.1 and LEED 2.2 WEc3.2) into a single notation for LEED 2009 (LEED 2009 WEc3). In LEED 2.2, the first credit was worth one point for a 20 percent reduction in potable water use and the second worth an additional point for a 30 percent reduction. In LEED 2009, the 20 percent reduction is now a requirement in the new prerequisite (LEED 2009 WEp1) and in the combined credit section (LEED WEc3), two, three, or four points are awarded for reductions of 30, 35, or 40 percent, respectively. Summaries of the credits and points for both versions can be found in Table 3.0.0. There are EP points under the Innovation and Design category available which relate to two items in the Water Efficiency category. These are available for exceeding each of the respective credit criteria to a minimum level as noted in the credit descriptions for WEc2 and WEc3. These are noted in Table 3.0.0.

In LEED 2.2, the Water Use Reduction subcategory could also be the basis for an additional ID credit. It was not called an exemplary performance point per se, since it is not exceeding the percentages in the credit, but it was given for indoor water use reductions for water fixtures that are not specifically covered in LEED 2.2 WEc3.1 and LEED 2.2 WEc3.2, although the basis for the calculations is dependent on the design indoor water rates for the fixtures as regulated in LEED 2.2 WEc3.1 and LEED 2.2 WEc3.2. It is based on a reduction in other fixture water uses (non-EPAct fixtures) equal to or greater than 10 percent of the design water use rate for the fixtures (EPAct fixtures) as calculated for LEED 2.2 WEc3.1 and LEED 2.2 WEc3.2. EPAct stands for the Energy Policy Act of 1992. This is a federal act that addresses energy and water use in commercial, institutional, and residential facilities. This innovative point is not mentioned in LEED 2009.

The second subcategory in the Water Efficiency category also has an EP point associated with it for achievement of a 100 percent reduction of sewage conveyance or treatment on-site in addition to a 50 percent reduction.

The items available for EP points in this category are shown in Table 3.0.0. Note that only a maximum total of four exemplary performance points are available for a project for LEED 2.2 and only a maximum of three are available in LEED 2009, and they may be from any of the noted exemplary performance point options in any of the SS, WE, EA, MR, or IEQ LEED categories (see Chap. 7).

Table 3.0.1 also has a column that notes the importance of the site boundary and site area in the credit calculations or verification. As mentioned in Sustainable Sites, these are variables that should be determined early in a project as they impact many of the credits. Some of the credits are easier to obtain if the site area is less, while others are easier to obtain if the site area is greater. Determination of the site area may sometimes be flexible, particularly for campus locations, but it should be reasonable and must be consistent throughout the LEED process. It must also be in agreement with other nearby project submissions. The site area is mentioned again in Chap. 8 as an important variable to analyze early and set in a project.

Water Efficiency Prerequisite

WE Prerequisite 1: Water Use Reduction Required

This prerequisite is new to LEED 2009 and is essentially what was worth one point under water use reduction in LEED 2.2. It requires that the design water use for the building, not including irrigation, is 20 percent less than a baseline. The calculations are different for commercial uses and residential uses and include only the following fixtures or fixture fittings, if included in the project: water closets, urinals, lavatory faucets, showers, kitchen sink faucets, and prerinse spray valves.

The values to use for the baseline and the applicable calculations are in the section of this book entitled 3.3 WE Credit Subcategory 3: *Water Use Reduction*.

3.1 WE Credit Subcategory 1: Water Efficient Landscaping

The intention of this credit subcategory is to reduce the use of potable water or natural surface waters or natural groundwater for outdoor irrigation. In LEED 2.2, WEc1.1 was worth one point, and WEc1.2 was worth one point *in addition to* the point for WEc1.1. In LEED 2009, these two are combined into a single WEc1 subcategory worth either two points or four points depending on the extent of compliance.

Potable water is water that is assumed to be acceptable for human consumption. Strategies include using more drought-tolerant plants in landscape beds near pavements and buildings, keeping native vegetation when possible, and using recycled waters for irrigation. Figure 3.1.1 shows a landscape bed where several drought-tolerant species

FIGURE 3.1.1 Landscape bed in a subdivision in Columbia, S.C., where several drought-tolerant species reduce irrigation needs. (*Photograph taken July 14, 2007.*)

(a)

(b)

Figure 3.1.2 (a) and (b) Shell Woodcreek office and multi-purpose building complex in Houston, TX views of native and adapted plantings. (*Photos taken November 11, 2009 by Heidi Brakewood.*)

were used next to a drive and parking area. Figure 3.1.2 depicts photos of native or adapted plantings at a new facility in Houston, Texas.

WE Credit 1: *Water Efficient Landscaping*

The former LEED 2.2 WEc1.1 and WEc1.2 (*Water Efficient Landscaping: Reduce by 50 percent and No Potable Water Use or No Irrigation*, respectively) is now combined into the single LEED WEc1: *Water Efficient Landscaping*.

USGBC Rating System WE Credit 1

LEED-NC 2.2 lists the Intent, Requirements, and Potential Technologies and Strategies for LEED 2.2 WEc1.1 and LEED 2.2 WEc1.2 as follows, with modifications to reflect the changes in LEED 2009:

Intent

Limit or eliminate the use of potable water, or other natural surface or subsurface water resources available on or near the project site, for landscape irrigation.

Requirements

(LEED 2009 Option 1 worth two points) Reduce potable water consumption for irrigation by 50% from a calculated mid-summer baseline case. Reductions shall be attributed to any combination of the following items:

- Plant species factor
- Irrigation efficiency
- Use of captured rainwater
- Use of recycled wastewater
- Use of water treated and conveyed by a public agency specifically for nonpotable uses

Per LEED 2009, groundwater seepage that is pumped away from building slabs and foundations may be used for irrigation needs, but only if it is demonstrated that this does not affect stormwater management systems.

(LEED 2009 Option 2 worth an additional two points) Achieve the requirements for Option 1 and:

LEED 2009 Path 1; Use only captured rainwater, recycled wastewater, recycled greywater, or water treated and conveyed by a public agency specifically for nonpotable uses for irrigation.

OR

LEED 2009 Path 2; Install landscaping that does not require permanent irrigation systems. Temporary irrigation systems used for plant establishment are allowed only if removed within one year of installation.

Potential Technologies and Strategies

Perform a soil/climate analysis to determine appropriate plant material and design the landscape with native or adapted plants to reduce or eliminate irrigation requirements. Where irrigation is required, use high-efficiency equipment and/or climate-based controllers.

LEED 2009 Calculations and Considerations (WE Credit)

All four credit points are automatically earned if there is no irrigation on the site. Otherwise, this credit subcagtegory usually requires a series of calculations. For two points, the calculations must show that the design water use of potable water is less than 50 percent of a baseline irrigation case. For four points, in addition to having none of the irrigation water potable, or coming from natural surface or groundwater sources near the site, the calculations must also show that the design water use of *any* water for irrigation is less than 50 percent of a baseline irrigation case.

The calculations are done in two main steps. First a calculation for a baseline (typical) landscape scenario is performed and then calculations for the design case are performed.

These are compared to determine the points available. The calculations are used to estimate the total water applied (TWA) for each of these two cases. Note that the total landscaped areas for both cases must be the same, and it is not appropriate to assume that the baseline is an entire site with only high water use such as turfgrass. The main calculation format is reviewed in the following section.

Calculating the Total Water Applied The TWA is determined using a series of landscape water usage factors and a reference evapotranspiration rate to first determine the evapotranspiration (ET) rates for both cases (baseline and design). These factors and the reference rate are as follows:

k_s The species factor (an indicator of water usage by vegetation species)

k_d The density factor (an indicator of water usage by how densely the vegetation is planted or how substantially the foliage covers the area)

k_{mc} The microclimate factor (an indicator of water usage with respect to the location of the vegetation on the site)

ET_0 The reference evapotranspiration rate ET_0 is specific to areas of the country and represents a typical evapotranspiration rate based on the local climate and a reference plant (grass or alfalfa). It is usually also based on conditions in that area for the month of July since often the greatest irrigation demand is in July. More information can be obtained from the American Society of Civil Engineers (ASCE) and the Irrigation Association (IA). Recently, a standardized reference evapotranspiration equation for the United States has been developed.

Each of the factors is divided into three main categories—low, average, and high water use—and further subdivided into five subcategories on the overall vegetation type or combination. The table for typical landscape factors (Table 3.1.1) is based on WE Credit 1 Table 1 from LEED-NC 2.2 and is also applicable to LEED 2009.

The species factor should be specific to the design species for the design case (k_{sD}), where, for instance, more succulent plants have low water use, etc. The species factor for the baseline case (k_{sB}) should be set to average values representative of conventional design practices.

	Factors	k_s low	k_s avg.	k_s high	k_d low	k_d avg.	k_d high	k_{mc} low	k_{mc} avg.	k_{mc} high	ET_0
Vegetation Type	Trees	0.2	0.5	0.9	0.5	1.0	1.3	0.5	1.0	1.4	Not Dependent on Vegetation Type
	Shrubs	0.2	0.5	0.7	0.5	1.0	1.1	0.5	1.0	1.3	
	Ground covers	0.2	0.5	0.7	0.5	1.0	1.1	0.5	1.0	1.2	
	Mixed: Trees/shrubs/ ground covers	0.2	0.5	0.9	0.6	1.1	1.3	0.5	1.0	1.4	
	Turfgrass	0.6	0.7	0.8	0.6	1.0	1.0	0.8	1.0	1.2	
Baseline		Conventional Design Practice			Conventional Design Practice			Use Low, Average, or High as in Design			Same as Design

TABLE 3.1.1 Landscape Factors and ET_0

Vegetation Type	Low	Average	High
Trees	<60	60–100	Layered
Shrubs	<90	90–100	Layered
Ground covers	<90	90–100	Layered

TABLE 3.1.2 Typical Density Factor Categories Based on Percent Ground Shading

The density factor is indicative of the number of plants planted and the total leaf area. More plants require more water. Some rules of thumb are based on ground shading and are shown in Table 3.1.2, but these cover only a few cases. The density factor for the design case (k_{dD}) should be set to that representative of the design plant spacings and densities. The density factor for the baseline case (k_{dB}) should be set to average values representative of conventional design practices.

The microclimate factor is based on where the landscape areas are with respect to other features on the site as well as site orientation. Some typical conditions that differentiate between low, average, and high as given in LEED-NC 2.2 and 2009 are shown in Table 3.1.3. The same microclimate levels (low, average, or high) are used in the design (k_{mcD}) and in the baseline calculations (k_{mcB}).

The reference evapotranspiration rate ET_0 for a project depends on the local climate. It is the same for both the design case and the baseline case. It can be found from irrigation references or websites and is usually given in inches per day. (It may also be given in inches, which usually means inches per the month of July.) Some typical values are given in Table 3.1.4. Table 3.1.4 is taken from the *Landscape Irrigation Scheduling and*

Low	Shaded areas and areas protected from wind (e.g., on north sides of buildings, courtyards, areas under wide building overhangs, and north sides of slopes)
Average	Evapotranspiration rate is unaffected by buildings, pavements, reflective surfaces, and slopes
High	Landscape area near heat-absorbing and reflective surfaces or landscape area exposed to particularly windy conditions (e.g., near parking lots, on west sides of buildings, on west or south sides of slopes, median, and wind tunnel effect areas)

TABLE 3.1.3 Typical Microclimate Factor Categories

Climate	Definition	ET_0 (Max.)(in/day)
Cool humid	<70°F >50% RH	0.10–0.15
Cool dry	<70°F <50% RH	0.15–0.20
Warm humid	70–90°F >50% RH	0.15–0.20
Warm dry	70–90°F <50% RH	0.20–0.25
Hot humid	>90°F >50% RH	0.20–0.30
Hot dry	>90°F <50% RH	0.30–0.45

TABLE 3.1.4 Some Example Reference ET_0 Values

Water Management Draft: Table 1-3: Mid-Summer Daily Reference ET_0 (Grass Reference) Based on Climate for Midday Mid-Summer Air Temperature and Relative Humidity on the IA (Irrigation Association) website, January 2006 (www.irrigation.org/about_et_list.htm). The values are based on typical July daily high temperatures (degrees Fahrenheit) and associated relative humidities (RHs).

These factors and ET_0 are used to determine the baseline and the design evapotranspiration rates for each landscape area (area i) on the site, ETB_{Li} and ETD_{Li}.

$$ETB_{Li} = (k_{sB})(k_{dB})(k_{mcB})(ET_0) \quad \text{for landscape area } i \tag{3.1.1}$$

$$ETD_{Li} = (k_{sD})(k_{dD})(k_{mcD})(ET_0) \quad \text{for landscape area } i \tag{3.1.2}$$

There are two other factors that influence irrigation and are included in the calculations. The first is the irrigation efficiency (IE) factor, which is representative of the type of irrigation system used. A drip irrigation system usually has an irrigation factor of 0.90 and is much more efficient at delivering the water more directly to the plants as needed than a typical sprinkler system which has an IE of around 0.65. For the baseline case, a "conventional irrigation" system type of the region should be used (*conventional irrigation* is defined as per LEED-NC 2.2 in App. B). The second factor, the controller efficiency (CE), has to do with using special sensors to control the sprinkler use based on conditions such as soil moisture. If no special sensors are used, then CE is equal to 1; otherwise the factor used should be supported by the manufacturer's literature and is less than 1, but significantly greater than 0. Some typical or representative values are given in Table 3.1.5.

The baseline and the design evapotranspiration rates for each landscape area A_i on the site (ETB_{Li} and ETD_{Li}) together with the baseline and design irrigation efficiencies (IEB_i and IED_i), and the controller efficiencies (CE_i) for each landscape area i, are used to calculate the baseline and design total water applied ($TWAB_i$ and $TWAD_i$) for each landscape area, and the baseline and design total water applied (TWAB and TWAD) for the project. Typically the controller efficiency for the baseline case remains at 1.0.

$$TWAB_i = (A_i)(ETB_{Li})(1.0/IEB_i) \quad \text{for landscape area } i \tag{3.1.3}$$

$$TWAD_i = (A_i)(ETD_{Li})(CE_i/IED_i) \quad \text{for landscape area } i \tag{3.1.4}$$

$$TWAB = \sum_i TWAB_i \tag{3.1.5}$$

$$TWAD = \sum_i TWAD_i \tag{3.1.6}$$

	Irrigation System	Irrigation Efficiency (IE)	Irrigation Controller	Controller Efficiency (CE)
Types	Drip irrigation	0.90	None	1
	Sprinkler	0.65	Per manufacturer	$0 \ll CE < 1$
Baseline	Conventional irrigation	0.65	None	1

TABLE 3.1.5 Other Irrigation Factors: Some Representative Values

Credit Determinations Now that the total water needs for the landscaping have been determined, the amount of potable water (or other special waters that are recommended not to be used for irrigation for environmental considerations) that will be applied in the design case (TPWA) can be determined from the total water applied for the design case (TWAD) less the reuse water (RW). The author has interpreted *reuse water* to be any waters from rainwater harvesting, reused graywaters, and municipal supplied "reuse" waters applied for irrigation that are not for potable uses. Other waters from the site such as pond water or nonpotable groundwater may not be subtracted out, except as noted for groundwater seepage pumped from slab and foundation areas. These other waters are considered to be important for other environmental reasons on the site. However, even with this clarification there are many definitions of graywaters. The WEc2 subcategory section on Innovative Wastewater Technologies gives some further insight into the definition of graywaters.

$$\mathrm{TPWA} = \mathrm{TWAD} - \mathrm{RW} \tag{3.1.7}$$

Note that the units must all cancel to give a unit for the TWAB and TWAD in volume per month or hour. Typically, the ET rates are given in inches per hour or month, the areas are given in square feet, and the other factors are dimensionless. In these cases, the conversion factor to obtain TWAB or TWAD in gallons per unit time is 0.6233 gal/($\mathrm{ft}^2 \cdot \mathrm{in}$), or inversely, 1.604 $\mathrm{ft}^2 \cdot \mathrm{in/gal}$.

The values for TWAB, TWAD, and TPWA can be used for the credit verifications. Two points for LEED 2009 WEc1 may be earned if the following inequality is true:

$$\frac{\mathrm{TPWA}}{\mathrm{TWAB}} \leq 0.50 \tag{3.1.8}$$

Two additional points can be given for LEED 2009 WEc1 if the following two equations are both true:

$$\frac{\mathrm{TWAD}}{\mathrm{TWAB}} \leq 0.50 \tag{3.1.9}$$

and

$$\mathrm{TPWA} = 0 \tag{3.1.10}$$

Special Circumstances and Exemplary Performance (WE Credit 1)

As mentioned previously, if there is no irrigation, then there are no calculations necessary to obtain both credits. If there is any irrigation, even if all is from reuse water, then calculations usually must be performed. There are some circumstances where it may be unclear whether the water used for watering is "irrigation" water per the intent of the credit or "process" water. One example is for the main field in a baseball stadium, where watering the main playing field may be considered irrigation or may be considered as a special water application (process water). Typical playing fields are irrigated, but main sports venue fields are sometimes looked at differently and have different requirements. These special exceptions should be reviewed through the CIR (Credit Interpretation Ruling) procedure.

There are many projects which do not provide any landscaping on-site and the LEED 2009 WEc1 credit points are not available for use on these projects, unless there are roof, courtyard garden spaces, or planters included on the site which cover at least 5 percent of the the total site area and these special garden spaces/planters fulfill the other requirements of the subcategory. In a similar fashion, it is not appropriate to substitute hardscape surfaces as the design cases against landscape areas for the baseline in the calculations. Hose bibs are not considered to be permanent irrigation facilities and can be used during periods of drought or for establishing plantings during the first year.

Note that there are many special health codes governing the use or storage of reuse waters. They vary by local or state jurisdiction. The designers, owners, and managers should become familiar with any special health-related requirements for use of reuse waters so that appropriate protective measures or treatments can be included in the design and operations. The submittals must also include narratives describing the various nonpotable water sources, local codes for their use, and descriptions of any special treatment of these waters and other related features.

An interesting item to note about these credits is that the calculated baseline total water applied for irrigation (TWAB) is based on the conventional practices in the region where the project is built. The author assumes that this implies that the conventional practices are those as of around 2005, when the LEED-NC 2.2 Reference Guide was adopted and there were not many facilities percentagewise built to the LEED standards. As the "green" movement expands, the author hopes that the new "conventional" practices will become greener and greener. The author suggests that the designers date their calculations, so that in future years, these earlier baseline criteria can possibly be used as the future baseline, even if LEED and other environmental water-saving programs become the standard or conventional practice in the area. Otherwise these credits will become harder to obtain and may be disregarded, which is not the intention of LEED. There have recently been significant strides made to adopt standards and develop protocols to evaluate products and processes for water efficiency. The Irrigation Association has recently started the Smart Water Applications Technologies (SWAT) program. Much of this ongoing work is being incorporated into the EPA's Water Sense program.

If there are special circumstances for these credits, a narrative must be submitted describing these circumstances. There is no exemplary performance point available for these credits in either version 2.2 or 2009.

3.2 WE Credit 2: *Innovative Wastewater Technologies*

USGBC Rating System

LEED-NC 2.2 lists the Intent, Requirements, and Potential Technologies and Strategies for WEc2 as follows, with the LEED 2009 Intent and Requirements essentially identical:

Intent

Reduce generation of wastewater and potable water demand, while increasing the local aquifer recharge.

Requirements

OPTION 1

Reduce potable water use for building sewage conveyance by 50% through the use of water conserving fixtures (water closets, urinals) or non-potable water (captured rainwater, recycled greywater, and on-site or municipally treated wastewater).

OR

OPTION 2

Treat 50% of wastewater on-site to tertiary standards. Treated water must be infiltrated or used on-site.

Potential Technologies and Strategies

Specify high-efficiency fixtures and dry fixtures such as composting toilet systems and non-water using urinals to reduce wastewater volumes. Consider reusing stormwater or greywater for sewage conveyance or on-site wastewater treatment systems (mechanical and/or natural). Options for on-site wastewater treatment include packaged biological nutrient removal systems, constructed wetlands, and high-efficiency filtration systems.

LEED 2009 Calculations and Considerations

First, the intent of these two options should be further explored as there are many definitions for *sewage* and *wastewater.* For the purposes of this credit, the volumes used for calculating either option should be based on "blackwater" volumes. As a default, blackwater is the water from toilets (water closets) and urinals. Some jurisdictions also include kitchen sinks and other wastewater sources. If these other sources are included, then they should be specifically noted in the calculations and the local or state definitions provided in the submittal. In fact, it is usually easier to obtain this credit if *blackwater* is defined as the wastewater from toilets and urinals only. In some senses, this is justified for Option 1 as kitchen sinks need to use potable water. Therefore, for the calculations as outlined in the following sections for this credit and for WEc3, it will be assumed that blackwater is the wastewater from water closets and urinals only. Again, any variances from this should be specifically outlined and noted in the submittals.

The calculations are based on the usage of the building with both a baseline case and a design case. The baseline for this credit is always based on the same usage of the building as the design case, but with water usage rates as established as the baseline in the Energy Policy Act (EPAct) of 1992.

The number of occupants should be differentiated by gender, by employee occupancies, and by transient occupancies. There are numerous variations of the occupancies based on various usages, but the typical assumptions for blackwater generation are as listed in Tables 3.2.1 and 3.2.2. These values have been taken from the LEED-NC 2.2 and 2009 Reference Guides. They refer to the daily usage for a full-time employee equivalent (FTE) on her or his shift, for the special definition of transient occupancy for water usage (TOW) which is the summation of *each and every* transient occupant in the building during a day (the transient uses have been further subdivided into student/visitor uses and retail customer uses) and for residential design occupancies. For residential uses these may be higher for certain uses, particularly for residential facilities where there are many children or retirees, as they may be home for more hours during the day. Obviously, for special residential uses such as retirement or nursing homes, the numbers for daily uses would need to be increased appropriately. Otherwise, the default values in Table 3.2.1 should be used.

Various special notes are needed for some of the blackwater generation rates listed in Table 3.2.2. Many of the typical values given for some fixtures in the LEED 2009 Reference Guide have changed significantly from those listed previously in the LEED 2.2 version and there are two new acronyms being used; HET for High Efficiency Toilet and HEU for High Efficiency Urinal. The values listed in Table 3.2.2 are from the 2009 version. In the calculations, actual flow values should be used for the fixtures as designed. Also, composting toilets and nonwater urinals do not use water for regular usage. They are not currently customary

Fixture Type	Gender	Variable Acronym	Daily Uses per Person: FTE* Uses	Transient: Student/ Visitor	Transient: Retail Customer	Resident†
Water closet use	Male	$WCUM_i$	1	0.1	0.1	5
Water closet use	Female	$WCUF_i$	3	0.5	0.2	5
Urinal use	Male	UUM_i	2	0.4	0.1	n/a
Urinal use	Female	UUF_i	0	0	0	n/a

*The students as listed in TOW represent transient students, such as those attending a college where they may take only a handful of classes in one building on certain days. Students in primary schools and other schools where they remain in the same building every "work" day for the entire school day are actually similar to full-time employees and should be treated as such. This should therefore be noted in the calculations and the submittals.

†These resident daily usage rates represent residents who on average also spend part of their days elsewhere, such as at school or work. It is recommended by the author that for residential facilities where there are usually 24-h occupancies, such as nursing homes, the resident rates for water closets be, at a minimum, 8, which is a sum of 3 (for FTE) and 5 (for residences). This should be noted in the calculations and the submittals.

Table 3.2.1 Default Typical Daily Blackwater Fixture Use by Gender

Specific Fixture Type	Gallons Generated per Use [Gallons per Flush (gpf)] (WCR_i or UR_i)
Baseline (conventional) water closet	1.6 (Blowout fixtures 3.5)
HET foam flush	0.05
HET single flush water closet	1.28 gravity 1.0 pressure assist
HET dual flush water closet	1.6 full flush 1.1 low flush 1.27 average*
Composting toilet	0
Baseline (conventional) urinal	1.0
HEU urinal	0.5
Nonwater urinal	0

*This average is as calculated for one-third usage with full flush.

Table 3.2.2 Typical Blackwater Generation by Fixture Type as Given in LEED 2009

FIGURE 3.2.1 Waterless urinal installed in the LEED certified public health building at the University of South Carolina, Columbia, S.C. (*Photograph taken July 25, 2007.*)

fixtures in the United States. Therefore, they have different maintenance and waste collection needs than more common water closets and urinals. The designers and owners should be very cognizant of the different maintenance needs, and it is recommended that they determine a way to meet these special needs early in the design phase of a project so that there are no additional concerns, costs, or changes later in the project. Figure 3.2.1 is a photo of a waterless urinal.

To perform the calculations, the non-gender-specific occupancies should be established as previously defined in Chap. 2. Occupancies must be consistent throughout a LEED submittal for a project. The calculations for WEc2 Option 1 are different for commercial and institutional uses than for residential uses and are summarized in the following sections.

Option 1: Commercial and Institutional Uses In a similar fashion to the calculations for occupancies in Chap. 2, the following definitions are given:

FTE_j Full-time equivalent building occupant during shift *j*

$FTE_{j,i}$ Full-time equivalent building occupancy of employee *i* during shift *j*. This is equal to 1 for a full-time employee and is equal to the normal hours worked (less than 8) divided by 8 for a part-time employee. Obviously, this will need to be modified if shifts are different from the standard 8 h.

The FTE_j is then determined by using the following set of equations from Chap. 2:

$$FTE_{j,i} = (\text{worker } i \text{ h})/8 \text{ h} \qquad \text{where } 0 < FTE_{j,i} \leq 1 \tag{2.4.1}$$

$$FTE_j = \Sigma\, FTE_{j,i} \qquad \text{for all employees in shift } j \tag{2.4.2}$$

Then the full-time employee equivalents need to be subdivided by gender. For each shift the numbers for male and female employees may not be equal. Let $FTEM_j$ and $FTEF_j$ be the male full-time employee equivalent and the female full-time employee equivalent for shift j, respectively. To be consistent throughout the LEED submittal, the following equality must hold for all shifts j:

$$FTE_j = FTEM_j + FTEF_j \qquad \text{for all shifts } j \tag{3.2.1}$$

In SS credit 4.2, the transient occupancies are estimated as average transient occupancies at any time over a shift, not as the total number of transient occupants who may go in or out of the building during the shift. This is done to estimate the peak number of bicycle rack spaces needed at any time. For example, if the building is a college building with faculty and staff offices and student classrooms, then in SSc4.2 the FTE_j portion of the calculations is based on the staff and faculty in the building during the typical workday, and the transient population is the average occupancy of the classrooms and study areas over this typical 8-h class time, not the total number of students who come in and out. The students may be in other campus buildings for other classes throughout the day. However, in WE credit 3 the LEED-NC 2.2 and 2009 Reference Guides base transient uses on the *total number of transients*, not the average over the shifts. To be consistent with these other water credits, let us define TOWM and TOWF to be the estimated total number of male and female transients, respectively, during a day. (TOW would be the summation of TOWM and TOWF.)

As mentioned previously, LEED-NC 2.2 and 2009 recommend that the typical daily fixture usage rates for water closets and urinals be separately analyzed for the different genders, building usages, and occupancies. Let $WCUM_i$ and $WCUF_i$ be the male water closet and female water closet usage rates, respectively, for various types of buildings or occupancies. Let UUM_i and UUF_i be the male and female urinal usage rates, respectively, for various building types and occupancies. Table 3.2.1 gives some typical default values for these daily fixture usage rates. It is important to also analyze the building usage on an annual basis. Let ND_{FTE} and ND_{TOW} be the number of days in a year that each type of occupancy (FTE or TOW) uses the building. ND_{FTE} is usually 260 days for office buildings and 365 days for residential or retail. For college buildings, ND_{FTE} may be 260 for staff and faculty, but ND_{TOW} may be much less for student occupancies. Summing the fixture usage rates times the occupancies over the various shifts as applicable for the entire year gives the total number of times water closets and urinals are typically used in a year (TWCU and TUU, respectively):

$$\begin{aligned} TWCU = {} & ND_{FTE}\,[\Sigma\,(FTEM_j \times WCUM_i) + \Sigma\,(FTEF_j \times WCUF_i)] \\ & + ND_{TOW}[(TOWM \times WCUM_i) + (TOWF \times WCUF_i)] \\ & \text{over all shifts } j \text{ and for each type of use } i \end{aligned} \tag{3.2.2}$$

$$\begin{aligned} TUU = {} & ND_{FTE}\,[\Sigma\,(FTEM_j \times UUM_i) + \Sigma\,(FTEF_j \times UUF_i)] \\ & + ND_{TOW}\,[(TOWM \times UUM_i) + (TOWF \times UUF_i)] \\ & \text{over all shifts } j \text{ and for each type of use } i \end{aligned} \tag{3.2.3}$$

Now both the baseline and the design blackwater annual generation rates can be determined. The blackwater baseline (BWB) generation rate, in gallons per year, can be calculated using the fixture flow rate numbers in Table 3.2.2 for the most typical conventional toilet as follows:

$$BWB = (1.6 \times TWCU) + (1.0 \times TUU) \quad gal/yr \tag{3.2.4}$$

To calculate the blackwater design (BWD) generation rate, the factors in Eq. (3.2.4) are replaced with those appropriate for the alternate fixture type i used such as WCR_i (the fixture rate for water closet type i) and UR_i (the fixture rate for urinal type i) from Table 3.2.2:

$$BWD = (WCR_i \times TWCU) + (UR_i \times TUU) \quad gal/yr \tag{3.2.5}$$

Note that for Option 1 the only differences between the base blackwater generation rate and the design blackwater generation rate are the fixture flow rates. However, Option 1 of WE credit 3.2 is based on a reduction in the wastewater generated, a reduction in potable water used for blackwater fixtures, or a combination of both. Therefore, the annual usage of other water sources, such as rainwater or graywater, for these fixtures can be subtracted from the total design blackwater generation rate. The designer should provide calculations to estimate the rate of these alternative waters available for use on an annual basis. For instance, rainwater availability is seasonal, so the totals available during the rainy seasons may be the only opportunity for potable water reduction.

Note also that there are local and state health and other regulatory codes that cover the use, storage, and possibly treatment of nonpotable waters in buildings. In addition, there is a large variation in the definitions used from locality to locality and from state to state on what constitutes graywater, similar to with what is included in blackwater. The designers, owners, and managers should become familiar with the appropriate local definitions and any special health-related requirements for the use of reuse waters or rainwaters so that any special protective measures or treatments can be included in the design and operations. The designer should include the appropriate local definitions of graywaters and other reuse waters, as well as the associated costs and systems in the design, and include a narrative of them in the submittal. Table 3.2.3 lists some of the various state interpretations of what may be included in graywater as taken from *Appendix 3: US State Regulations* compiled in 1999 on the www.weblife.org website, humanure section.

Let GW be the annual graywater or other alternative nonpotable water source usage in the wastewater fixtures; then WE credit 2 is obtained via Option 1 if the following is true:

$$\frac{BWD - GW}{BWB} \leq 0.50 \tag{3.2.6}$$

Option 1: Residential Uses The building designer should declare the design occupancy (DO) for the buildings. Note that design occupancies are not the same as maximum occupancies. LEED 2009 gives some recommended residential design occupancies as two persons per one bedroom unit and one additional person per additional bedroom. In addition, usually residential facilities do not have urinals and do have equal gender distributions as noted in Table 3.2.1. Equation (3.2.2) can be used to estimate the total water closet usage. In this case, the transient occupancies are zero, there are no urinals, and there are no shifts. Using a water closet rate of 5 uses per day as given in Table 3.2.1,

State	Shower, Bath, Bathroom Sink Wastewaters	Kitchen Sink, Dishwater Wastewaters	Laundry Washing Machine Wastewaters
Alabama	Yes	Yes*	Yes
Alaska	Yes	Yes	Yes
Arizona	Yes	Yes*	Yes
California	Yes	—	Yes
Colorado	Yes	Yes	Yes
Connecticut	Yes	Yes	Yes
Florida	Yes	—	Yes
Georgia	Yes	—	Yes
Hawaii	Yes	Yes	Yes
Idaho	Yes	—	Yes*
Kentucky	Yes	Yes*	Yes
Maine	Yes	Yes*	Yes
Maryland	Yes	—	Yes
Massachusetts	Yes	Yes*	Yes
Michigan	Yes	Yes	Yes
Minnesota	Yes	Yes	Yes
Missouri	Yes	—	Yes
Nebraska	Yes	—	Yes
Nevada	Yes	—	Yes
New Jersey	Yes	—	Yes
New Mexico	Yes	—	Yes
New York	Yes	Yes	Yes
Oregon	Yes	Yes	Yes
Rhode Island	Yes	Yes*	Yes
South Dakota	Yes	Yes	Yes
Texas	Yes	—	Yes
Utah	Yes	—	Yes
Washington	Yes	Yes	Yes

*With some exclusions.

Source: The information is from *Appendix 3: US State Regulations* compiled in 1999 by J. C. Jenkins on the www.weblife.org website, humanure section, as interpreted by Wu et al. (2007) with Permission from College Publishing.

Table 3.2.3 Waters Considered to Be Graywaters by Various States

and substituting DO for FTE, Eq. (3.2.2) simplifies to Eq. (3.2.7) such that the estimated total annual number of water closet usages is

$$\text{TWCU} = 365 \times 5 \times \text{DO} \qquad \text{For residential uses} \tag{3.2.7}$$

Equations (3.2.4) through (3.2.6) can again be used for the Option 1 calculations relating to these residential uses. In cases where there will be no alternative nonpotable water used for the water closets, the calculations can be further reduced to the following simple equation for obtaining Option 1 WE credit 2:

$$\frac{\text{WCR}_i}{1.6} \leq 0.50 \qquad \text{for WCR}_i \text{ in gallons per flush, for typical mixed-gender residences with only potable water to water closets} \tag{3.2.8}$$

Option 2 Option 2 of WE credit 2 is for providing on-site wastewater treatment that is treated to tertiary standards and infiltrated and/or reused on-site. The calculations use the design wastewater generation rate (BWD) as calculated in Eq. (3.2.5) (or as applicable for residential facilities). Let TBW be the annual volume of blackwater treated and infiltrated and/or reused on-site. Option 2 of WE credit 2 may then be obtained if the following is true:

$$\frac{\text{TBW}}{\text{BWD}} \geq 0.50 \tag{3.2.9}$$

The calculations for Option 2 do not include nonpotable waters used for flushing toilets in the numerator unless these nonpotable waters are part of the treated and reused blackwater generated (i.e., part of the TBW). In addition to the calculations, the submittals must include a narrative and additional information on how, and how much of, the blackwaters are treated, infiltrated, and/or reused on-site.

Special Circumstances and Exemplary Performance (WE Credit 3)

As mentioned previously, if the local jurisdiction defines blackwater to include other wastewaters, then the volumes of these wastewaters generated should be included in the total baseline and design blackwater generation rates (BWB and BWD) for both options. This should be appropriately noted in the submittals. If there are special circumstances for this credit, a narrative must be submitted describing these circumstances.

There is an EP point available for this credit for each option based on the following criteria:

- Option 1: There is no potable water used for sewage conveyance (BWD = GW).
- Option 2: 100 percent of these wastewaters are treated to tertiary standards and infiltrated or reused on-site (TBW = BWD).

3.3 WE Credit Subcategory 3: *Water Use Reduction*

The intention of this credit subcategory is to reduce the use of potable water for indoor use. In LEED 2.2, WEc3.1 is worth one point for a 20 percent reduction and WEc3.2 is one point *in addition to* the point for WEc3.1 for a 30 percent reduction. In LEED 2009, overall building water use reduction is addressed under the required WE prerequisite 1

(WEp1) for the 20 percent reduction, and then the WE credit 3 (WEc3) for two, three, and four points for 30, 35, and 40 percent reductions, respectively.

WE Credit 3: *Water Use Reduction*

USGBC Rating System WE Credit 3

LEED-NC 2.2 lists the Intent, Requirements, and Potential Technologies and Strategies for WEc3 as follows, and as modified to be applicable to LEED 2009.

Intent

Maximize water efficiency within buildings to reduce the burden on municipal water supply and wastewater systems.

Requirements

Employ strategies that in aggregate use less water than the water use baseline calculated for the building (not including irrigation) after meeting the Energy Policy Act of 1992 fixture performance requirements (or as modified to the 2006 UPC and IPC in LEED 2009). Calculations are based on estimated occupant usage and shall include only the following fixtures (as applicable to the building): water closets, urinals, lavatory faucets, showers and kitchen sinks (with prerinse spray valves also included in LEED 2009).

Potential Technologies and Strategies

Use high-efficiency fixtures, dry fixtures such as composting toilet systems and non-water using urinals, and occupant sensors to reduce the potable water demand. Consider reuse of stormwater and greywater for non-potable applications such as toilet and urinal flushing and custodial uses.

Calculations and Considerations (WE Prerequisite 1 and WE Credit 3) LEED 2009

The calculations for WE prerequisite 1 and WE credit 3 are very similar to the calculations for Option 1 of WE credit 2, with four additional types of fixtures added: lavatory faucets, showers, prerinse spray valves, and kitchen sinks. Bathtubs are not addressed, but the author assumes that all residential facilities that have bathtubs are modeled as if they have a shower head, and the calculations are based on the shower head flow. The intention of the credit is not to reduce the volume of water as needed for proper bathing or food preparation and cleanup, but to limit the wasted water from unused faucet flow. Also, not addressed in these credits are dishwashers and clothes washers.

Many water usage calculations for other purposes are based on the number of fixtures in a building. The calculations for WE prerequisite 1 and WE credit 3 are not based on the number of fixtures, but rather on the number of users, as was done in the WE credit 2 calculations and will also differentiate between the number of male and female users in case they are not equal. However, in most cases, the default assumption is that the facility is gender-neutral, with one-half of the users male and the one-half female.

Commercial and Institutional Uses The total annual water closet and urinal usage rates (in number of times used per year) are again estimated using the FTE estimates as developed in SSc4.2, defining the following:

FTE_j Full-time equivalent building occupant during shift j.

$FTE_{j,i}$ Full-time equivalent building occupancy of employee i during shift j. This is equal to 1 for a full-time employee and is equal to the normal hours worked (less than 8) divided by 8 for a part-time employee. Obviously, this will need to be modified if shifts are different from the standard 8 h.

The FTE_j is then determined using the following set of equations:

$$FTE_{j,i} = (\text{worker } i \text{ h})/8 \text{ h} \qquad \text{where } 0 < FTE_{j,i} \leq 1 \tag{2.4.1}$$

$$FTE_j = \Sigma\, FTE_{j,i} \qquad \text{for all employees in shift } j \tag{2.4.2}$$

Then the full-time employee equivalents need to be subdivided by gender. For each shift the numbers for male and female employees may not be equal. Let $FTEM_j$ and $FTEF_j$ be the male full-time employee equivalent and the female full-time employee equivalent for shift j, respectively. To be consistent throughout the LEED submittal, the following equality must hold for all shifts j:

$$FTE_j = FTEM_j + FTEF_j \qquad \text{for all shifts } j \tag{3.2.1}$$

In SS credit 4.2, the transient occupancies are estimated as average transient occupancies at any one time over a shift, not as the total number of transient occupants who may go in or out of the building during the shift. This is done to estimate the peak number of bicycle rack spaces needed at any time. However, for WE prerequisite 1 and credit 3, the LEED-NC 2009 Reference Guide bases transient uses on the *total number of transients*, not the average over the shifts. To be consistent with this, let us define TOWM and TOWF to be the estimated total number of male and female transients, respectively, during a day. TOW, as previously described, would be the summation of TOWM and TOWF.

As mentioned previously, LEED-NC 2009 recommends that the typical daily fixture usage rates for water closets and urinals be separately analyzed for the different genders, building usages, and occupancies. Let $WCUM_i$ and $WCUF_i$ be the male water closet and female water closet usage rates, respectively, for various types of buildings or occupancies. Let UUM_i and UUF_i be the male and female urinal usage rates, respectively, for various building types and occupancies. Table 3.3.1 gives some typical default values for these daily fixture usage rates. It is important to also analyze the building usage on an annual basis. Let ND_{FTE} and ND_{TOW} be the number of days in a year that each type of occupancy (FTE or TOW) uses the building. Usually ND_{FTE} is 260 days for office buildings and 365 days for residential or retail. For college buildings, ND_{FTE} may be 260 days for staff and faculty, but ND_{TOW} may be much less for student occupancies. Just as in WEc2, summing the fixture usage rates times the occupancies over the various shifts as applicable for the entire year gives the total number of times water closets and urinals are typically used in a year (TWCU and TUU, respectively). These were previously given in Eqs. (3.2.2) and (3.2.3) and are reiterated here:

$$\begin{aligned} TWCU = {} & ND_{FTE} \times (\Sigma\, (FTEM_j \times WCUM_i) + \Sigma\, (FTEF_j \times WCUF_i)) \\ & + ND_{TOW} \times [(TOWM \times WCUM_i) + (TOWF \times WCUF_i)] \\ & \text{over all shifts } j \text{ and for each type of use } i \end{aligned} \tag{3.2.2}$$

$$\begin{aligned} TUU = {} & ND_{FTE} \times (\Sigma\, (FTEM_j \times UUM_i) + \Sigma\, (FTEF_j \times UUF_i)) \\ & + ND_{TOW} \times [(TOWM \times UUM_i) + (TOWF \times UUF_i)] \\ & \text{over all shifts } j \text{ and for each type of use } i \end{aligned} \tag{3.2.3}$$

Lavatory faucets, showers, prerinse spray, and kitchen sink usage rates are usually not considered to be gender-specific. It is typically assumed that the lavatory faucet is used for every use of a water closet and/or urinal, which combined should be the same for

Fixture Type	Gender	Daily Uses per Person				
		Variable Acronym	FTE* Uses	TOW: Student/ Visitor	TOW: Retail Customer	Resident†
Water closet use	Male	$WCUM_i$	1	0.1	0.1	5
Water closet use	Female	$WCUF_i$	3	0.5	0.2	5
Urinal use	Male	UUM_i	2	0.4	0.1	n/a
Urinal use	Female	UUF_i	0	0	0	n/a
Commercial lavatory faucet (duration 15 s for nonmetered)	n/a	LFU_i	3	0.5	0.2	n/a
Residential lavatory faucet (duration 60 s)	n/a	LFU_i	n/a	n/a	n/a	5
Shower (commercial duration 300 s and residential duration 480 s)	n/a	SU_i	0.1	0	0	1
Kitchen sink (commercial duration 15 s and residential duration 60 s)	n/a	KSU_i	1	0	0	4

*The students as listed in TOW represent transient students, such as those attending a college where they may take only a handful of classes in one building on certain days. Students in primary schools and other schools where they remain in the same building every "work" day for the entire school day are actually similar to full-time employees and should be treated as such. This should be noted in the calculations and the submittals.

†These resident daily usage rates represent residents who on average also spend part of many of their days elsewhere, such as at school or work. It is recommended by the author that for residential facilities where there are usually 24-h occupancies, such as nursing homes, the resident rates for water closets and lavatories be, at a minimum, 8, which is a sum of 3 (for full-time employees) and 5 (for residences). This should be noted in the calculations and the submittals.

Table 3.3.1 Default Typical Daily Fixture Use by Gender

either male or female occupants. The LEED-NC 2009 Reference Guide also gives default baseline values for lavatory, shower, and kitchen sink (separated for residential and nonresidential uses) which can be found with the previously listed blackwater fixture daily use rates in Table 3.3.1. No information is provided on estimating the annual prerinse spray usages (TPRU) and it is therefore assumed that this will be calculated by the project team. These values are based on EPAct 1992.

Using these variables, the equations for total lavatory faucet use (TLFU), total kitchen sink use (TKSU), and total shower use (TSU) are

$$TLFU = [ND_{FTE} \Sigma (FTE_j \times LFU_i)] + [ND_{TOW} (TOW \times LFU_i)]$$
over all shifts j and for each type of use i (3.3.1)

$$TKSU = [ND_{FTE} \Sigma (FTE_j \times KSU_i)] + [ND_{TOW} (TOW \times KSU_i)]$$
over all shifts j and for each type of use i (3.3.2)

$$TSU = [ND_{FTE} \Sigma (FTE_j \times SU_i)] + [ND_{TOW} (TOW \times SU_i)]$$
over all shifts j and for each type of use i (3.3.3)

The equations for water volume usage based on lavatory faucet, shower, prerinse spray, and kitchen sink uses have an additional variable that was not included in the equations for blackwater generation. Unlike toilets and urinals, where the uses are calculated on a volume per flush or use, sink and shower volumes per use are often based on both flow rates from the fixture and a time or duration that the fixture is kept turned on. The applicable durations are given in Table 3.3.1. Note that Table 3.3.1 gives different durations for residential and commercial uses. The baseline EPAct 1992 (conventional) flow rates as updated for the 2006 Uniform Plumbing Code (UPC) and the 2006 International Plumbing Code (IPC) revisions and water-efficient flow rates from the LEED-NC 2009 Reference Guide for lavatory faucets, kitchen sinks, prerinse sprays, and showers (LFR_i, KSR_i, PRR_i, and SR_i, respectively) are given in Table 3.3.2, as are the previously given baseline values for the water closets and urinals from Table 3.2.2.

A special note is again needed for some of the blackwater generation rates in Table 3.2.2. Composting toilets and nonwater urinals do not use water for regular usage. They are not currently customary fixtures in the United States. Therefore, they have different maintenance and waste collection needs than more common water closets and urinals. The designers and owners should be very cognizant of the different maintenance needs, and it is recommended that they determine a way to meet these special needs early in the design phase of a project so that there are no additional concerns, costs, or changes later in the project.

Now both the baseline and the design annual water volume usage rates for the fixtures applicable to WE prerequisite 1 and WE credit 3 can be determined. Durations for typical prerinse spray usages will need to be provided by the project team. The following equations also do not address private (hotel or hospital room) lavatory use in commercial facilities and these would also need to be added in. The applicable indoor water usage baseline (IWUB), in gallons per year, can be calculated using the durations in Table 3.3.1 and the fixture flow rate numbers in Table 3.3.2 as follows:

$$IWUB_{no\text{-}auto} = (1.6 \times TWCU) + (1.0 \times TUU) + [0.5(0.25 \text{ min}) \times TLFU] + [2.2(0.25 \text{ min}) \times TKSU] + [2.5(5 \text{ min}) \times TSU] + [1.6(? \text{ min}) \times TPRU]$$
for lavatory faucets *without* autocontrols, in gallons per year, no private applications (3.3.4)

Specific Fixture Type	Gallons Generated per Use [Gallons per Flush or Cycle (gpf)] (WCR_i, LCF_i or UR_i)	Gallons Generated per Minute (gpm) (LFR_i, KSR_i, PRR_i or SR_i)
Baseline water closet (commercial and residential)	1.6	n/a
Baseline urinal	1.0	n/a
Baseline commercial lavatory faucet (not private applications)	0.25 metered	0.5 not metered
Baseline commercial lavatory faucet (private applications such as hotel rooms)	n/a	2.2
Baseline residential lavatory faucet	n/a	2.2
(Baseline Commercial lavatory faucet LEED 2.2)	(n/a)	(2.5)
Baseline prerinse spray valves	n/a	1.6
Baseline kitchen sink	n/a	~~2.5~~ 2.2 gpm
Low-flow kitchen sink	n/a	1.8
Baseline shower	n/a	2.5
Low-flow shower	n/a	1.8

*The project team may add other fixture types to this list as applicable. See Table 3.2.2 for typical water closet and urinal options. LEED 2009 also allows for additional showerhead flows for shower areas >2500 in^2 and provides a listing of EPA Watersense options for design including 1.28 gpf WC, 1.5–2 gpm showerheads, and 1.5 gpm private lavatory fixtures.

TABLE 3.3.2 Baseline Flow Rates by Fixture Type and Some Typical Low-flow Kitchen and Shower Rates*

or

$$\begin{aligned} IWUB_{auto} = & (1.6 \times TWCU) + (1.0 \times TUU) + [0.25 \times TLFU] \\ & + [2.2(0.25 \text{ min}) \times TKSU] + [2.5(5 \text{ min}) \times TSU] + [1.6(? \text{ min}) \\ & \times TPRU] \text{ for lavatory faucets } \textit{with} \text{ autocontrols (metered),} \\ & \text{in gallons per year, no private applications} \end{aligned} \tag{3.3.5}$$

To calculate the applicable design indoor water usage (IWUD), the factors in Eqs. (3.3.4) and (3.3.5) are replaced with those appropriate for the alternate fixture type *i* used such as LFR_i, KSR_i, and SR_i from the design (see Table 3.3.2 for some typical values). Note that the prerinse spray rates and the metered lavatory faucet volumes have not been modified and will need to be reduced in these equations if the design values are lower.

$$\begin{aligned} IWUD_{no\text{-}auto} = & (WCR_i \times TWCU) + (UR_i \times TUU) + [LFR_i\,(0.25 \text{ min}) \times TLFU] \\ & + [KSR_i\,(0.25 \text{ min}) \times TKSU] + [SR_i\,(5 \text{ min}) \times TSU] + [1.6\,(? \text{ min}) \\ & \times TPRU] \text{ for lavatory faucets } \textit{without} \text{ autocontrols, in gallons} \\ & \text{per year, no private applications} \end{aligned} \tag{3.3.6}$$

or

$$
\begin{aligned}
IWUD_{auto} = {} & (WCR_i \times TWCU) + (UR_i \times TUU) + [0.25 \times TLFU] \\
& + [KSR_i\,(0.25\text{ min}) \times TKSU] + [SR_i\,(5\text{ min}) \times TSU] \\
& + [1.6(?\text{ min}) \times TPRU] \text{ for lavatory faucets } \textit{with} \text{ autocontrols} \\
& \text{(metered), in gallons per year, no private applications}
\end{aligned} \tag{3.3.7}
$$

The only differences between the base water usage rate and the design water usage rate are the fixture flow rates. If autocontrols are used in the design case, then autocontrols are assumed in the baseline case. This is different from LEED-NC 2.1, and designs which meet the version 2.1 criterion might not meet the version 2.2 criterion. Also, in LEED 2009, the baseline lavatory faucet rates for commercial uses (nonprivate rooms) are significantly reduced from the LEED 2.2 values. In fact, they are now similar to what were previously considered to be ultra-low flow rates. In previous versions of LEED, commercial lavatory faucet use counted for a substantial portion of the calculated water use reductions in many cases. With this baseline change from the 2006 UPC and 2006 IPC, lavatory faucets will no longer be as important toward credit compliance.

As in WE credit 3.2, a reduction in potable water used for blackwater fixtures can be from a reduction in the flows, the alternate use of nonpotable water in these fixtures, or a combination of both. Therefore, the annual usage of other water sources, such as rainwater or graywater, for these fixtures can be subtracted from the total design water usage rate. The designer should provide calculations to estimate the rate of these alternative waters available for use on an annual basis. For instance, rainwater availability is seasonal, so the totals available during the rainy seasons may be the only opportunity for potable water reduction. Note also that there are local and state health and other regulatory codes that cover the use, storage, and possibly treatment of nonpotable waters in buildings. The designers, owners, and managers should become familiar with any special health-related requirements for use of reuse waters or rainwaters so that any special protective measures or treatments can be included in the design and operations. The designer should include the associated costs and systems in the design and include a narrative of them in the submittal. Let GW be the annual graywater or other alternative water source usage in the wastewater fixtures; then WE prerequisite (WEp1) is obtained if the following is true:

$$
\frac{IWUD - GW}{IWUB} \le 0.80 \tag{3.3.8}
$$

In like manner, two, three, or four points for WE credit 3 are obtained if the following are also true, respectively:

$$
\frac{IWUD - GW}{IWUB} \le 0.70 \text{ or} \le 0.65 \text{ or} \le 0.60 \tag{3.3.9}
$$

Residential Uses The building designer should declare the design occupancy (DO) for the buildings. LEED 2009 gives some recommended residential design occupancies as two persons per one bedroom unit and one additional person per additional bedroom. Usually, residential facilities do not have urinals, and they typically have equal gender distributions. Equation (3.2.2) can again be used to estimate the total water closet usage. In this case, the transient occupancies are zero, there are no urinals, and there are no shifts. By using a water closet rate of 5 uses per day as given in Table 3.2.1, and substituting DO for FTE, Eq. (3.2.2) simplifies to Eq. (3.3.10) such that the estimated total annual number of water closet usages per year is

$$\text{TWCU} = 365 \times \text{DO} \times 5 \qquad \text{for residential uses in number of uses per year} \qquad (3.3.10)$$

In like manner, Eqs. (3.3.1) to (3.3.3) can be modified for annual lavatory, kitchen sink, and shower usage rates in number of uses per year to

$$\text{TLFU} = 365 \times \text{DO} \times 5 \qquad \text{for residential uses in number of uses per year} \qquad (3.3.11)$$

$$\text{TKSU} = 365 \times \text{DO} \times 4 \qquad \text{for residential uses in number of uses per year} \qquad (3.3.12)$$

$$\text{TSU} = 365 \times \text{DO} \qquad \text{for residential uses in number of uses per year} \qquad (3.3.13)$$

Prerinse spray valves and metered lavatory faucets are typically not applicable to residential uses and are therefore not included. Similar to the commercial and institutional derivation, now both the baseline and the design annual water volume usage rates for the fixtures applicable to WE prerequisite 1 and WE credit 3 can be determined. The applicable indoor water usage baseline (IWUB), in gallons per year, can be calculated using the durations in Table 3.3.1 and the fixture flow rate numbers in Table 3.3.2:

$$\begin{aligned}\text{IWUB} = {} & (1.6 \times \text{TWCU}) + [2.2(1.0 \text{ min}) \times \text{TLFU}] \\ & + [2.2(0.25 \text{ min}) \times \text{TKSU}] + [2.5(8 \text{ min}) \times \text{TSU}] \\ & \text{for residences, gal/yr} \qquad (3.3.14)\end{aligned}$$

To calculate the applicable design indoor water usage (IWUD), the factors in Eq. (3.3.14) are replaced with those appropriate for the alternate fixture type i used such as LFR_i, KSR_i, and SR_i from the design (see Table 3.3.2 for some typical values):

$$\begin{aligned}\text{IWUD} = {} & (\text{WCR}_i \times \text{TWCU}) + [\text{LFR}_i(1.0 \text{ min}) \times \text{TLFU}] \\ & + [\text{KSR}_i(0.25 \text{ min}) \times \text{TKSU}] + [\text{SR}_i(8 \text{ min}) \times \text{TSU}] \\ & \text{for residences, gal/yr} \qquad (3.3.15)\end{aligned}$$

Note that the only differences between the base water usage rate and the design water usage rate are the fixture flow rates.

Similar to the cases for commercial/institutional buildings, a reduction in potable water used for blackwater fixtures can be due to a reduction in the flows, the alternate use of nonpotable water in these fixtures, or a combination of both. Therefore, Eqs. (3.3.8) and (3.3.9) can be used for WE prerequisite 1 and WE credit 3 compliance calculations. However, there will rarely be a case in which reuse waters or rainwaters (GW) are used in water closets for residential uses. Thus, the required WE prerequisite 1 is fulfilled when there are no reuse waters or rainwaters used for typical indoor fixtures if the following is true:

$$\frac{\text{IWUD}}{\text{IWUB}} \leq 0.80 \qquad \text{for no reuse waters or rainwaters (GW) used in indoor residential fixtures} \qquad (3.3.16)$$

In like manner, two, three, or four points for WE credit 3 are obtained if the following are also true, respectively:

$$\frac{IWUD}{IWUB} \leq 0.70 \text{ or } \leq 0.65 \text{ or } \leq 0.60 \quad \text{for no reuse waters or rainwaters (GW) used in residential indoor fixtures} \tag{3.3.17}$$

Special Circumstances and Exemplary Performance

If there are special circumstances for these credits, a narrative must be submitted describing these circumstances.

In LEED 2.2, there were two different Innovation and Design (ID) points which could be obtained related to the WE credit 3 subcategory. One was an EP point, and the other was based on non-EPAct regulated fixtures.

The EP point available in the ID category for Water Use Reduction as per LEED-NC 2.2 may be awarded by decreasing the water usages as defined in LEED 2.2 WEc3.1 (20 percent) and LEED 2.2 WEc3.2 (30 percent) by 40 percent. It is related to the overall subcategory and brings the total LEED 2.2 points available for overall indoor potable water use reduction of the EPAct regulated fixtures to three if the following inequality is true:

$$\frac{IWUD - GW}{IWUB} \leq 0.60 \text{ LEED 2.2} \tag{3.3.18}$$

In LEED 2.2, project teams were also encouraged to earn an ID credit for reduction of water usage in non-EPAct regulated and/or process water (non-EPAct) consuming fixtures. These might include dishwashers and clothes washers for residential or applicable retail and commercial uses, or cooling towers for applicable uses. There are still some options available in LEED 2009 for schools. The criterion for LEED 2.2 was based on a reduction from these changes to non-EPAct regulated or process water consuming fixtures, which is at least equal to 10 percent of the total design regulated water use (10 percent of IWUD). Therefore, for the applicable non-EPAct fixtures involved, the baseline water usage (non-EPActB) and the design water usage (non-EPActD) should be calculated. An additional LEED 2.2 ID credit might have been obtained if the following was true:

$$\frac{\text{non-EPActB} - \text{non-EPActD}}{IWUD} \geq 0.10 \text{ LEED 2.2} \tag{3.3.19}$$

LEED 2009 grants an EP point for WEc3 if the following is true:

$$\frac{IWUD - GW}{IWUB} \leq 0.55 \tag{3.3.20}$$

3.4 Discussion and Overview

Care should be taken to evaluate the site area for the project for the WEc1 credit subcategory and many of the Sustainable Sites subcategories. Maximizing naturally vegetated open spaces and designing the project around them can aid in stormwater management, reducing irrigation needs, lowering urban heat island impacts, and encouraging ecological diversity and sustainability. In addition, full-time, transient, and design occupancies are important in the calculations for the WEc2, WEc3, and SSc4 subcategories. These should be established early in the project phase and used consistently throughout.

References

Allen, R. G., et al. (Eds.) (2005), *Standardized Reference Evapotranspiration Equation*, The American Society of Civil Engineers, Reston, VA.

EPA ORD (2002), *Onsite Wastewater Treatment Systems Manual*, U.S. Environmental Protection Agency, Office of Research and Development, website accessed July 10, 2007, http://www.epa.gov/ORD/NRMRL/Pubs/625R00008/625R00008.htm.

EPA OW (2007), http://www.epa.gov/water/, U.S. Environmental Protection Agency, Office of Water, website accessed July 10, 2007.

EPA OW (2007), http://www.epa.gov/watersense/, U.S. Environmental Protection Agency, Office of Water, Water Sense; Efficiency Made Easy, website accessed July 10, 2007.

EPA OW (2007), "National Efficiency Standards and Specifications for Residential and Commercial Water-Using Fixtures and Appliances," U.S. Environmental Protection Agency, Office of Water, Water Sense Document, http://www.epa.gov/watersense/docs/matrix508.pdf, accessed July 10, 2007.

IA (2006), http://www.irrigation.org/about_et_list.htm, evapotranspiration rates on the Irrigation Association website, accessed January 2006.

IA (2007), http://www.irrigation.org/SIM/default.aspx?pg=consumer_resources.htm&id=234, website for irrigation resources, Irrigation Association, accessed June 14, 2007.

IAPMO (2006), ANSI UPC 1-2006, Uniform Plumbing Code 2006, Section 402.0, *Water Conserving Fixtures and Fittings*, International Association of Plumbing and Mechanical Officials.

ICC (2006), International Plumbing Code 2006, Section 604, *Design of Water Distribution Systems*, International Code Council.

Jenkins, J. C. (1999), http://www.weblife.org/humanure/, website accessed July 26, 2007 with information from *The Humanure Handbook*. Jenkins Publishing, Grove City, PA.

Poremba, S. M. (2007), "Smart Solutions," *Water Efficiency, The Journal for Water Conservation*, July/August, 2(4), Forester Communications, Santa Barbara, Calif.

Rainbarrel Guide (2007), http://www.rainbarrelguide.com/, Rainbarrel Guide webpage accessed July 12, 2007.

Rainbird (2007), http://www.rainbird.com/, website for Rain Bird Corporation accessed July 10, 2007.

RMI (2007), http://www.rmi.org/, Rocky Mountain Institute webpage, accessed July 10, 2007.

US Code of Federal Regulations (1992), Energy Policy Act of 1992 (EPAct 1992).

US Code of Federal Regulations (2005), Energy Policy Act of 2005 (EPAct 2005), passed July 29, 2005.

USGBC (2003), *LEED-NC for New Construction, Reference Guide*, Version 2.1, 2d ed., May, U.S. Green Building Council, Washington, D.C.

USGBC (2005–2007), *LEED-NC for New Construction, Reference Guide*, Version 2.2, 1st ed., U.S. Green Building Council, Washington, D.C., October 2005 with errata posted through Spring 2007.

USGBC (2009), *LEED Reference Guide for Green Building Design and Construction*, 2009 Edition, U.S. Green Building Council, Washington, D.C., April 2009.

Wu, X., B. Akinci, and C. I. Davidson (2007), "Modeling Graywater in Residences: Using Shower Effluent in the Toilet Reservoir," *Journal of Green Building*, Spring, 2(2): 111–120.

Exercises

1. You plan to plant 24,000 ft^2 of area with vegetation on a site in Columbia, S.C.
   - **A.** You plan to use a typical sprinkler system. What is your expected TPWA for the following combination of plantings?
     - 8000 ft^2 will be north of the planned building and will be mixed tree and shrub species with average water needs, densely planted.
     - 8000 ft^2 will be average density, low-water-use lawn on a sloped area west of the building.
     - The remaining will be a shrubbed landscape bed, with species of average water needs and spaced around normally.
   - **B.** You then decide to try to get two points for LEED 2009 WE credit 1 by keeping all the areas natural with the existing woods that require no irrigation except for the lawned area mentioned. Will you meet the credit requirement with only these changes? You can assume that the design in part A is a typical practice.
   - **C.** Is this also eligible for an RP point? (See Chap. 7.)

2. You plan to plant 24,000 ft^2 of area with vegetation on a site in East Hampton on Long Island, N.Y.
   - **A.** You plan to use a typical sprinkler system. What is your expected TPWA for the following combination of plantings?
     - 7000 ft^2 will be north of the planned building and will be mixed tree and shrub species with average water needs, densely planted.
     - 9000 ft^2 will be average density, low-water-use lawn on a sloped area west of the building.
     - The remaining will be a shrubbed landscape bed, with species of average water needs and spaced around normally.
   - **B.** You then decide to try to get two points for LEED 2009 WE credit 1 by keeping all the areas natural with the existing woods that require no irrigation except for the lawned area mentioned. Will you meet the credit requirement with only these changes? You can assume that the design in part A is a typical practice.
   - **C.** Is this also eligible for an RP point? (See Chap. 7.)

3. You plan to plant 20,000 ft^2 of area with vegetation on a site in San Francisco.
   - **A.** You plan to use a typical sprinkler system. What is your expected TPWA for the following plantings?
     - 7000 ft^2 will be north of the planned building and will be mixed tree and shrub species with average water needs, densely planted.
     - 5000 ft^2 will be average density, low-water-use lawn on a sloped area west of the building.
     - The remaining will be a shrubbed landscape bed, with species of high water needs and the plants spaced around normally.
   - **B.** You then decide to try to get two points for LEED 2009 WE credit 1 by using low-water-need plants and a drip irrigation system in the landscape bed. You can also extend the modified landscape bed into the lawn area. How much of the lawn area would need to be converted to obtain this credit? You can assume that the design in part A is a typical practice.
   - **C.** Is this also eligible for an RP point? (See Chap. 7.)

4. Your client really would like an expansive, thick lawn in front of her new office building and would like to obtain at least two WEc1 points by using rainwater harvesting. The building is in Austin, TX. The building could be oriented with the front façade facing either north or west. You are deciding between a low-water and an average water-need lawn. Determine how much water is needed to irrigate the lawn with a sprinkler system per square foot for the four possible options for the month of July.

5. You are building an office building for 150 men and 100 women. Shower facilities will be provided for those who exercise before work or during lunch. You do not want to use automatic controls on the lavatory faucets.

- **A.** Calculate a baseline water usage.
- **B.** To be more green, you decide to go with the HET single flush gravity water closets, waterless urinals, and low-flow showers. Perform calculations to determine the points you might obtain for LEED 2009 WEc2 and WEc3.

6. You are building an office building for 150 men and 100 women. Shower facilities will be provided for those who exercise before work or during lunch. You decide to use automatic controls on the lavatory faucets.

- **A.** Calculate a baseline water usage.
- **B.** To be more green, you decide to go with the HET single flush gravity water closets, waterless urinals, and low-flow showers. Perform calculations to determine the points you might obtain for LEED 2009 WEc2 and WEc3.

7. You are building an apartment complex with 80 three-bedroom units near an elementary school. The facility will include a playground and pool complex. The designers estimate that there will be on average 4.0 occupants in each apartment.

- **A.** Calculate a baseline water usage.
- **B.** To be more green, you decide to go with the dual flush water closets, low-flow kitchen sinks, and low-flow showers. Perform calculations to determine the points you might obtain for LEED 2009 WEc2 and WEc3.

8. You are building an apartment complex with 80 three-bedroom units near the downtown office district. The designers estimate that there will be on average 3.5 occupants in each apartment.

- **A.** Calculate a baseline water usage.
- **B.** To be more green, you decide to go with HET single flush gravity water closets, low-flow showers, and low-flow kitchen sinks. Perform calculations to determine the points you might obtain for LEED 2009 WEc2 and WEc3.

9. You are building an industrial building for a mixed workforce. The day shift will have about 150 employees, and about 20 employees will be needed to work both the swing and the graveyard shifts to keep the machines running. Shower facilities will be provided. You originally do not want to use automatic controls (metering) on the lavatory faucets.

- **A.** Calculate a baseline water usage.
- **B.** You have heard about the new dual flush water closets and decide to use them in the women's rooms. You also decide to go with low-flow urinals. Which LEED 2009 WE points might you earn?
- **C.** What additional changes might need to be made to earn the exemplary performance point for LEED 2009 WEc3?

10. You are building an industrial building for a mixed workforce. The day shift will have about 200 employees, and about 20 employees will be needed to work both the swing and the graveyard shifts to keep the machines running. Shower facilities will be provided. You originally want to use automatic controls (metering) on the lavatory faucets.

- **A.** Calculate a baseline water usage.
- **B.** Your design will be using HET single flush pressure assist water closets, low-flow urinals, and low-flow showers. What LEED 2009 WE points might you earn?
- **C.** You would like to earn the EP point for LEED 2009 WEc2 by plumbing graywater or rainwater to the water closets and urinals. What minimum average monthly volume of these waters will you need to provide in order to accomplish this?

11. Rewrite Eqs. (3.3.16) and (3.3.17) for residential uses if there are reuse waters (GW) used.
12. Comment on the health and other protective measures which might need to be taken if reuse waters are used for indoor water closets or urinals in the jurisdiction where your work or school is located.
13. Comment on the health and other protective measures which might need to be taken if reuse waters are used for irrigation needs in the jurisdiction of your work or school.
14. Compare the TPWA for planting the landscape areas in Exercise 1A to planting them in your hometown.
15. Compare the TPWA for planting the landscape areas in Exercise 2A to planting them at your work or school location.
16. Compare the TPWA for planting the landscape areas in Exercise 3A to planting them in Kansas City, MO.
17. Which, if any, of the WE credits are available for Regional Priority points in the zip code of your work or school? (See Chap. 7.)
18. Which, if any, of the WE credits are available for Regional Priority points in the following major cities? (See Chap. 7.)
    - **A.** New York City
    - **B.** Houston, TX
    - **C.** Los Angeles, CA
    - **D.** Portland, OR
    - **E.** Miami, FL
    - **F.** Kansas City, MO

CHAPTER 4

LEED Energy and Atmosphere

The Energy and Atmosphere category of the U.S. Green Building Council (USGBC) rating system for new construction version 2.2 (LEED-NC 2.2) and version 2009 (LEED-NC 2009) consists of three prerequisite subcategories and six credit subcategories that together may earn a possible 17 credit points maximum for version 2.2 and 35 possible points for version 2009 not including exemplary performance points. The notation format for the prerequisites and credits is, for example, EAp1 for Energy and Atmosphere prerequisite 1 and EAc1 for Energy and Atmosphere credit 1. To facilitate cross-referencing between this text and the USGBC rating system, the second digit in a section heading, equation, table, or figure number represents the credit subcategory number for sections that deal directly with a USGBC LEED-NC 2.2 or LEED-NC 2009 credit subcategory.

The Energy and Atmosphere (EA) portion deals with practices and policies that reduce the use of energy at the site, reduce the use of nonrenewable energy both at the site and at the energy source, and reduce the impact on the global climate, atmosphere, and environment from both activities at the site and energy sources off-site. Many of the subcategories in the Sustainable Sites category also deal with practices that may have an impact on the climate, but usually on a local scale.

The reason why Energy and Atmosphere are combined is that a significant portion of the air pollution and global climate impacts come from energy sources. Therefore, reducing or changing these energy sources has a large impact on the atmosphere, particularly on a more regional or global scale. For example, fossil-based energy sources, such as fuel oil and coal used in the production of energy, may also release many pollutants to the air and release additional carbon dioxide into the atmosphere. Carbon dioxide is considered to be a greenhouse gas, and it is considered to be nonrenewable to use these types of fuels, both from an energy source perspective and from a global carbon cycle perspective. Carbon dioxide emissions can be from the production of materials and other practices related to construction, but since most of the energy used in the United States is fossil fuel–based, reducing energy usage is advantageous in a resource and economic sense as well as in an atmospheric and air pollution prevention sense. To be inclusive, other credits related to regional or global climate issues are also included in the Energy and Atmosphere category, even though they are not specific only to energy issues. Also, one of the credits for enhanced commissioning is not limited solely to energy systems, but since a large proportion of the systems being commissioned are energy-related, the entire enhanced commissioning subcategory is included under Energy and Atmosphere.

In both versions, the three prerequisite subcategories are EA Prerequisite 1: *Fundamental Commissioning of the Building Energy Systems* (EAp1), EA Prerequisite 2: *Minimum Energy Performance* (EAp2), and EA Prerequisite 3: *Fundamental Refrigerant Management* (EAp3). As it is typical for prerequisites, there are no points associated with any of these three.

The points for the credits in the Energy and Atmosphere category are not necessarily individually related to a specific credit in LEED-NC 2.2 and all in LEED-NC 2009. Table 4.0.0 is a summary of the prerequisites and credits in both versions (2.2 and 2009) with a comparison of the points available and a summary of many of the major differences.

The prerequisites and credits in the Energy and Atmosphere category can be grouped into the following four overall sets of goals:

- EAp1 and EAc3 provide, for commissioning, a design and construction service that aids in improving various system efficiencies and reliabilities at a site.
- EAp2, EAc1, and EAc5 help optimize the energy performance and reduce energy usage at the site.
- EAp3 and EAc4 reduce the amount of pollutants at the site that are related to the ozone hole in the stratosphere. These pollutants are ranked by their ozone-depleting potential (ODP). EAc4 takes into consideration those ODP pollutants that might also contribute to an index referred to as the *global warming potential* (GWP).
- EAc2 and EAc6 promote the use of renewable or "green" energy sources. EAc2 promotes its use on-site, and EAc6 promotes its use off-site.

Some of the three prerequisites and six credit subcategories in the Energy and Atmosphere category were previously referenced with different names in the older version 2.1 as follows:

- **EA Prerequisite 1** (EAp1): *Fundamental Commissioning of the Building Energy Systems* (previously referred to as *Fundamental Building Systems Commissioning* in LEED-NC 2.1)
- **EA Prerequisite 2** (EAp2): *Minimum Energy Performance*
- **EA Prerequisite 3** (EAp3): *Fundamental Refrigerant Management* (previously referred to as *CFC Reduction in HVAC&R Equipment* in LEED-NC 2.1)
- **EA Credit 1** (EAc1): *Optimize Energy Performance* (EB)
- **EA Credit 2** (EAc2): *On-Site Renewable Energy* (previously referred to as *Renewable Energy* in LEED-NC 2.1)
- **EA Credit 3** (EAc3): *Enhanced Commissioning* (previously referred to as *Additional Commissioning* in LEED-NC 2.1) (EB)
- **EA Credit 4** (EAc4): *Enhanced Refrigerant Management* (previously referred to as *Ozone Depletion* in LEED-NC 2.1)
- **EA Credit 5** (EAc5): *Measurement and Verification* (EB)
- **EA Credit 6** (EAc6): *Green Power*

Subcategory 2009	Subcategory 2.2	Pts 2009	Pts 2.2	EP Pts 2009	EP Pts 2.2	Major Comments
EAp1: Fundamental Commissioning of Building Energy Systems	EAp1: Fundamental Commissioning of Building Energy Systems	na	na	na	na	–
EAp2: Minimum Energy Performance	EAp2: Minimum Energy Performance	na	na	na	na	*LEED 2009 EAp2 is similar to LEED 2.2 EAc1 for minimal points. Points are also obtained for this under EAc1 for 2009 Options 2 and 3.
EAp3: Fundamental Refrigerant Management	EAp3: Fundamental Refrigerant Management	na	na	na	na	†Exemption for small units (<0.5 lbs refrigerant) not specifically noted, but implied in LEED 2009.
EAc1: Optimize Energy Performance	EAc1: Optimize Energy Performance	1–19	1–10 2 req'd	1 pt 50% (new) or 46% (exist)	–	• *Option 1: 1–19 pts (2009) or 1–10 pts (2.2) • Option 2: 1 pt (2009) or 4 pts (2.2) • Option 3: 1–3 pts (2009) or 2–5 pts (2.2 updated) or 1 pt (2.2 obsolete)
EAc2: On-site Renewable Energy	EAc2: On-site Renewable Energy	1–7	1–3	1 (15%)	–	LEED 2009 awards more points than LEED 2.2.
EAc3: Enhanced Commissioning	EAc3: Enhanced Commissioning	2	1	1	–	2009 EP is for Comprehensive Envelope Commissioning.
EAc4: Enhanced Refrigerant Management	EAc4: Enhanced Refrigerant Management	2	1	–	–	• LEED 2009 includes not operating fire suppression systems with CFCs, HCFCs, or Halons. • Equipment life defaults vary. • Exemption for small units (<0.5 lb refrigerant) in 2009 and 2.2.

Table 4.0.0 EA Summary Table of Major Differences between Versions 2.2 and 2009 and Associated Points (Pts) (*Continued*)

Subcategory 2009	Subcategory 2.2	Pts 2009	Pts 2.2	EP Pts 2009	EP Pts 2.2	Major Comments
EAc5: Measurement and Verification	EAc5: Measurement and Verification	3	1	–	–	†LEED 2009: Provide a process for corrective action if necessary.
EAc6: Green Power (35% – 2 yr)	EAc6: Green Power (35% – 2 yr)	2	1	1 (70%)	1 (70% or 4 yr)	Option 1 basis in LEED 2.1 includes onsite renewable energy if use EAc1, implies same in 2.2 and 2009.
	Total	1	1	1	1	

*• LEED 2009 requires 10% or 5% improvement in building energy performance for new and existing buildings, respectively in the prerequisite based on ASHRAE 90.1-2007 Appendix G while LEED 2.2 EAc1 had the first 2 points required for 14% and 7% reductions, respectively based on ASHRAE 90.1-2004 Appendix G, both using a Whole Building Energy Simulation (Option 1).

• The prescriptive paths for LEED 2.2 EAc1 are now also a part of the prerequisite and may also earn some EAc1 points. Of note:

Option 2 LEED 2009, in addition to being required if chosen, additionally includes a path for Small Retail Buildings (<20,000 sf) and Small Warehouses and Self Storage Buildings (<50,000 sf) based on the ASHRAE Advanced Energy Design Guides for the same, 2006 and 2008, respectively. A project also earns 1 point under LEED 2009 EAc1 for following these required prescriptive options based on the Advanced Energy Guides, previously worth 4 points in LEED 2.2.

Option 3 LEED 2009, in addition to being required if chosen, 1 point is given for the required portions (complying with Sections 1 and 2 of the Advanced Buildings™ Core Performance Guide) and an additional point is given for every three strategies implemented from Section 3 (with exceptions), for a maximum of 2 additional points. Note that the original version of LEED 2.2 had a 1 point prescriptive Option 3, which could no longer be used after 6/26/2007 when a minimum of 2 EAc1 points was required (it became an Option 4 which is now obsolete). The aforementioned comments are based on the LEED 2.2 revision with the 2–5 point Option 3 based on the Advanced Buildings™ Core Performance Guide for buildings under 100,000 sf. It allotted 3 points for office, schools, public assembly, and retail projects, and 2 points for all other projects except warehouse, laboratory, and health care projects which complied with the Sections 1 and 2. It gave a point for every three strategies implemented from Section 3 (with exceptions), for a maximum of 2 additional points. https://www.usgbc.org/showfile.aspx?documentID=3198

• Certain Mandatory Measures are still required for all projects in LEED 2009 and are part of ASHRAE 90.1-2007.

• Option 1 subtracts on-site renewable energy from the energy used (proposed performance) in both LEED 2.2 and LEED 2009.

†Projects with a district energy system can now find guidance at www.usgbc.org/projecttools for LEED 2009.

Table 4.0.0 EA Summary Table of Major Differences between Versions 2.2 and 2009 and Associated Points (Pts)

The LEED-NC 2.2 and 2009 Reference Guides provide tables that summarize some of these prerequisite and credit characteristic, including the main phase of the project for submittal document preparation (Design or Construction).

Even though the entire LEED process encompasses all phases of a project, the prerequisites or credits that are noted as Construction Submittal credits usually have major actions going on through the construction phase that, unless documented through that time, may not be able to be verified from preconstruction documents or the built project. Sometimes these credits have activities associated with them that cannot or need not be performed until the beginning of, or subsequent to, the construction phase. EAp1, EAc3, EAc5, and EAc6 have been noted in the construction submittal column in Table 4.0.1.

Three of the credit subcategories listed in Table 4.0.1 have the icon EB noted after their title (EAc1, EAc3, and EAc5). This icon is a tool to help those who wish to proceed with the continuing LEED-EB certification (Existing Building) after certification of LEED-NC (New Construction or Major Renovation) is obtained. Those credits noted with this icon are usually significantly more cost-effective to implement during the construction of the building than later during its operation. They are also shown in Table 4.0.1. There

Prerequisite or Credit	Construction Submittal	EB Icon	Regional Variations May Be Applicable in LEED 2009	Location of the Site Boundary and the Site Area Important in the Calculations	Related Prerequisite
EAp1: Fundamental Commissioning of the Building Energy Systems	*	n/a	Yes		n/a
EAp2: Minimum Energy Performance		n/a	Yes		n/a
EAp3: Fundamental Refrigerant Management		n/a	No		n/a
EAc1: Optimize Energy Performance		Yes	Yes	No	EAp2
EAc2: On-Site Renewable Energy		No	Yes	Yes	n/a
EAc3: Enhanced Commissioning	*	Yes	Yes	No	EAp1
EAc4: Enhanced Refrigerant Management		No	No	No	EAp3
EAc5: Measurement and Verification	*	Yes	No	No	n/a
EAc6: Green Power	*	No	Yes	Not usually	n/a

TABLE 4.0.1 EA EB Icon, Regional Variation Applicability, and Miscellaneous LEED 2009

are many prerequisites or credits which might have variations applicable to the design or implementation of the project related to regional conditions in the United States. All the prerequisites and credits for which regional variations might be applicable are noted in the LEED 2009 Reference Guide and are so listed in Table 4.0.1 for the Energy and Atmosphere category.

There are EP points under the Innovation and Design category available that relate to several items in other categories. These are available for exceeding each of the respective credit criteria to a minimum level as noted in the respective credit descriptions. The available EP points for both versions 2.2 and 2009 are listed in Table 4.0.0. Note that only a maximum total of four EP points are available in total for a project in LEED 2.2 and a maximum of three in LEED 2009 and they may be from any of the noted EP options in any of the SS, WE, EA, MR, or IEQ LEED categories (see Chap. 7).

Two of the variables which must be determined for each project and which must be kept consistent throughout the credits in a LEED submittal are the size and location of the project site. The location of the site boundary is particularly important to credit EAc2, since the locations of renewable energy sources on-site are the basis of this credit. If renewable sources are located off-site, but used on-site, then they may only be applied to EAc6. There may be exceptions to this for special circumstance, such as campus settings. Also note that there is no rule that project areas are always contiguous, so it might be possible to include portions of some campus renewable energy sources in the project; however, this might then impact other credits, such as SSc2. Therefore, there might be a need for narratives describing any such variations and how the project still meets the intent of the credit. Usually, however, separate campus facilities for renewable energy sources are counted as green power for EAc6. Any renewable energy that is counted for EAc6 cannot also be counted toward EAc2. The importance of the LEED boundary to EAc2 is also noted in Table 4.0.1.

Three of the credit subcategories are actually optimizations or enhancements of EA prerequisites. The final column in Table 4.0.1 lists the prerequisites that are in this way related to the credits.

In summary, most of the Energy and Atmosphere credits address problems associated with energy use and production by a combination of methods. They aim to reduce energy usage and to use forms of energy that have a less harmful impact on the environment. Although several of the credits in the Energy and Atmosphere category might include a capital cost investment, many provide a means to potentially reduce the associated utility costs over the life of the building. The three prerequisite subcategories *must* be adhered to and verified for each project.

Energy and Atmosphere Prerequisites

EA Prerequisite 1: *Fundamental Commissioning of the Building Energy Systems*

This prerequisite is intended to aid in the efficiency and verification of the energy systems in a building.

USGBC Rating System

LEED-NC 2.2 lists the Intent, Benefits of Commissioning, Requirements, Commissioned Systems, and Potential Technologies and Strategies for this credit as follows, most of which are essentially the same in LEED 2009:

Intent

Verify that the building's energy related systems are installed, calibrated, and perform according to the owner's project requirements, basis of design, and construction documents.

Benefits of Commissioning

Benefits of commissioning include reduced energy use, lower operating costs, reduced contractor callbacks, better building documentation, improved occupant productivity, and verification that the systems perform in accordance with the owner's project requirements.

Requirements

The following commissioning process activities shall be completed by the commissioning team.

1) Designate an individual as the Commissioning Authority (CxA) to lead, review and oversee the completion of the commissioning process activities.
   a) The CxA shall have documented commissioning authority experience in at least two building projects.
   b) The individual serving as the CxA shall be independent of the project's design and construction management, though they may be employees of the firms providing those services. The CxA may be a qualified employee or consultant of the Owner.
   c) The CxA shall report results, findings and recommendations directly to the Owner.
   d) For projects smaller than 50,000 gross square feet, the CxA may include qualified persons on the design or construction teams who have the required experience.
2) The owner shall document the Owner's Project Requirements (OPR). The design team shall develop the Basis of Design (BOD). The CxA shall review these documents for clarity and completeness. The Owner and design team shall be responsible for updates to their respective documents.
3) Develop and incorporate commissioning requirements into the construction documents.
4) Develop and implement a commissioning plan.
5) Verify the installation and performance of the systems to be commissioned.
6) Complete a summary commissioning report.

Commissioned Systems

Commissioning process activities shall be completed for the following energy related systems, at a minimum:

- Heating, ventilating, air conditioning and refrigeration (HVAC&R) systems (mechanical and passive) and associated controls
- Lighting and daylighting controls
- Domestic hot water systems
- Renewable energy systems (wind, solar etc.)

Potential Technologies and Strategies

Owners are encouraged to seek out qualified individuals to lead the commissioning process. Qualified individuals are identified as those who possess a high level of experience in the following areas:

- Energy systems design, installation and operation
- Commissioning planning and process management
- Hands-on field experience with energy systems performance, interaction, start up, balancing, testing, troubleshooting, operation, and maintenance procedures
- Energy systems automation control knowledge

Owners are encouraged to consider including water-using systems, building envelope systems, and other systems in the scope of the commissioning plan as appropriate. The building envelope is an important component of a facility which impacts energy consumption, occupant comfort and indoor air quality. While it is not required to be commissioned by LEED, an owner can receive significant financial savings and reduced risk of poor indoor air quality by including building envelope commissioning.

The LEED-NC 2.2 and 2009 Reference Guides provide guidance on the rigor expected for this prerequisite for the following:

- Owner's project requirements
- Basis of design
- Commissioning plan
- Performance verification documentation
- Commissioning report

Calculations and Considerations

There are no referenced standards for this prerequisite, but the narrative given by the USGBC is very detailed as to the expectations. Previously in LEED-NC 2.1, this prerequisite also included some systems other than energy systems, such as water systems. These other systems are no longer in the prerequisite, but many are included in EAc3: *Enhanced Commissioning*. This prerequisite has a detailed list of overall project-related documentation that must be provided and gives a detailed overview of the project as previously listed in the bulleted list. There are two main steps early on in this commissioning process:

- A commissioning authority (CxA) is designated.
- The two documents called the Owner's Project Requirements (OPR) and the Basis of Design (BOD) are developed, and then reviewed by the CxA. (Usually these documents are developed by the owner and the design team with the aid of the commissioning authority.)

The LEED 2009 Reference Guide provides good summaries of many of the items which should be included in the OPR and the BOD, such as:

Owner and user requirements (OPR)
Environmental and sustainability goals (OPR)
Energy efficiency goals (OPR)
Indoor environmental quality requirements (OPR)
Equipment and systems expectations (OPR)
Building occupant and operating and maintenance (O&M) personnel requirements (OPR)
Primary design assumptions (BOD)
Standards, codes, regulations (BOD)
Narrative descriptions of performance criteria (BOD)

As the design phase of the project proceeds, in addition to periodically updating the noted documents, the following two steps are performed:

- A commissioning plan is developed and implemented.
- Commissioning requirements are incorporated into the construction documents.

Various members of the project team assume responsibility for different portions of the commissioning plan, with the general contractor assuming responsibility for most of the requirements outlined in the construction documents. Finally, during and after construction the following two main items are done:

- The installation and performance of the listed commissioned systems are verified.
- A summary commissioning report is completed.

The CxA should have experience with at least two projects of similar magnitude and complexity as the project in question. The USGBC expects this commissioning authority to be familiar with these steps. Further details can be found in the LEED-NC 2.2 and 2009 Reference Guides and will not be included here. However, it is very important that the commissioning steps be well documented and verified during the construction phase. They are a part of the construction phase submittal. The following list is a good outline from the LEED-NC 2.2 and 2009 Reference Guides of the items that should be covered for the contractor in the construction documents so that his or her construction engineers and managers can adequately fulfill their portion of the fundamental commissioning (Cx) requirements:

- Commissioning team involvement
- Contractor's responsibilities
- Submittals and submittal review procedures for Cx process/systems
- Operations and maintenance documentation, system manuals
- Meetings
- Construction verification procedures
- Start-up plan development and implementation/Functional performance testing/ Acceptance and closeout
- Training
- Warranty review site visit

There is an additional commissioning opportunity that can earn a credit point in 2.2 and 2 credit points in the 2009 version, EAc3: *Enhanced Commissioning*. A summary table of the additional steps in the commissioning process that are needed for enhanced commissioning as compared to fundamental commissioning has been included in the LEED-NC 2009 Reference Guide and is included here as Table 4.0.2.

Various definitions have also been included in the LEED-NC 2009 Reference Guide. These are presented in App. B and cover the following terms: *basis of design* (BOD), *commissioning* (Cx), *commissioning plan, commissioning report, commission specification, commissioning team, installation inspection, owner's project requirements* (OPR), and *systems performance testing*.

EA Prerequisite 2: *Minimum Energy Performance*

The intention of this prerequisite is to establish and require a minimum energy efficiency in the building and many of its systems.

USGBC Rating System LEED 2002

LEED-NC 2.2 lists the Intent, Requirements, and Potential Technologies and Strategies for EAp2 as follows:

Tasks	Responsibilities	
	For EAp1 Only	For EAp1 and EAc3
Designate commissioning authority (CxA)	Owner or project team	Owner or project team
Document Owner's Project Requirements	Owner or CxA design team	Owner or CxA design team
Develop Basis of Design	Owner or CxA design team	Owner or CxA design team or CxA design team
Review OPR and BOD	CxA	CxA
Incorporate commissioning requirements into the construction documents	Project team or CxA	Project team or CxA
Conduct commissioning design review prior to midconstruction documents	n/a	CxA
Develop and implement a commissioning plan	Project team or CxA	Project team or CxA
Review contractor submittals applicable to systems being commissioned	n/a	CxA
Verify the installation and performance of commissioned systems	CxA	CxA
Develop a systems manual for the commissioned systems	n/a	Project team or CxA
Verify that the requirements for training are completed	n/a	Project team or CxA
Complete a summary commissioning report	CxA	CxA
Review building operation within 10 months after substantial completion	n/a	CxA

TABLE 4.0.2 Tasks and Primary Responsibilities for Commissioning per LEED-NC 2009 Reference Guide

Intent

Establish the minimum level of energy efficiency for the proposed building and systems.

Requirements

Design the building project to comply with both

- The mandatory provisions (Sections 5.4, 6.4, 7.4, 8.4, 9.4 and 10.4) of ASHRAE/IESNA Standard 90.1-2004

And

- The prescriptive requirements (Sections 5.5, 6.5, 7.5 and 9.5) or performance requirements (Section 11) of ASHRAE/IESNA Standard 90.1-2004

Potential Technologies and Strategies

Design the building envelope, HVAC, lighting, and other systems to maximize energy performance. The ASHRAE 90.1-2004 User's Manual contains worksheets that can be used to document compliance with this prerequisite. For projects pursuing points under EA Credit 1, the computer simulation model may be used to confirm satisfaction of this prerequisite. If a local code has demonstrated quantitative and textual equivalence following, at a minimum, the U.S. Department of Energy standard process for commercial energy code determination, then it may be used to satisfy this prerequisite in lieu of ASHRAE 90.1-2004. Details on the DOE process for commercial energy code determination can be found at www.energycodes.gov/implement/determinations_com.stm.

USGBC Rating System LEED 2009

This prerequisite has been substantially changed in LEED-NC 2009 and is now similar to the first few points in EAc1, Optimize Energy Performance from LEED-NC 2.2. Refer to the three options and calculation pathways in the updated LEED-NC 2009 EAc1 for detailed calculations. A summary of the requirements are outlined in Table 4.0.0.

Option 1 for EAp2 in LEED 2009 is based on a whole building energy simulation and requires a 10 percent improvement in the proposed building performance for new buildings and a 5 percent improvement for major renovations to existing buildings based on Appendix G of ASHRAE Standard 90.1-2007 (with errata but without addenda). All the energy costs associated with the project must be included, with the default process energy costs assumed at 25 percent of the total energy costs for the baseline building similar to the LEED 2.2 EAc1 requirements. The process energy remains at this default value for the design case, except if the team uses the exceptional calculation method (ECM) in Appendix G to document applicable process energy reductions. The process energy cost items include office and general miscellaneous equipment, computers, elevators and escalators, kitchen cooking and refrigeration, laundry washing and drying, lighting exempt from the lighting power allowance such as medical equipment lighting, etc. The regulated (or non-process) energy is the main focus of the energy reduction. It includes interior, parking, facade or exterior lighting (with exceptions), heating, ventilation and air conditioning (HVAC) (such as for space heating and cooling, fans, pumps, toilet exhaust, parking garage ventilation, kitchen hood exhaust) and service water heating for domestic use or heating. Option 1 in LEED 2009 EAp2 also requires compliance with mandatory provisions listed in Sections 5.4, 6.4, 7.4, 8.4, 9.4, 10.4 in Standard 90.1-2007 (with errata but without addenda). The prescriptive requirements in Sections 5.5, 6.5, 7.5, and 9.5 of ASHRAE Standard 90.1-2007 are not specifically noted in LEED 2009 EAp2, but are referred to in the LEED 2009 reference guide and therefore compliance is implied. (There is also an alternative code that may be used for California projects based on California Title 24-2005, Part 6 for Option 1.)

Option 2 for LEED 2009 EAp2 is titled the Prescriptive Compliance Path: ASHRAE Advanced Energy Design Guide, while Option 3 is titled Prescriptive Compliance Path: Advanced Buildings Core Performance (TM) Guide based on a guide developed by the New Buildings Institute (NBI). Options 2 and 3 both have restrictions on the size of the project and also have specific requirements for certain uses. Summaries of Options 2 and 3 for LEED 2009 EAp2 are listed in Table 4.0.3.

Calculations and Considerations LEED 2009

This prerequisite uses the ASHRAE/IESNA 90.1-2007: Energy Standard for Buildings except Low-Rise Residential. This standard was sponsored by the American Society of Heating, Refrigerating and Air-Conditioning Engineers (ASHRAE) and the Illuminating Engineering Society of North America (IESNA) under an American National Standards Institute (ANSI) process. It sets minimum energy requirements and energy efficiency requirements. It does not apply to buildings that use neither electricity nor fossil fuel, manufactured houses, or portions of buildings that use energy primarily for industrial, manufacturing, or commercial processes.

Option and Path	Guidance	Requirements
Option 2 Path 1	ASHRAE Advanced Energy Design Guide for Small Office Buildings 2004	• Building less than 20,000 sf GFA • Office occupancy
Option 2 Path 2	ASHRAE Advanced Energy Design Guide for Small Retail Buildings 2006	• Building less than 20,000 sf GFA • Retail occupancy
Option 2 Path 3	ASHRAE Advanced Energy Design Guide for Small Warehouses and Self-Storage Buildings 2008	• Building less than 50,000 sf GFA • Warehouse or self-storage occupancy
Option 3	Advance Buildings™ Core Performance™ Guide	• Building less than 100,000 sf GFA • Health care, warehouse, and laboratory facilities are not eligible • Office, school, public assembly, and retail projects comply with Section 1: Design Process Strategies and Section 2: Core Performance Requirements • Other project types comply with the basic requirements • Additional requirements for EAc1 points are: The building must have a window-to-wall ratio less than 0.40 1 point given for compliance with Sections 1 and 2 1 point additional for every 3 strategies implemented from Section 3 for a maximum total of 3 points (strategies 3.1–Cool Roofs, 3.8–Night Venting, and 3.13–Additional Commissioning not included)

Table 4.0.3 Summary of Options 2 and 3 in LEED 2009 EAp2: Minimum Energy Performance and EAc1 Optimize Energy Performance

This prerequisite requires that certain mandatory provisions in the energy standard be met. In addition, there are other items that may be met by either adhering to some prescriptive requirements in the standard or alternatively meeting the intent of these prescriptive requirements by meeting certain overall design-based performance criteria. Prescriptive requirements in standards generally refer to actions that are specifically met, whereas, with design-based performance criteria, there may be many design options that will fulfill a similar goal. For instance, a prescriptive requirement might state that all windows must be double-paned, whereas a performance requirement might state that the overall energy loss from the windows must not exceed a certain value.

Energy-Related Component	
Section 5	Building envelope (including semiheated spaces such as warehouses)
Section 6	Heating, ventilating, and air conditioning (including parking garage ventilation, freeze protection, exhaust air energy recovery, and condenser heat recovery for service water heating)
Section 7	Service water heating (including swimming pools)
Section 8	Power (including all building power distribution systems)
Section 9	Lighting (including lighting for exit signs, building exterior, grounds, and parking garage)
Section 10	Other equipment (including all permanently wired electric motors)

Source: Courtesy LEED-NC 2.2 Reference Guide.

TABLE 4.0.4 ASHRAE/IESNA 90.1-2004 and 2007 Components Applicable to EAp2 and EAc1

The general energy-related components of the building that are covered by this prerequisite have been summarized by the LEED-NC 2.2 and 2009 Reference Guides and are listed in Table 4.0.4.

A review or summary of the extensive calculations needed for the energy performance is not intended to be a part of this book. There are many energy models used regularly by energy experts to determine the energy requirements of the building and system. However, there are a few items that the LEED engineer may find useful in understanding this prerequisite and its implication to other credits. Six of these are listed as follows:

- There are certain mandatory provisions that must be met in each section listed in Table 4.0.4. In addition, as previously mentioned, either certain prescriptive requirements or certain performance requirements must be met.
- If EAc1 Option 1 is sought, the alternative energy simulation documentation referred to as the Performance Rating Method option in App. G of this same standard shall be used to document these items of variance from the prescriptive requirements.
- The United States is segregated into eight climate zones, and there are different prescriptive requirements for many sections in each of these zones.
- The building envelope requirements in Section 5 of Table 4.0.4 cover many mandatory items such as insulation, doors, and windows. If the prescriptive method is used, then there are maximum allowed window and skylight areas (40 percent of the gross wall area for windows and 5 percent of the roof area for skylights in LEED 2009). This is important to note as windows and skylights impact other credits in the indoor Environmental Quality (EQ) category of LEED. Otherwise, either the cost budget method or the performance rating method can be used, and the performance requirements have greater flexibility.

Mandatory measures (all projects)	• Building Envelope Compliance Documentation (Part I)—Mandatory Provisions Checklist • HVAC Compliance Documentation (Part II)—Mandatory Provisions Checklist • Service Water Heating Compliance Documentation (Part I)—Mandatory Provisions Checklist • Lighting Compliance Documentation (Part I)—Mandatory Provisions Checklist
Prescriptive requirements (projects using prescriptive compliance approach)	• Building Envelope Compliance Documentation (Part II) • HVAC Compliance Documentation—Part I for buildings <25,000 ft² using the simplified approach or Part III for all other buildings • Service Water Heating Compliance Documentation
Performance requirements (projects using performance compliance approach and when EA Credit 1 Option 1 is being sought)	• Performance Rating Report • Table documenting energy-related features included in the design, and including all energy features that differ between the baseline design and proposed design models

Source: Courtesy LEED-NC 2.2 Reference Guide.

Table 4.0.5 EAp2 Compliance Forms Based on ASHRAE Standard 90.1-2007 and Updated for LEED 2009

- There are exceptions to Section 6 (HVAC) for projects served by existing HVAC systems such as central energy plants. For example, existing systems that service the project, but are outside of it, do not have to conform to all the requirements of this prerequisite.
- Section 9 covers lighting and includes both indoor and exterior lighting. The reduction in exterior lighting from this prerequisite is also a part of SSc8.
- Credit for natural ventilation may be obtained in the energy modeling on a case-by-case basis with additional documentation.

Table 4.0.5 is based on the LEED-NC 2009 Reference Guide and gives a good summary of some of the documentation for the various paths chosen for this prerequisite. Note that most LEED-NC projects do seek the EA credit 1, and therefore, only the first and third items in this table are applicable.

EA Prerequisite 3: *Fundamental Refrigerant Management*

The intent of this prerequisite is to help maintain the ozone levels in the stratosphere. The chemical and physical makeup of the stratosphere is different from the atmosphere in the troposphere where we live and breathe. The stratosphere is located above the troposphere and is where many jets fly. Over the past century, many chemicals were developed as effective means for refrigeration and fire suppression. When these chemicals leaked or were expelled into the troposphere, the effects seemed to be fairly benign, and therefore, they became very commonly used. However, scientists have recently shown that these same chemicals react differently when they enter an atmospheric environment similar to that found in the stratosphere. There, many of these same chemical compounds apparently serve as catalysts that destroy the amount of ozone (O_3) in the stratosphere, converting it into oxygen molecules (O_2). In the troposphere, we usually regard O_2 as a "good" molecule, since this is the form of oxygen that we use to breathe, and we usually regard O_3 as a "bad" molecule, as it can cause air pollution problems and adversely impact health. However, in the stratosphere, the O_3 molecule advantageously aids in filtering out some of the ultraviolet radiation from the sun that might otherwise reach the surface of the earth and negatively impact health and property. As is well known, places in the stratosphere where the concentrations of O_3 have been reduced are commonly referred to as "ozone holes." Therefore, there is an effort to internationally reduce the use of certain refrigerants and fire suppression chemicals that might have the greatest negative impact on the amount of ozone in the stratosphere.

EA prerequisite 3 deals with a group of common refrigerants collectively referred to as chlorofluorocarbons (CFCs). They are organic carbon–based compounds in which the hydrogen atoms have been replaced by chlorine and fluorine atoms. The chlorine atoms are thought to be some of the chemicals responsible for problems in ozone depletion in the stratosphere. There are alternative compounds which can be used. Many of these are discussed under EA credit 4. Many of the CFCs also go by the DuPont trademark name *Freon*. The intent of this prerequisite is to ban or limit the use of CFCs in new projects and major renovations.

USGBC Rating System

LEED-NC 2.2 lists the Intent, Requirements, and Potential Technologies and Strategies for EAp3 as follows, and the requirements in LEED 2009 are essentially identical:

Intent

Reduce ozone depletion.

Requirements

Zero use of CFC-based refrigerants in new base building HVAC&R systems. When reusing existing base building HVAC equipment, complete a comprehensive CFC phase out conversion prior to project completion. Phase-out plans extending beyond the project completion date will be considered on their merits. (Small HVAC units (defined as containing less than 0.5 lbs of refrigerant), and other equipment such as standard refrigerators, small water coolers, and any other cooling equipment that contains less than 0.5 lbs of refrigerant, are not considered part of the "base building" system and are not subject to the requirements of this credit.)

Potential Technologies and Strategies

When reusing existing HVAC systems, conduct an inventory to identify equipment that uses CFC refrigerants and provide a replacement schedule for these refrigerants. For new buildings, specify new HVAC equipment in the base building that uses no CFC refrigerants.

Calculations and Considerations

This prerequisite is quite simple. There will be no CFCs used for heating, ventilating, air conditioning, and refrigeration (HVAC&R) base building systems in new buildings, and they must be taken out of existing buildings if economically feasible for which there must be a detailed phase-out plan. There must be a third-party audit proving that this is not economically feasible to replace or convert the system based on a 10 year simple payback if existing systems are to remain.

4.1 EA Credit 1: *Optimize Energy Performance*

The intention of this credit is to give credit for minimizing energy usage on-site, and also to encourage minimizing the use of off-site power and other off-site energy sources.

USGBC Rating System

LEED-NC 2.2 lists the Intent for EAc1 as follows, which is essentially identical to the LEED 2009 Intent. The requirements from LEED 2.2 have been updated to reflect the LEED 2009 changes.

Intent

Achieve increasing levels of energy performance above the baseline in the prerequisite standard to reduce environmental and economic impacts associated with excessive energy use.

Requirements The requirements have changed significantly with the conversion to the LEED 2009 version, with the most significant change being the point values given, and minimum energy reductions from the baseline now part of the prerequisite EAp2. The baseline standard is now the 2007 version of ASHRAE/IESNA Standard 90.1 Appendix G (with errata but without addenda) for Option 1. The 2009 version allows the project team to select one of the three compliance path options described. Project teams documenting achievement using any of the three options are assumed to be in compliance with EA Prerequisite 2.

OPTION 1 — WHOLE BUILDING ENERGY SIMULATION

This option requires that the project team demonstrate a percentage improvement in the proposed building performance rating compared to the baseline building performance rating by a whole building project simulation using the Building Performance Rating Method in Appendix G of the Standard. The points available for EA credit 1 for both the 2.2 and the 2009 versions of LEED New Construction and Major Renovations are displayed in Table 4.1.1 with the percent reduction from a baseline energy cost noted for each.

Appendix G requires that the energy analysis done for the Building Performance Rating Method include ALL of the energy costs within and associated with the building project. To achieve points using this credit, the proposed design must also adhere to the requirements as listed in EAp2 which include that the project:

- Must comply with the mandatory provisions (Sections 5.4, 6.4, 7.4, 8.4, 9.4 and 10.4);
- Must include all the energy costs within and associated with the building project; and
- Must be compared against a baseline building that complies with Appendix G. The default process energy cost is 25% of the total energy cost for the baseline building. For buildings where the process energy cost is less than 25% of the baseline building energy cost, the LEED submittal must include supporting documentation substantiating that process energy inputs are appropriate. Process energy and regulated (nonprocess) energy are as defined in EAp2 and process loads shall be identical for both the baseline building performance rating and for the proposed building performance rating. Project teams may follow the Exceptional Calculation Method in ASHRAE 90.1 2007 to document measures that reduce process loads. As in LEED 2.2, documentation of process load energy savings shall include a list of the assumptions made for both the base and proposed design, and theoretical or empirical information supporting these assumptions. Projects in California may again use the Title 24-2005, Part 6 standard in place of the ASHRAE standard as stated in EAp2 for EAc1 Option 1.

Points	LEED 2009 Percent Reduction		LEED 2.2 Percent Reduction	
	New Buildings	Existing Building	New Buildings	Existing Building
Required (EAp2)	10	5	0	0
1	12	8	10.5 (required)	3.5 (required)
2	14	10	14 (required)	7 (required)
3	16	12	17.5	10.5
4	18	14	21	14
5	20	16	24.5	17.5
6	22	18	28	21
7	24	20	31.5	24.5
8	26	22	35	28
9	28	24	38.5	31.5
10	30	26	42	35
11	32	28	–	–
12	34	30	–	–
13	36	32	–	–
14	38	34	–	–
15	40	36	–	–
16	42	38	–	–
17	44	40	–	–
18	46	42	–	–
19	48	44	–	–
19+1 EP	50	46	–	–

TABLE 4.1.1 Table of Point Differences between Version 2.2 and Version 2009, EAc1 Option 1 Based on Percent Reductions in Energy Uses from a Baseline

OR

OPTION 2 — PRESCRIPTIVE COMPLIANCE PATH (4 points in version 2.2, 1 point in version 2009)

Option 2 is based on ASHRAE Advanced Energy Design Guides for various uses. In version 2.2 only the Small Office Building pathway was available, but version 2009 includes two additional pathways for Small Retail Buildings (Path 2) and Small Warehouses and Self Storage Buildings (Path 3), based on the 2006 and 2008 applicable ASHRAE Advanced Energy Design Guide, respectively. The requirements are listed in Table 4.0.0 and are both required as the EAp2 Minimum Energy Performance prerequisite and available as one point for EAc1 in the 2009 version of LEED-NC. For small office buildings LEED 2.2 required to:

Comply with the prescriptive measures of the ASHRAE Advanced Energy Design Guide for Small Office Buildings 2004. The following restrictions apply:

- Buildings must be under 20,000 square feet
- Buildings must be office occupancy
- Project teams must fully comply with all applicable criteria as established in the Advanced Energy Design Guide for the climate zone in which the building is located

OR

OPTION 3 — PRESCRIPTIVE COMPLIANCE PATH (1–3 points for version 2009)

The two versions are based on different standards. LEED 2.2 originally was worth 1 point and required the project to comply as follows, but this was replaced prior to LEED 2009 as the minimum 2 point requirement for EAc1 took effect. Comply with the Basic Criteria and Prescriptive Measures of the Advanced Buildings Benchmark Version 1.1 with the exception of the following sections: 1.7 Monitoring and Trend-logging, 1.11 Indoor Air Quality, and 1.14 Networked Computer Monitor Control. The following restrictions apply:

- Project teams must fully comply with all applicable criteria as established in Advanced Buildings Benchmark for the climate zone in which the building is located.

LEED 2009 requires this prescriptive Option 3 to comply with the prescriptive measures in the Advanced Buildings (TM) Core Performance (TM) Guide by the New Buildings Institute (NBI). The requirements and available points are listed in Table 4.0.3. Note that one point is automatic for EAc1 as its requirements are part of the prerequisite EAp2. Up to two additional points can be earned by also applying a minimum of 3 strategies for each extra point from the Enhanced Performance Section 3 strategy list from this guide, with the exception of Cool Roof, Night Venting and Additional Commissioning strategies, as these are addressed elsewhere in LEED 2009. The eleven enhanced performance strategy categories which can be considered are listed in Table 4.1.2.

Section	Strategy
3.2	Daylighting and controls
3.3	Additional lighting power reductions
3.4	Plug loads, appliance efficiency
3.5	Supply air temperature reset (VAV)
3.6	Indirect evaporative cooling
3.7	Heat recovery
3.9	Premium economizer performance
3.10	Variable speed control
3.11	Demand responsive buildings (peak power reduction)
3.12	On-site supply of renewable energy
3.14	Fault detection and diagnostics

TABLE 4.1.2 EAc1 Option 3: Enhanced Performance Strategies Available from the Advanced Buildings™ Core Performance™ Guide (LEED 2009)

Calculations and Considerations LEED 2009

EA credit 1 can be worth 1 to 19 points for version 2009 depending on the option chosen and the energy performance obtained. There are three options. Option 1 is based on a performance model approach that simulates the whole building. Option 1 is very calculation-intensive, and from 1 to 19 points can be obtained, with an additional Exemplary Performance (EP) point for version 2009. (Point values available for both versions are listed in Table 4.1.1.). The three overall strategies that will aid in obtaining many of these available points are to reduce energy demand, harvest free energy, and improve the efficiency of the energy systems.

Reducing demand includes items such as occupancy sensors so that many of the systems operate only when needed, and using colors or building orientation to decrease potential loads. Figure 4.1.1 shows architectural shading features installed on a building. These help reduce the air conditioning load.

Harvesting free energy includes such items as taking advantage of natural lighting, as with light tubes, light shelves, building orientation, and clerestories. Examples of some of these are also shown in Chap. 6, where natural lighting is encouraged for indoor environmental quality, in addition to energy reduction.

Improving efficiencies refers to individual items such as more efficient equipment and higher-performance lighting. It also includes looking at the overall systems. For example, a heating system could have more effectively sized ductwork, or distributed smaller heaters could be used to reduce line losses. Figure 4.1.2 depicts a Trane TRAQ damper with built-in, temperature-compensated flow measurement. This damper allows the building controls to bring in the right amount of outside air to meet indoor air quality standards while eliminating excess outside air intake, which can save on

Figure 4.1.1 Architectural shading features on the LEED certified Public Health Building at the University of South Carolina, Columbia, S.C. (*Photograph taken July 25, 2007.*)

Figure 4.1.2 Trane TRAQ damper which balances air intake for both indoor air quality and energy performance. (*Photograph Courtesy of Trane.*)

energy. Figure 4.1.3 depicts a Trane energy wheel. This wheel is part of an HVAC system and is used to reduce energy demand. It can recover energy from the conditioned exhaust air and uses it to pretreat the incoming outside air. Another developing concept in improving efficiencies is the idea of thermal energy storage (TES), where cooling energy is transferred into storage via mechanisms such as chilled water or ice during off-peak hours and is then used during peak hours. This technique reduces the demand on the equipment and the energy network. These are just a few of the examples of how energy efficiencies can be improved.

Options 2 and 3 are prescriptive-based options. They have requirements where certain specific actions must be taken which are the opposite of design-based performance requirements, where many different actions may result in the same level of performance. (The definition of prescriptive requirements is further explained in EAp2.) Options 2 and 3 are summarized in Table 4.0.3. There are no calculations for Option 2 or Option 3. Details of the prescriptive measures are not a part of this book but can be found in the LEED-NC 2.2 and 2009 Reference Guides.

Option 1 The performance modeling procedures and various standard energy models used for Option 1 are not a detailed part of this book. However, there are a few items relating to

FIGURE 4.1.3 Trane energy wheel for pretreating incoming air to the HVAC system. (*Photograph Courtesy of Trane.*)

the approach and calculations that the project team and engineers other than the energy modeler may find useful in acquiring a general understanding of LEED energy credits, as well as how this credit can interrelate with other credits. Some of these items are as follows:

- Both an annual baseline and an annual design energy model are established. They are based on annual energy costs and do not include any building construction or *first costs* (design, material, or installation costs) for the systems. The dollar values are based on either the local energy utility rates or, if these are not available, the default state average rates as published annually by the US Department of Energy (DOE) Energy Information Administration and available at www.eia.doe.gov. The 2003 Commercial Sector Average Energy Costs by State (CSAECS) on the U.S. Department of Energy (DOE) website were compiled in the version 2.2 Reference Guide with $CSAECSE_i$ and $CSAECSF_i$ representing the electrical and natural gas costs, respectively, for state *i*. These values are given in Table 4.1.3.
- The baseline and design models must be based on the ASHRAE/IESNA Standard 90.1-2007 App. G, Building Performance Rating Method. This is the method used to quantify exceedance of Standard 90.1. The energy cost budget (ECB) method in Section 11 of this same standard was accepted as a rating method in earlier versions of LEED but is no longer accepted.
- The numerical results of the baseline annual energy model are referred to as the baseline building performance (BBP), and the design annual energy costs as determined by this model are referred to as the proposed design energy model (PDEM). The energy model values include both regulated and process

	Electrical Rate ($/kWh) (default value of CSAECSE$_i$)*	Natural Gas ($/kBtu) (default value of CSAECSF$_i$)*
Alabama	0.0682	0.00938
Alaska	0.1646	0.00355
Arizona	0.0670	0.00758
Arkansas	0.0526	0.00668
California	0.1171	0.00843
Colorado	0.0597	0.00476
Connecticut	0.0900	0.01101
Delaware	0.0693	0.00840
District of Columbia	0.0645	0.01266
Florida	0.0678	0.01083
Georgia	0.0669	0.00957
Hawaii	0.1502	0.01926
Idaho	0.0601	0.00612
Illinois	0.0758	0.00794
Indiana	0.0585	0.00844
Iowa	0.0602	0.00750
Kansas	0.0611	0.00753
Kentucky	0.0520	0.00760
Louisiana	0.0664	0.00861
Maine	0.1019	0.01086
Maryland	0.0659	0.00807
Massachusetts	0.0848	0.01071
Michigan	0.0701	0.00631
Minnesota	0.0546	0.00778
Mississippi	0.0721	n/a
Missouri	0.0505	0.00796
Montana	0.0601	0.00623
Nebraska	0.0500	0.00698
Nevada	0.0955	0.00723
New Hampshire	0.0973	0.00917
New Jersey	0.0835	0.00835
New Mexico	0.0737	0.00659

Table 4.1.3 Default Energy Costs per State/District *i* [DOE Energy Information Administration 2003 Commercial Sector Average Energy Costs by State (CSAECS)]

	Electrical Rate ($/kWh) (default value of $CSAECSE_i$)*	**Natural Gas ($/kBtu) (default value of $CSAECSF_i$)***
New York	0.1113	0.00895
North Carolina	0.0641	0.00863
North Dakota	0.0547	0.00682
Ohio	0.0723	0.00789
Oklahoma	0.0571	0.00755
Oregon	0.0657	0.00775
Pennsylvania	0.0819	0.00898
Rhode Island	0.0834	0.00964
South Carolina	0.0652	0.00992
South Dakota	0.0605	0.00963
Tennessee	0.0631	0.00832
Texas	0.0695	0.00757
Utah	0.0538	0.00539
Vermont	0.1087	0.00778
Virginia	0.0572	0.00920
Washington	0.0624	0.00669
West Virginia	0.0545	0.00734
Wisconsin	0.0645	0.00822
Wyoming	0.0548	0.00469

*Use current values from www.eia.doe.gov for $CSAECSE_i$ or $CSAECSF_i$ if current local values are not available.

TABLE 4.1.3 (*Continued*)

(nonregulated) loads. The default is to assume that the process loads are approximately 25 percent of the baseline regulated loads. If this is not the case, then information should be provided to verify the difference.

- Both models (BBP and PDEM) are based on all the energy loads including process (or nonregulated) loads as previously defined. These types of loads have been summarized in the Requirements section. Usually only regulated loads are reduced by the energy performance methods in the design. If some applicable process loads are reduced in the design, then a special procedure in the ASHRAE/IESNA Standard 90.1-2007 App. G called the *exceptional calculation method* (ECM) can be used to document and quantify these savings into the design model (PDEM).
- The potential energy savings represented as the difference between the base model (BBP) and the proposed design (PDEM) can be realized in many different ways. These may include reduced loads, more efficient equipment, and the

recovery of potential energy losses by various methods, such as water preheating from waste streams, and the use of some natural sources, such as daylighting. For the purposes of the calculations presented in the following section, all these as just listed are included in the determination of the annual PDEM costs. If the energy model used cannot adequately include some of these recovery savings, then these can be calculated separately by using the ECM method and then incorporated into the PDEM.

- The savings from on-site renewable energy sources that produce power are calculated separately and are referred to as the *renewable energy cost* (REC). They are defined in greater detail in EAc2. The anticipated annual costs of off-site energy usage as represented by the PDEM can be offset by the on-site renewable energy source cost savings (REC). The on-site renewable energy usage values (RECs) can be subtracted from the PDEM as if the power they provide to the project were free. As in the PDEM calculations, the actual capital costs of the REC are not included. The REC savings are virtual savings based on the typical off-site utility rates.
- The baseline building is different from the proposed building in many ways. The LEED-NC 2.2 and 2009 Reference Guides list many of these differences in detail. Some of them include the following:
  1. The proposed design is modeled with the actual building orientation on the lot. The baseline design is the average of this orientation and the other three orthogonal orientations.
  2. Many building schedules, such as lighting or occupancy schedules, can differ between the two models if the differences can be defended. (See Chap. 8 for more information on schedules with respect to design and construction.)
  3. The baseline building is based on a specific assembly for floors, walls, and roofs as outlined in App. G of the standard.
  4. The total exterior wall area occupied by vertical fenestrations in the base building must be the same percent of the total exterior wall area as in the design case, unless the design case percentage is greater than 40 percent and then the base building is restricted to 40 percent.
  5. The vertical fenestrations in the baseline case are distributed uniformly on all sides in version 2.2, but distributed in the same proportion as the design in version 2009. In version 2009, the baseline case has these fenestrations flush to the outer wall without any shading features.
  6. The baseline building can be based on a different HVAC system, but again, this variation must be defensible.
  7. In version 2009, roof surfaces are modeled with a solar reflectivity of 0.30 in the base case.
  8. In version 2009, projects with natural ventilation may be able to include these savings in the design, but will be evaluated on a case-by-case basis.
  9. Combined heat and power (CHP) systems capture waste heat from electricity for heating. ASHRAE 90.1 2007 App. G with the performance rating method does not address how to incorporate the benefits of district CHP systems into the design model and additional insight into how these energy savings can be realized for the EAc1 Option 1 calculations is available in the LEED 2009 Reference Guide.
  10. In version 2009, the baseline case does not take credit for automatic lighting controls.

In summary, Option 1 is calculation-intensive. There are a total of five simulations that need to be performed, including one of the design case and four of the baseline case, one in each of the four orthogonal orientations. The four baseline case orientations are then averaged to obtain the baseline.

The annual energy percent cost savings (AECS) can be calculated based on a *proposed building performance* (PBP) and the *base building performance* (BBP). The use of on-site renewable energy as calculated as the REC may increase the points obtained from Option 1 of this credit, as the REC is used as a reduction in the calculation of the PBP, offsetting this amount of energy use as if it were free. The PBP is calculated as follows:

$$\text{PBP} = \text{PDEM} - \text{REC} \tag{4.1.1}$$

The annual energy percent cost savings (AECS) as used in the point determinations for EAc1 can then be calculated from either of the following two equations:

$$\text{AECS} = 100\left(1 - \frac{\text{PBP}}{\text{BBP}}\right) \tag{4.1.2}$$

or alternatively

$$\text{AECS} = 100\left(1 - \frac{\text{PDEM}}{\text{BBP}} + \frac{\text{REC}}{\text{BBP}}\right) \tag{4.1.3}$$

Equation (4.1.3) shows the positive impact that using on-site renewable energy has on the point potential for EAc1, when using Option 1.

LEED-NC 2.2 and 2009 include both regulated and process loads in assessing Option 1. Since much of the energy efficiency focus is on regulated loads, it is also useful to be able to segregate out these two parts of the energy loads for both the base building and the proposed design. This is readily done by defining BBPR and BBPP as the regulated BBP load and the process BBP load, respectively, and by defining PDEMR and PDEMP as the regulated PDEM load and the process PDEM load, respectively. By definition,

$$\text{BBP} = \text{BBPR} + \text{BBPP} \tag{4.1.4}$$

and

$$\text{PDEM} = \text{PDEMR} + \text{PDEMP} \tag{4.1.5}$$

It is usually easier to assess the regulated loads (BBPR and PDEMR) in the evaluations, as these are the ones most appropriately considered for energy efficiency measures. In most cases, the base process load is considered to be 25 percent of the base and does not change from the base to the proposed design. When this is the case, Eqs. (4.1.4) and (4.1.5) can be simplified to the following:

$$\text{BBP} = \text{BBPR}/0.75 \qquad \text{base process is 25\% of base load} \tag{4.1.6}$$

and

$$\text{PDEM} = \text{PDEMR} + \text{BBPR}/3 \qquad \text{base process is 25\% of base load and does not change in design} \tag{4.1.7}$$

Substituting this into Eq. (4.1.3) results in the following simplification based only on regulated loads:

$$AECS = 75\left(1 - \frac{PDEMR}{BBPR} + \frac{REC}{BBPR}\right) \quad \text{base process is 25\% of base load and does not change in design} \quad (4.1.8)$$

Equations (4.1.1) through (4.1.8) are for Option 1. A summary of the symbols as used for calculating EAc1 Option 1 can be found in Table 4.1.4.

Symbol	Description	Comments
BBP	Base building performance ($)	• Total includes both regulated and process loads. • Average of four model orientations.
BBPP	Base building performance process loads ($)	Default is BBPP = 25% of BBP
BBPR	Base building performance regulated loads ($)	
PDEM	Proposed design energy model ($)	Total includes both regulated and process loads.
PDEMP	Proposed design energy model process loads ($)	Process is same for base and proposed unless specially addressed by the exceptional calculation method, but usually only regulated loads are reduced.
PDEMR	Proposed design energy model regulated loads ($)	
REC	On-site renewable energy cost "savings" ($)	Equivalent "savings" of off-site utility costs for using on-site renewable energies.
PBP	Proposed building performance ($)	Proposed design costs less rec savings: PBP = PDEM – REC
AECS	Annual energy cost savings (%)	$AECS = 100\left(1 - \frac{PBP}{BBP}\right)$ or $AECS = 100\left(1 - \frac{PDEM}{BBP} + \frac{REC}{BBP}\right)$
ECM	Exceptional calculation method	Method to calculate changes to the proposed design for some process load changes or some energy cost savings for measures not readily included in the energy model used.

TABLE 4.1.4 Summary of Symbols and Criteria Equations for EAc1 Option 1

Options 2 and 3 Options 2 and 3 are not model based and therefore do not have a method for applying on-site renewable energy sources to the point total for this credit. Their prescriptive requirements are well outlined in the LEED-NC 2.2 and 2009 Reference Guides.

Special Circumstances and Exemplary Performance

EAc1 is also listed with the EB icon, primarily because the construction assembly of much of the exterior of the building impacts energy losses. In addition, the orientation of the building is a significant portion of Option 1. Envelope treatments for energy efficiency are different based on orientation and climate region of the country. Figures 4.1.4, 4.1.5, and 4.1.6 are all of a new office and multipurpose building complex in Houston, Texas. Figures 4.1.4 and 4.1.5 depict western exposure envelope treatments, whereas Fig. 4.1.6 is of a southern exposure shading feature.

The Indoor Environmental Quality category promotes good indoor air quality, and one of the strategies used to do this is increased ventilation, which often increases the energy load. Ventilation strategies should be carefully evaluated in conjunction with energy efficiencies and impacts to maintain an optimum balance between these two goals.

There were no exemplary performance points associated with this credit in version 2.2, but version 2009 allows one EP point for Option 1 as summarized for exemplary energy performance reductions as noted in Table 4.1.1 (50 percent and 46 percent for new and existing buildings, respectively.)

FIGURE 4.1.4 Shell Woodcreek office and multipurpose building complex in Houston, TX as featured on the cover: Western exposure envelope treatments. (*Photo taken November 11, 2009 by Heidi Brakewood.*)

Figure 4.1.5 Shell Woodcreek office and multipurpose building complex in Houston, TX: Western exposure envelope treatments close-up view. (*Photo taken November 11, 2009 by Heidi Brakewood.*)

Figure 4.1.6 Shell Woodcreek office and multipurpose building complex in Houston, TX: Southern exposure shading features. (*Photo taken November 11, 2009 by Heidi Brakewood.*)

4.2 EA Credit 2: *On-Site Renewable Energy*

The intention of EAc2 is to encourage the production of renewable energy sources on-site. This will serve to decrease the load on the off-site energy resources and also improve energy distribution efficiencies, as the potential energy losses that can be allocated to transport to the site are decreased. The energy offsets can earn one to seven points in LEED 2009.

USGBC Rating System

LEED-NC 2.2 lists the Intent, Requirements, and Potential Technologies and Strategies for EAc2 for which the following are also valid in LEED 2009 as follows:

Intent

Encourage and recognize increasing levels of on-site renewable energy self-supply in order to reduce environmental and economic impacts associated with fossil fuel energy use.

Requirements

Use on-site renewable energy systems to offset building energy cost. Calculate project performance by expressing the energy produced by the renewable systems as a percentage of the building annual energy cost. Use the building annual energy cost calculated in EA Credit 1 or use the Department of Energy (DOE) Commercial Buildings Energy Consumption Survey (CBECS) database to determine the estimated electricity use.

Potential Technologies and Strategies

Assess the project for nonpolluting and renewable energy potential including solar, wind, geothermal, low-impact hydro, biomass and bio-gas strategies. When applying these strategies, take advantage of net metering with the local utility. (The two versions of LEED give different point values for various levels of on-site renewable energy as a percentage of the building energy costs. These values are listed in Table 4.2.1.)

Points	LEED 2009 On-Site Renewable Energy as Percent of Proposed Energy Use*	LEED 2.2 On-Site Renewable Energy as Percent of Proposed Energy Use*
1	1	2.5
2	3	7.5
3	5	12.5
4	7	–
5	9	–
6	11	–
7	13	–
7 + 1 EP	15	–

*For both versions, use the design energy usage (on-site renewable energy not excluded) as per EAc1 Option 1 if applied. Otherwise use the Department of Energy (DOE) Commercial Buildings Energy Consumptions Survey (CBECS) database for estimated energy requirements based on proposed building use.

Table 4.2.1 EAc2 Point Differences between Versions 2.2 and 2009

Calculations and Considerations

The project boundary is an important variable in the calculations since this credit requires these renewable energy sources to be provided on-site. There may be some exceptions to this if the building is part of a campus setting where on-campus renewable energies are shared, but usually large, facility wide renewable energy sources are considered to be a part of EAc6: *Green Power*, and are not counted as part of EAc2. There is no rule stating that the project area must always be contiguous, but justifications for the area being separated and other special circumstances should be appropriately documented and approved by the USGBC.

The Reference Guides specifically list the systems that are eligible for applying to EAc2 as on-site renewable energy (REC). In general, the system must produce electrical or thermal energy on-site. Other on-site systems that are not included can usually be used as energy offsets to EAc1, Option 1. If the renewable energy sources come from off-site, then they are not a part of EAc2, but may be a part of EAc 6: *Green Power*. The lists of eligible and noneligible on-site energy sources from the LEED-NC 2.2 and 2009 Reference Guides are summarized in Table 4.2.2. Note that hydrogen fuel cells are not addressed. This is so because hydrogen fuel cells are not energy sources, but rather are energy storage devices. Their applicability would be dependent on whether the source for producing the hydrogen used in the fuel cell came from a renewable or nonrenewable source.

Energy Source	Systems Eligible for EAc2	Systems Not Eligible for EAc2
Solar	Photovoltaic systems	Passive solar strategies
	Solar thermal systems	Daylighting
		Architectural features
Biofuel energy systems with the following fuels	Untreated wood wastes	Wood coated with paints, plastics, or formica or more than 1% treated with halogen or arsensic-based preservatives
	Wood mill residues	Other forestry biomass waste
	Agricultural crops or wastes	Municipal solid waste
	Animal wastes and other organic wastes	
	Landfill gas	
Ground source	Geothermal heating	Geoexchange heat pumps
	Geothermal electric	
Surface water	Low-impact hydroelectric	
	Wave and tidal power	
Wind	Wind power	

Source: LEED-NC 2.2 and 2009 Reference Guides.

TABLE 4.2.2 Summary of On-Site Renewable Energy Systems

The credit gives points based on the percent of an anticipated on-site *energy usage* that is provided by the eligible on-site renewable sources as listed in Table 4.2.1. As noted in EAc1, these energy cost savings are referred to as the renewable energy costs (RECs) and are given in dollars per year. They are virtual savings using the on-site energy generated times the applicable local or state average energy cost rate, as defined in EAc1.

The denominator in the eligibility equations is more complex. If Option 1 of EAc1 is sought, then the denominator is the proposed design energy model (PDEM). (It was the PBP for EAc1 Option 1 cases originally in earlier versions.) If not, then the denominator is the *building annual energy cost* (BAEC) and is calculated using the local energy rates ($CSAECSE_i$ and $CSAECSF_i$), or if current local values are not available, the U.S. Department of Energy (DOE) Energy Information Administration (EIA) Commercial Sector Average Energy Costs by State available from www.eia.doe.gov (the 2003 values are listed in Table 4.1.2 in the EAc1 section) along with default energy consumption of both electricity and nonelectrical fuel usage rates per square foot of the building, based on typical building usages. The applicable default usage rates have been compiled by the USGBC from median data in the DOE EIA 1999 Commercial Buildings Energy Consumption Survey (CBECS) and are also tabulated in Table 4.2.3.

For the calculations of the building annual energy cost (BAEC), the U.S. Department of Energy (DOE) Energy Information Administration (EIA) 2003 Commercial Sector

Building Use	Median Electrical Use [kWh/(ft^2 · yr)] ($CBECSE_j$)	Median Fuel Use [kBtu/(ft^2 · yr)] ($CBECSF_j$)
Education	6.6	52.5
Food sales	58.9	143.3
Food service	28.7	137.8
Health care inpatient	21.5	50.2
Health care outpatient	9.7	56.5
Lodging	12.6	39.2
Retail (other than mall)	8.0	18.0
Enclosed and strip malls	14.5	50.6
Office	11.7	58.5
Public assembly	6.8	72.9
Public order and safety	4.1	23.7
Religious worship	2.5	103.6
Service	6.1	33.8
Warehouse and storage	3.0	96.9
Other	13.8	52.5

TABLE 4.2.3 Default Energy Consumption Intensities by Building Usage *j* [*Based on DOE EIA 1999 Commercial Buildings Energy Consumption Survey (CBECS)*]

Average Energy Costs by State (CSAECS) and the median data in the DOE EIA 1999 Commercial Buildings Energy Consumption Survey for electrical ($CBECSE_j$) and for fuel use ($CBECSF_j$) for usage *j* are usually used as default values, although it is preferred that actual local energy cost rates be used. Additional exceptions to these default values should be documented, and explanations should be given in the submittals with supporting data. So for all projects that do not seek Option 1 of EAc1, the design energy cost is this BAEC, whose default value is calculated by using Eq. (4.2.1). The GFA in Eq. (4.2.1) is the gross floor area of the building as defined in Chap. 2. (If the project is made up of several buildings or building portions with different uses as listed in Table 4.2.3, then the commercial building energy consumptions should be separately analyzed for each portion, multiplied by the gross floor area in that building portion, and summed to obtain the BAEC.)

$$\text{BAEC} = [(\text{CSAECSE}_i \times \text{CBECSE}_j) + (\text{CSAECSF}_i \times \text{CBECSF}_j)](\text{GFA})$$
$$\text{for state/district } i \text{ and use } j \qquad \text{(if EAc1 Option 1 is not sought)} \qquad (4.2.1)$$

Now the PDEM for projects using Option 1 of EAc1 and the BAEC for all other projects are used to determine the percent renewable energy (PREPDEM or PREBAEC) as a percent of the PDEM or BAEC, respectively, to determine the credit points listed in the Requirements section for EAc2. For projects that seek points via Option 1 of EAc1:

$$\text{PREPDEM} = 100 \times \frac{\text{REC}}{\text{PDEM}} \qquad \text{if EAc1 Option 1 is sought} \qquad (4.2.2)$$

and for all other projects

$$\text{PREBAEC} = 100 \times \frac{\text{REC}}{\text{BAEC}} \qquad \text{if EAc1 Option 1 is not sought} \qquad (4.2.3)$$

Special Circumstances and Exemplary Performance

Environmental attributes which are associated with the on-site renewable energy for EAc2 cannot be sold; they must be retained or retired. More information on current renewable energy options can be found on the websites of the DOE National Renewable Energy Laboratory (NREL), the DOE Energy Information Administration (EIA), *Distributed Energy—The Journal for Onsite Power Solutions,* and the EPA Clean Energy program. LEED 2009 also allows for sales of RECs but with restrictions to encourage on-site RECs over off-site RECs. These restrictions are listed in the LEED 2009 Reference Guide.

There were no exemplary performance points associated with this credit for version 2.2, whereas version 2009 allows for an EP point for a 15% reduction as noted in Table 4.2.1.

4.3 EA Credit 3: *Enhanced Commissioning*

The intent of EAc3 is to gain even more benefits for system efficiencies and reliabilities by enhancing the commissioning process. LEED 2009 awards two points for this credit (LEED 2.2 awarded one point).

USGBC Rating System

LEED-NC 2.2 lists the Intent and Requirements for EAc3 as follows, and the requirements in LEED 2009 are essentially the same except that the 2009 Guide is referred to:

Intent

Begin the commissioning process early during the design process and execute additional activities after systems performance verification is completed.

Requirements

Implement, or have a contract in place to implement, the following additional commissioning process activities in addition to the requirements of EA Prerequisite 1 and in accordance with the LEED-NC 2.2 (or LEED 2009 as applicable) Reference Guide:

1. Prior to the start of the construction documents phase, designate an independent Commissioning Authority (CxA) to lead, review, and oversee the completion of all commissioning process activities. The CxA shall, at a minimum, perform Tasks 2, 3 and 6. Other team members may perform Tasks 4 and 5.
   a. The CxA shall have documented commissioning authority experience in at least two building projects.
   b. The individual serving as the CxA shall be:
      i. independent of the work of design and construction;
      ii. not an employee of the design firm, though they may be contracted through them;
      iii. not an employee of, or contracted through, a contractor or construction manager holding construction contracts; and
      iv. (can be) a qualified employee or consultant of the Owner.
   c. The CxA shall report results, findings and recommendations directly to the Owner.
   d. This requirement has no deviation for project size.
2. The CxA shall conduct, at a minimum, one commissioning design review of the Owner's Project Requirements (OPR), Basis of Design (BOD), and design documents prior to mid-construction documents phase and back-check the review comments in the subsequent design submission.
3. The CxA shall review contractor submittals applicable to systems being commissioned for compliance with the OPR and BOD. This review shall be concurrent with A/E reviews and submitted to the design team and the Owner.
4. Develop a systems manual that provides future operating staff the information needed to understand and optimally operate the commissioned systems.
5. Verify that the requirements for training operating personnel and building occupants are completed.
6. Ensure the involvement by the CxA in reviewing building operation within 10 months after substantial completion with O&M staff and occupants. Include a plan for resolution of outstanding commissioning-related issues.

Calculations and Considerations

EAc3 requires a more comprehensive commissioning process in addition to the fundamental commissioning requirements as outlined in EAp1. It is important that the commissioning authority be familiar with the LEED requirements. A summary table of the additional steps as compared to fundamental commissioning has been included as Table 4.0.2. Definitions of CxA, BOD, and OPR are listed in App. B.

Special Circumstances and Exemplary Performance

EAc3 has been listed as a construction phase submittal, primarily because many of the activities cannot be performed or verified until during or after substantial completion of the project. It is also listed with the EB icon. The potential for many benefits associated with a commissioning-type review process for system efficiencies that do not necessitate substantial investment of additional monies for future retrofits is lost if not initiated in the early phases of the initial project. There were no exemplary performance points associated with this credit in LEED 2.2. In LEED 2009, an EP point may be obtained based on comprehensive envelope commissioning being included.

4.4 EA Credit 4: *Enhanced Refrigerant Management*

As mentioned in EAp3, there is an effort internationally to reduce the use of certain refrigerants and fire suppression chemicals that have negatively impacted the amount of ozone in the stratosphere. The Montreal Protocol, as mentioned in the credit intent, is the international agreement that sets timetables for the elimination of production and reduction in use of many of these chemicals. Its full title is the Montreal Protocol on Substances that Deplete the Ozone Layer. There is an international panel related to the Montreal Protocol called the Technology and Assessment Panel (TEAP).

The intention of EA credit 4 is to reduce the use of refrigeration and fire suppression chemicals that have a large ozone depletion potential (ODP), including many as listed in the Montreal Protocol, while also considering the global warming potential (GWP) of the alternatives. The global warming potential of a substance is an index that was established by the International Panel on Climate Change (IPCC). LEED 2009 awards two points for EAc4, whereas LEED 2.2 awarded one point.

USGBC Rating System

LEED-NC 2.2, as modified for verbiage in LEED 2009, lists the Intent, Requirements, and Potential Technologies and Strategies for EAc4 as follows:

Intent

Reduce ozone depletion and support early compliance with the Montreal Protocol while minimizing direct contributions to climate change.

Requirements

OPTION 1

Do not use refrigerants.

OR

OPTION 2

Select refrigerants and HVAC&R that minimize or eliminate the emission of compounds that contribute to ozone depletion and global climate change. The base building HVAC&R equipment shall comply with the following formula, which sets a maximum threshold for the combined contributions to ozone depletion and global warming potential:

$$\text{LCGWP} = \text{LCODP} \times 100{,}000 \leq 100$$

Where:
$\text{LCODP} = [\text{ODPr} \times (\text{Lr} \times \text{Life} + \text{Mr}) \times \text{Rc}]/\text{Life}$
$\text{LCGWP} = [\text{GWPr} \times (\text{Lr} \times \text{Life} + \text{Mr}) \times \text{Rc}]/\text{Life}$
LCODP: Lifecycle Ozone Depletion Potential (lbCFC_{11}/Ton-Year)

LCGWP: Lifecycle Direct Global Warming Potential ($lbCO_2$/Ton-Year)
GWPr: Global Warming Potential of Refrigerant (0 to 12,000 $lbCO_2$/lbr)
ODPr: Ozone Depletion Potential of Refrigerant (0 to 0.2 $lbCFC_{11}$/lbr)
Lr: Refrigerant Leakage Rate (0.5% to 2.0%; default of 2% unless otherwise demonstrated)
Mr: End-of-life Refrigerant Loss (2% to 10%; default of 10% unless otherwise demonstrated)
Rc: Refrigerant Charge (0.5 to 5.0 lb of refrigerant per ton of gross ARI rated cooling capacity)
Life: Equipment Life (default based on equipment type, unless otherwise demonstrated)

For multiple types of equipment, a weighted average of all base building level HVAC&R equipment shall be applied using the following formula:

$$[\Sigma\ (LCGWP + LCODP \times 100{,}000) \times Qunit\ ]/Qtotal \le 100$$

Where:
Qunit = Gross ARI rated cooling capacity of an individual HVAC or refrigeration unit (Tons)
Qtotal = Total gross ARI rated cooling capacity of all HVAC or refrigeration
ALL OPTIONS

Small HVAC units (defined as containing less than 0.5 lbs of refrigerant), and other equipment such as standard refrigerators, small water coolers, and any other cooling equipment that contains less than 0.5 lbs of refrigerant, are not considered part of the "base building" system and are not subject to the requirements of this credit.

AND

Do not operate or install fire suppression systems that contain ozone-depleting substances (CFCs, HCFCs or halons).

Potential Technologies and Strategies

Design and operate the facility without mechanical cooling and refrigeration equipment. Where mechanical cooling is used, utilize base building HVAC and refrigeration systems for the refrigeration cycle that minimize direct impact on ozone depletion and climate change. Select HVAC&R equipment with reduced refrigerant charge and increased equipment life. Maintain equipment to prevent leakage of refrigerant to the atmosphere. Utilize fire suppression systems that do not contain HCFCs or halons.

Calculations and Considerations

Two issues must be addressed to fulfill the requirements of this credit. The first deals with the refrigerant processes in heating, ventilation, air conditioning, and refrigeration (HVAC&R) systems, and the second deals with fire suppression. The two options listed in the credit are the two options for HVAC&R alternatives, and one of these must be met. The last item listed in the Requirements section is for fire suppression systems and must be abided by in all cases.

As mentioned in EA prerequisite 3, CFCs (chlorofluorocarbons) are being phased out due to their potential impact on the ozone levels in the stratosphere. (Many of these CFCs and other refrigerants commonly go by the DuPont trademark name Freon.) However, some of the alternative compounds may also impact the stratospheric ozone, but not necessarily to the extent that the CFCs do. Two examples of alternatives are the groups of chemicals referred to as the hydrochlorofluorocarbons (HCFCs) and hydrofluorocarbons (HFCs). HCFCs are similar to CFCs but also contain some hydrogen atoms instead of chlorine atoms and therefore have a lower potential for ozone depletion than do the CFCs. HFCs are also similar to HCFCs and CFCs but with no chlorine atoms, and therefore the ozone depletion potential of HFCs is negligible.

In addition, other chemicals used for fire suppression may impact the ozone in the stratosphere. Of particular concern are halons, which are organic compounds that contain bromine atoms in addition to chlorine and fluorine. A bromine atom can have an even greater impact than a chlorine atom has on the concentration of ozone in the stratosphere.

To compare the impact of the various refrigerants and fire suppression chemicals on the ozone in the stratosphere, a scale has been developed called the *ozone depletion potential* (ODP). One of the refrigerants typically referred to as CFC-11 has been given the ranking of 1 on the ODP scale, and the other chemicals are compared to CFC-11 on a mass basis. Most compounds are ranked with a value less than 1, except for those with bromine atoms, which may be higher.

There is concern that some of the alternatives to CFCs and other chemicals used for refrigeration and fire suppression may have an impact on another global atmospheric phenomenon, specifically global warming. Global climate change is of major interest, as changes in the global climate may cause weather and ocean level variations that can adversely impact humans and the environment. Carbon dioxide (CO_2) is a common compound in the troposphere that most living creatures expire. Scientists have shown that its concentration in the troposphere has been increasing in the last few centuries, and there is concern that the additional CO_2 and other chemicals in the atmosphere that may come from anthropogenic processes may have a blanketing effect, causing average temperatures to rise. There is much debate over the importance of CO_2 and these other gases in overall climate change.

Carbon dioxide has been given a global warming potential (GWP) of 1. Several refrigerants and fire suppression chemicals have also been given a GWP ranking based on the GWP of CO_2 on a mass basis. Many different sources list many different values for ODPs and GWPs as their values are estimated from very complex models and scenarios. The GWPs and ODPs of many refrigerants are given in the LEED-NC 2.2 and 2009 Reference Guides. These and several other compounds are listed in Table 4.4.1 and are based on Tables 1-5 and 1-6 of *The Scientific Assessment of Ozone Depletion, 2002*, a report of the World Meteorological Association's Global Ozone Research and Monitoring Project; those that are not updated in this 2002 report are from *The Scientific Assessment of Ozone Depletion, 1998*.

The ODPs and GWPs are based on the mass of the noted refrigerant or fire suppression chemical with respect to the mass of the associated base chemical, CFC-11 and CO_2, respectively. In the case of the GWP, note that there are significantly larger quantities of CO_2 emitted on a mass basis into the atmosphere than there are of most other refrigerants. In addition, there is much scientific debate on the relative impact that water vapor has on the global climate. Regardless, the other refrigerants and fire suppression chemicals have been ranked with a GWP as compared to that of CO_2, so that designers and regulators can make informed decisions as to the use of these chemicals from both an ODP and a GWP perspective. More information on these compounds and other ozone-depleting compounds can be found on the U.S. Environmental Protection Agency Ozone Depletion website: http://www.epa.gov/ozone/strathome.html.

The ODPs and GWPs are both sets of comparative environmental scales of certain chemicals. If chemicals are going to be chosen based on both of these environmental concerns, then there also needs to be a comparative environmental ranking between the scales. The USGBC has developed a formula for use in LEED certification that gives a comparative ranking of 1 to 100,000 between the relative importance of the mass of CO_2 to the mass of CFC-11 for global warming and ozone depletion potential, respectively. The relative multiplier 100,000 lb_{CO2}/lb_{CFC11} will be used in the overall certification equation presented later in this section [Eq. (4.4.5)].

Refrigerant or Fire Suppression Compound (Chemical Formula)	Also Referred to as	ODP_r^*	GWP_r^*	Common Building Application
Chlorofluorocarbons				
CFC-11 (CCl_3F)		1	4,680	Centrifugal chillers
CFC-12 (CCl_2F_2)		1	10,720	Refrigerators, chillers
CFC-114 ($ClF_2C\text{-}CClF_2$)		0.94	9,800	Centrifugal chillers
CFC-500		0.605	7,900	Centrifugal chillers, humidifiers
CFC-502		0.221	4,600	Low-temperature refrigeration
Halons				
Halon 1211 ($CBrClF_2$)		6	1,860	Fire suppression
Halon 1301		12	7,030	Fire suppression
Halon 2402		<8.6	1,620	Fire suppression
Hydrochlorofluoro-carbons				
HCFC-22 ($CHClF_2$)	R-22	0.04	1,780	Air conditioning, chillers
HCFC-123	R-123	0.02	76	CFC-11 replacement
Hydrofluorocarbons				
HFC-23 (CHF_3)		~0	12,240	Ultralow-temperature refrigeration
HFC-134a (CF_3CFH_2)	R-134a	~0	1,320	CFC-12 or HCFC-22 replacement
HFC-245fa ($CF_3CH_2CHF_2$)	R-245fa	~0	1,020	Insulation agent, centrifugal chillers
HFC-404a		~0	3,900	Low-temperature refrigeration
HFC-407c	R-407c	~0	1,700	HCFC-22 replacement
HFC-410a	R-410a	~0	1,890	Air conditioning
HFC-507a (mix of CF_3CF_2H and CF_3CH_3)		~0	3,900	Low-temperature refrigeration
Natural refrigerants				
Carbon dioxide (CO_2)		~0	1	
Ammonia (NH_3)		~0	0	
Propane		~0	3	

*The units for ODP_r are lb_{CFC11}/lb_r, and the units for GWP_r are lb_{CO2}/lb_r for each refrigerant compound *r*, where lb_r represents the mass (in pounds) of the respective refrigerant or fire suppression chemical, lb_{CFC11} represents the mass (in pounds) of CFC-11, and lb_{CO2} represents the mass (in pounds) of CO_2.
The values for the refrigerants are also valid for LEED 2009. Some representative chemical formulas are also shown.
Source: Courtesy LEED-NC 2.2 Reference Guide.

TABLE 4.4.1 Some Relative Ozone Depletion Potentials (ODPs) and Global Warming Potentials (GWPs).

To obtain this credit, the project must comply with either Option 1 or Option 2 relating to refrigeration and HVAC systems, *and* the project must also comply with the requirement that the fire suppression systems do not contain ozone-depleting substances (CFCs, HCFCs, or halons).

Option 1 and Option 2 only apply to systems or units that contain more than 0.5 lb of refrigerant. Items such as standard household refrigerators and small water coolers are excluded.

Option 1 Option 1, which relates to refrigeration and HVAC systems, is very simple. If the facility uses no refrigerants except for those small units which are exempted, then this option is fulfilled. In addition, if there are only natural refrigerants used for the applicable systems (CO_2, NH_3, water, or propane), then this option is fulfilled. No calculations are needed.

Option 2 Option 2 is more complex. It requires a series of calculations on the base building HVAC systems using the following 10 variables:

GWP_r Global warming potential of refrigerant *r* (These are listed in Table 4.4.1 or can be obtained from the EPA website. They typically range from 0 to 13,000 lb_{CO2}/lb_r.)

$LCGWP_i$ The equivalent annual life-cycle-based global warming potential for system *i* in lb_{CO2}/ton of gross ARI-rated cooling capacity of system *i* per year [$lb_{CO2}/(ton_{ac} \cdot yr)$]

$LCODP_i$ The equivalent annual life-cycle-based ozone depleting potential for system *i* in lb_{CFC11}/ton of gross ARI-rated cooling capacity of system *i* per year [$lb_{CFC11}/(ton_{ac} \cdot yr)$]

$Life_i$ Equipment life of system *i* (The values as listed in the LEED-NC 2.2 and 2009 Reference Guides are listed in Table 4.4.2. Other values may be substituted if appropriately documented by manufacturers' data and warranties.)

Type of Equipment	LEED 2009 $Life_i$ (yr)*	LEED 2.2 $Life_i$ (yr)*
Room or window air-conditioning units and heat pumps	10	10
Unitary, split and packaged air-conditioning units, and heat pumps	15	15
Reciprocating compressors and reciprocating chillers	20	20
Scroll compressors	20	–
Centrifugal chillers	25	23
Screw or absorption chillers	23	23
Water-cooled packaged air conditioners	24	–
All other HVAC&R equipment	15	15

*Or as updated from Abramson (see the ASHRAE Applications Handbook)

Table 4.4.2 Equipment Life as Given in LEED-NC 2.2 and 2009 Reference Guides

Lr_i Annual refrigerant leakage rate of system i (The values range from 0.005 to 0.02 lb refrigerant leaked per pound of refrigerant in the unit per year with a default of 0.02 unless otherwise shown, such as with manufacturers' test data. This is based on an annual leakage of 0.5 to 2.0 percent of the total system refrigerant. Zero leakage may not be claimed.)

Mr_i End-of-life refrigerant loss of system i (The values range from 0.02 to 0.10 lb refrigerant lost per pound of refrigerant in the unit with a default of 0.10 unless otherwise shown, such as with manufacturers' test data. This is based on the assumption that 2 to 10 percent of the total system refrigerant will not be recovered when the system life is over, based on either installation or removal losses. Zero loss may not be claimed.)

ODP_r Ozone depletion potential of refrigerant r (These are listed in Table 4.4.1 or can be obtained from the EPA website. For refrigerants other than the CFCs, they typically range from 0 to 0.2 lb_{CFC11}/lb_r.)

$Q_{unit\ i}$ The gross ARI rated cooling capacity of an individual HVAC or refrigeration unit i in ton_{ac} s (A ton_{ac} of cooling capacity is based on an older system of rating refrigeration when tons of ice were delivered for iceboxes. The tonnage given is actually in tons of ice melted in 24 h. One ton_{ac} of cooling can be directly converted to 12,000 Btu/h.)

Q_{total} The total gross ARI-rated cooling capacity of all HVAC or refrigeration in ton_{ac} s ($\Sigma Q_{unit\ i}$ over all systems i)

Rc_i Refrigerant charge of system i (This usually ranges from 0.5 to 5.0 lb of refrigerant per ton_{ac} of gross ARI-rated cooling capacity.)

$LCODP_i$ and $LCGWP_i$, the equivalent annual life-cycle-based ozone depleting potential and the equivalent annual life-cycle-based global warming potential for system i, are calculated by Eqs. (4.4.1) and (4.4.2). In each equation, the variable to the left side of the addition sign represents the contributions from annual leakage and the variable to the right side of the addition sign represents an annualized contribution from installation of the unit and disposing of the unit at the end of its useful life.

$$LCODP_i = [Lr_i + (Mr_i/Life_i)]\,(ODP_r \times Rc_i) \tag{4.4.1}$$

$$LCGWP_i = [Lr_i + (Mr_i/Life_i)]\,(GWP_r \times Rc_i) \tag{4.4.2}$$

For systems with the default values of leakage and loss and a 10-year life, these equations simplify to Eqs. (4.4.3) and (4.4.4). Similar simplified equations can be written for the various default values and equipment lives.

$$LCODP_i = 0.03(ODP_r \times Rc_i) \quad \text{for default leakage and loss, 10-year life} \tag{4.4.3}$$

$$LCGWP_i = 0.03(GWP_r \times Rc_i) \quad \text{for default leakage and loss, 10-year life} \tag{4.4.4}$$

Finally, a weighted average of all base building level HVAC&R equipment systems shall be used over all systems i such that

$$\{\Sigma_i\,[LCGWP_i + (LCODP_i \times 100{,}000\ lb_{CO2}/lb_{CFC11})]Q_{unit\ i}\}/Q_{total} \le 100\ lb_{CO2}/(ton_{ac} \cdot yr) \tag{4.4.5}$$

For projects with multiple systems, each individual system does not need to meet the criteria based on the inequality in Eq. (4.4.5), but the overall project average must meet the criteria. The LEED-NC 2.2 Reference Guide has a table in the EAc4 section that gives default maximum allowable equipment refrigerant charges in pounds of refrigerant per ton of cooling for several refrigerants based on the default losses and leaks and applicable equipment life. These default maxima are for the individual pieces of equipment. These default maxima do not have to be met if the overall project complies with Eq. (4.4.5). There is also technical guidance for adapting this credit to projects with district energy systems on the USGBC Registered Project Tools page found at http://www.usgbc.org/projecttools.

It is important that the refrigerant also be chosen based on compliance with the Montreal Protocol, as many refrigerants are being phased out over the next two decades. It is recommended that project designers refer to the EPA ozone depletion and global warming potential lists on its website for information on new or alternative refrigerants. In addition, in 2004 several major corporations formed an initiative called *Refrigerants, Naturally!* The intent of this initiative is to promote and develop refrigeration technologies which are HFC-free. Its website contains information on many alternative systems, such as those with carbon dioxide and solar cooling. Tri-State Generation and Transmission Association, Inc., also has a Web-based *Energy Library* which gives information on many alternatives.

Special Circumstances and Exemplary Performance

In summary, two points will be awarded in this category for LEED 2009 (one for LEED 2.2) if no CFCs, HCFCs, or halons are used for fire suppression systems, *and* either the criteria in Eq. (4.4.5) are met or there are no refrigerants used other than natural refrigerants in the base building HVAC&R systems. There is no EP point available for this credit.

4.5 EA Credit 5: *Measurement and Verification*

Designing a system to be more efficient and the system actually performing in that manner in various situations with different occupancies are two separate issues. Therefore, credit is also given if it can be proved that the systems are operating as intended. A side benefit of this credit is that it gives the owners the knowledge of which systems may not be operating as efficiently as desired and, therefore, the opportunity for modification and improvement. LEED 2009 awards three points for EAc5, whereas LEED 2.2 awarded one point.

USGBC Rating System

LEED-NC 2.2 lists the Intent, Requirements, and Potential Technologies and Strategies for EAc5 as follows, with modifications to incorporate additional LEED 2009 verbiage, particularly with numbering the credit options and providing a process for corrective action if appropriate:

Intent

Provide for the ongoing accountability of building energy consumption over time.

Requirements

- Option 1: Develop and implement a Measurement & Verification (M&V) Plan consistent with Option D: Calibrated Simulation (Savings Estimation Method 2), as specified in the *International Performance Measurement & Verification Protocol (IPMVP) Volume III: Concepts and Options for Determining Energy Savings in New Construction, April, 2003.*

Or

- Option 2: Option B: Energy Conservation Measure Isolation, as specified in the *International Performance Measurement & Verification Protocol (IPMVP) Volume III: Concepts and Options for Determining Energy Savings in New Construction, April, 2003.*

- For both options: the M&V period shall cover a period of no less than one year of post-construction occupancy and a process for corrective action must be provided if the results of the M&V Plan indicate that energy savings are not being achieved.

Potential Technologies and Strategies

Develop an M&V Plan to evaluate building and/or energy system performance. Characterize the building and/or energy systems through energy simulation or engineering analysis. Install the necessary metering equipment to measure energy use. Track performance by comparing predicted performance to actual performance, broken down by component or system as appropriate. Evaluate energy efficiency by comparing actual performance to baseline performance. While the IPMVP describes specific actions for verifying savings associated with energy conservation measures (ECMs) and strategies, this LEED credit expands upon typical IPMVP M&V objectives. M&V activities should not necessarily be confined to energy systems where ECMs or energy conservation strategies have been implemented. The IPMVP provides guidance on M&V strategies and their appropriate applications for various situations. These strategies should be used in conjunction with monitoring and trend logging of significant energy systems to provide for the ongoing accountability of building energy performance.

Calculations and Considerations

EAc5 gives an additional avenue for enhanced energy performance by implementing a program that measures and verifies (M&V) energy performance in the built environment. If this is done, then there is a potential for detecting and modifying energy inefficiencies over the lifetime of a building. The credit is given if the Measurement and Verification (M&V) Plan is developed and implemented through at least one year of postconstruction occupancy. The engineering details for obtaining this credit are varied, and it is not the intention of this book to go into these details. However, there are a few items useful for all participants in a green building project to understand.

This credit is worth three points in LEED 2009 and is based on documents developed by the Efficiency Valuation Organization (EVO), which is a nonprofit organization devoted to global energy efficiency (www.evo-world.org/ipmvp.php). The relevant standard is the *International Performance Measurement and Verification Protocol (IPMVP), Volume III: Concepts and Options for Determining Energy Savings in New Construction, April 2003.* There are two different approaches that can be taken. For smaller or less complex buildings where it is relatively easy to isolate the separate energy-saving and energy systems, the credit may be obtained by following the Energy Conservation Measure (ECM) Isolation method (Option B) of the IPMVP Volume III. For more complex energy systems, the Whole Building Calibration Simulation method (Option D) of the IPMVP Volume III may be chosen. In

Building Energy System Types	IPMVP Vol. III Option	Method Name
Simple, easily isolated (LEED 2009 Option 2)	Option B	Energy Conservation Measure (ECM) Isolation
Complex (LEED 2009 Option 1)	Option D	Whole Building Calibration Simulation

Table 4.5.1 Summary of IPMVP Approaches for EAc5

addition to the development of an M&V Plan, there may be a substantial amount of metering which may be needed on various building systems to measure the energy usages and savings. Table 4.5.1 summarizes these two approaches.

It is important that the submittals include the IPMVP option chosen, a copy of the M&V Plan as written according to the requirements in IPMVP Volume III, appropriate models and calculations based on the data collected, and a narrative if there are any special circumstances or explanations of calculation methods. This submittal must cover at least a full year of postconstruction occupancy, with this minimum one year time period beginning after the building has reached a reasonable occupancy and use level. Therefore, it is not usually reasonable to schedule the submittal exactly one year from the certificate of occupancy (CofO), which may be issued by the local building official or building authority. Instead, the project team should estimate an occupancy schedule, keep track of the building occupancies, and have the one-year period begin after most of the uses are established. In the case where one group plans to be the sole occupant of the new building and move the entire workforce within a month of issuance of the CofO, then the one-year period may start relatively quickly after construction ends. In the case where an office building is constructed that has multiple occupants, there may be some floors or areas that may take a year or so after construction to find tenants, and the verification of this credit may take longer. If the expected energy savings are not being experienced, then a process for corrective action must also be submitted and implemented. There is also technical guidance for adapting this credit to projects with district energy systems on the USGBC Registered Project Tools page found at http://www.usgbc.org/projecttools.

Special Circumstance and Exemplary Performance

EAc5 has been given an EB notation by the USGBC. It is also designated as a credit that needs a construction phase submittal, as much of the activity is during or postconstruction. There is no EP point available for this credit.

4.6 EA Credit 6: *Green Power*

The intent of EAc6 is to encourage the use of power from a power grid that has been made from renewable energy sources. Projects were awarded one point for this credit in LEED 2.2 and will be awarded two points in LEED 2009.

USGBC Rating System

LEED-NC 2.2 lists the Intent, Requirements, and Potential Technologies and Strategies for EAc2 as follows, with additional verbiage from LEED 2009 to number the options:

Intent

Encourage the development and use of grid-source, renewable energy technologies on a net zero pollution basis.

Requirements

Provide at least 35% of the building's electricity from renewable sources by engaging in at least a two-year renewable energy contract. Renewable sources are as defined by the Center for Resource Solutions (CRS) Green-e products certification requirements.

DETERMINE THE BASELINE ELECTRICITY USE

Option 1: Use the annual electricity consumption from the results of EA Credit 1.

OR

Option 2: Use the Department of Energy (DOE) Commercial Buildings Energy Consumption Survey (CBECS) database to determine the estimated electricity use.

Potential Technologies and Strategies

Determine the energy needs of the building and investigate opportunities to engage in a green power contract. Green power is derived from solar, wind, geothermal, biomass or low impact hydro sources. Visit www.green-e.org for details about the Green-e program. The power product purchased to comply with credit requirements need not be Green-e certified. Other sources of green power are eligible if they satisfy the Green-e program's technical requirements. Renewable energy certificates (RECs), green tags and other forms of green power that comply with Green-e's technical requirements can be used to document compliance with EA Credit 6 requirements.

Calculations and Considerations (LEED 2009)

EAc6 is based on either purchasing or trading electricity from renewable off-site power sources for at least 35 percent of the defined electricity needs for the project during two years. The definition of renewable power sources is per the Center for Resource Solutions' Green-e Product Certification Requirements, which can be found at www.green-e.org. The electricity can be purchased from the local power provider, if available. If the "green" electricity is not directly available, but the area provider has a green power program, then the contract can include enrollment into this program with its additional premium to facilitate green power usage elsewhere. If the area provider cannot offer either of these options, then there is also the opportunity to purchase Green-e renewable energy certificates (RECs), In this case, power is purchased locally, and the RECs for the calculated amount are purchased. The RECs represent the premium needed for purchasing renewable power elsewhere in the country and facilitate its widespread use.

The threshold value of green power purchased is based on the electrical power needs portion of the energy model if the EAc1 Option 1 is sought. Otherwise, it is based on the electrical power needs portion of the building annual energy cost (BAEC) calculation as outlined in EAc2. Either way, electricity cost rates do not have to be evaluated since only the electrical power needs are being analyzed, and there is no need to have a cost-based bridge for adding different power usage impacts. The sets of calculations based on the BAEC method and the EAc1 Option 1 energy model method are described in the following paragraphs.

Based on the BAEC method when the EAc1 Option 1 is not sought, the building annual energy usage of electricity (BAEE) can be found by using the median annual electrical intensities for the particular building type ($CBECSE_i$) as listed in Table 4.2.3 under EAc2. The calculation for BAEE is given in Eq. (4.6.1). [This is the electricity portion of Eq. (4.2.1) and does not include cost rates.] The GFA in Eq. (4.6.1) is the gross floor area of the building as defined in Chap. 2. If the project is made up of several buildings or building portions with different uses as listed in Table 4.2.3, then the commercial building energy consumptions ($CBECSE_j$) for electricity should be separately analyzed for each portion, multiplied by the gross floor area in that building portion, and summed to obtain the BAEE.

$$BAEE = (CBECSE_j)(GFA) \qquad \text{if EAc1 Option 1 is not sought} \qquad (4.6.1)$$

On the other hand, the basis for the electric usage when EAc1 Option 1 is sought is a portion of the Proposed Building Performance (PBP) as previously calculated in Eq. (4.1.1). In LEED-NC 2.1, a point was earned in EAc6 if the green power purchased was at least 50 percent of the difference between all regulated electricity loads and the electricity provided by on-site renewal energy sources. For this earlier version, the EAc1 calculations were all based on regulated loads. In LEED-NC 2.2, the calculations for EAc1 Option 1 are based on both regulated and nonregulated loads for which the nonregulated loads are assumed to be approximately 25 percent of the regulated loads. (Nonregulated loads are sometimes referred to as *process loads*.) Therefore, the threshold in LEED-NC 2.2 has been reduced to 35 percent to take into account the calculation changes. In LEED-NC 2.1, the modeled energy electrical use was further modified to account for on-site renewable energy production. Keeping this modification in LEED-NC 2.2 and carried forward to LEED 2009, the author has interpreted that the calculation appropriate for the first compliance path (EAc1 Option 1) should again include this reduction in the denominator of the on-site renewable energy, so that the percent green power purchased is truly only the percent of the demand from the off-site grid.

Keeping this modification in mind, let PBE (proposed building electrical performance) be the electrical energy portion of the PBP as defined in EAc1 Option 1. If PDEE is the proposed design electrical energy portion (in electrical usage units) as given from the proposed design energy model (PDEM), then the PBE is the difference in the electrical usage portion of the PDEM in units of electrical power per year (PDEE) and the electrical portion of the annual on-site renewable energy (REC) in units of electrical power per year (REE).

$$PBE = PDEE - REE \qquad \text{if EAc1 Option 1 is sought} \qquad (4.6.2)$$

Using the annual amount of green electricity (GE) purchased or traded from off-site, and the definitions for proposed building energy needs, PBE and BAEE, based on the EAc1 Option 1 compliance pathway and other energy calculation pathways, respectively, the percent green electricity (PGE) can be calculated for each as

$$PGEPB = 100 \times \frac{GE}{PBE} \qquad \text{if EAc1 Option 1 is sought} \qquad (4.6.3)$$

or alternatively

$$PGEPB = 100 \times \frac{GE}{PDEE - REE} \qquad \text{if EAc1 Option 1 is sought} \qquad (4.6.4)$$

and

$$PGEBAE = 100 \times \frac{GE}{BAEE} \quad \text{if EAc1 Option 1 is not sought} \tag{4.6.5}$$

PGEPB or PGEBAE must be greater than or equal to 35 for the two points (LEED 2009) or one point (LEED 2.2) to be awarded in EAc6. Table 4.6.1 gives a summary of the symbols and equations.

There is also technical guidance for adapting this credit to projects with district energy systems on the USGBC Registered Project Tools page found at http://www.usgbc.org/projecttools.

Symbol	Description	Comments	Relevant Equations	Restrictions
GE	Annual amount of green electricity from off-site	Can be purchased or traded		
BAEE	Building annual energy usage of electricity	Electrical energy part of BAEC, the building annual energy cost method	$BAEE = (CBECSE_j)(GFA)$	Use if EAc1 Option 1 is not sought
PGEBAE	Percent green electricity based on the BAEE	Based on the BAEC method	$PGEBAE = 100 \times \frac{GE}{BAEE}$	Use if EAc1 Option 1 is not sought
REE	Annual on-site renewable electrical energy provided	Electrical energy portion of the annual on-site renewable energy (REC)		
PDEE	Annual proposed design electrical energy	Electrical energy part of the proposed design energy model (PDEM)		Use if EAc1 Option 1 is sought
PBE	Annual proposed building electrical performance	Electrical energy part of the proposed building performance (PBP)	$PBE = PDEE - REE$	Use if EAc1 Option 1 is sought
PGEPB	Percent green electricity based on the PBE	Based on the proposed design energy model (PDEM)	$PGEPB = 100 \times \frac{GE}{PBE}$ or $PGEPB = 100 \times \frac{GE}{PDEE - REE}$	Use if EAc1 Option 1 is sought

TABLE 4.6.1 Green Power Credit Summary of Symbols and Equations

Special Circumstances and Exemplary Performance

Environmental attributes which are associated with the green power purchased cannot be sold; they must be retained or retired. EAc6 has been designated as a construction phase submittal, mainly because the utility contracts are usually not fully developed until construction is nearly complete. If there are special circumstances for these credits, a narrative must be submitted describing these circumstances.

In LEED 2.2, exemplary performance (EP) credit can be achieved by doubling the requirement, either by doubling the amount of green electricity purchased to 70 percent or by doubling the length of the contract to 4 years. In LEED 2009, this EP point is awarded if 70 percent of the power is purchased from renewable sources. No mention is made of extending the contract.

4.7 Discussion and Overview

The Energy and Atmosphere category seeks to reduce energy usage and the impacts of energy usage on the environment. There are also additional comments which can be made about the multiple benefits or effects of some of the measures used to obtain these goals. Two examples relate to the promotion of on-site renewable energy sources and to the location of the "breathing zone" within the interior spaces. Also, it might be useful to address the use of hydrogen fuel cells and how that may fit in with the energy goals. These are briefly discussed in the following sections.

On-Site Renewable Energy As previously discussed, EAc1, EAc2, and EAc6 all seek to reduce the use of nonrenewable energy sources for both environmental and economic reasons. Combined, the three credit subcategories give extra credit if on-site renewable energy sources are used in this reduction. There are many reasons for this, including the reliability of the overall power options when there are multiple sources, the importance of distributing the use of energy sources so that they are not as concentrated and therefore do not have such large, concentrated impacts on the environment in only a few areas, and security issues in the energy network. Not only are there points available in EAc2 for on-site renewable energy sources, but also use of these on-site sources makes it possible to earn more points in EAc1 Option 1 and to reach the 35 percent threshold in EAc6 for the same building energy needs.

For example, let us look at new building project X using LEED 2009, where the project has equal cost values of electrical and natural gas needs. In the proposed design if both the electrical and natural gas loads are reduced 22 percent from the baseline design and there is no on-site renewable electricity production, then six points may be earned based on EAc1 Option 1 as shown in Eq. (4.7.1).

$$\text{AECS} = 100 \times \left(1 - \frac{\text{PBP}}{\text{BBP}}\right) = 22 \tag{4.7.1}$$

So PBP/BBP is 0.78, but since PBP equals PDEM less the REC, and REC is zero, then PDEM/BBP is 0.78.

Let us then hypothetically reevaluate the situation if 24 percent of the electrical needs (12 percent of the total cost value of the total project energy needs or REC = 0.12PDEM) can be generated by renewable sources on-site. Then the following two equations apply:

$$\text{AECS} = 100 \times \left(1 - \frac{\text{PDEM}}{\text{BBP}} + \frac{\text{REC}}{\text{BBP}}\right) = 31.4 \tag{4.7.2}$$

and

$$PREPDEM = 100 \times \left(\frac{REC}{PDEM} \right) = 12 \qquad (4.7.3)$$

The total points available now from EAc1 and EAc2 are 16 (ten and six for each, respectively).

In addition, without the on-site renewable energy sources, the threshold of the percent of purchased green power needed based on the design electrical use decreases from 35 percent to just under 27 percent.

Initially, without including on-site renewable energy sources, to obtain the point for EAc6, the following must be true:

$$35 \leq 100(GE/PBE) \qquad (4.7.4)$$

but if REE = 0.24PBE, then

$$35 \leq 100[GE/(PBE - 0.24PBE)] = 0.76[100(GE/PBE)] \qquad (4.7.5)$$

or

$$26.6 \leq 100\,(GE/PBE) \qquad (4.7.6)$$

In summary, the incorporation of on-site renewable energy into a project scope can facilitate obtaining many additional points in a project, several more than if just EAc2 is evaluated. This helps justify additional first costs that may be realized for installing these on-site systems.

Hydrogen Fuel Cells Hydrogen fuel cells are currently under development to help meet our energy needs. They are not specifically addressed in the LEED-NC 2009 categories, but may be considered for additional innovative points under certain circumstances. The hydrogen in the fuel cells by itself is not an energy source, but rather an intermediary energy storage fuel, and the hydrogen is made from other compounds such as H_2O or CH_4 with energy used in the formation process. Since there are many options for the energy used in the hydrogen production process and the source compounds for the hydrogen, the use of hydrogen fuel cells may or may not be considered renewable. However, there are many other potential environmental benefits of using hydrogen fuel cells on a project. One is the decrease in the on-site emissions of many criteria air pollutants, such as nitrogen oxides (NO_x), carbon monoxide (CO), and volatile organic compounds (VOCs), as compared to many alternative-fuel choices. If hydrogen fuel cells are used on the site, then several issues should be considered by the designers to maintain the effectiveness of the system. The hydrogen fuel cells usually intake ambient air as the oxygen (O_2) source to produce the main product, water. Similar to the air intakes for makeup air in a building, care must be taken in the location of the air intakes for the fuel cells so that the source air is not contaminated with pollutants. Air intakes located near fossil fuel motor vehicle loading areas or drives may direct exhausts into the cells that can foul the membranes. Figure 4.7.1 shows a fuel cell on a LEED-NC certified building in Columbia, S.C. In addition to heating water in the building, the waste energy in the fuel cell coolant system is used to preheat the water.

Figure 4.7.1 Fuel Cell at West Quad on the University of South Carolina Campus in Columbia, S.C.

Breathing Zone The designers of the energy systems should be cognizant of many parameters that affect other credits. One is the breathing zone. This is used in the Indoor Environmental Quality category for EQc1 if any of the options for CO_2 monitoring are used, and for EQc2 if the air quality testing option is sought. The *breathing zone* is an area defined as being between two imaginary planes that are located 3 and 6 ft above the floor, and are bound vertically by planes that are around 2 ft off from any wall or fixed air unit. The air quality parameter measurements are taken within the breathing zone as specified. LEED 2009 allows for these measurements outside of the vertical planes if necessary. If the breathing zone represents the volumes that are typically important for good air quality, then designing the energy systems in the building such that improved air quality is focused in these volumes would be very important for a green building.

References

Andrepoint, J. (2006), "Developments in Thermal Energy Storage: Large Applications, Low Temps, High Efficiency, and Capital Savings," *Energy Engineering,* 103(4): 7–18.

ASHRAE (2004), *90.1-2004: Energy Standard for Buildings Except Low-Rise Residential Building,* American Society of Heating, Refrigerating and Air-Conditioning Engineers, Atlanta, GA.

ASHRAE (2004), *Advanced Energy Design Guide for Small Office Buildings,* American Society of Heating, Refrigerating and Air-Conditioning Engineers, Atlanta, GA.

ASHRAE (2006), *Advanced Energy Design Guide for Small Warehouses and Self Storage Buildings,* American Society of Heating, Refrigerating and Air-Conditioning Engineers, Atlanta, GA.

ASHRAE (2007), *90.1-2007: Energy Standard for Buildings Except Low-Rise Residential Building*, American Society of Heating, Refrigerating and Air-Conditioning Engineers, Atlanta, GA.

ASHRAE (2007), *ASHRAE HVAC Applications Handbook*, American Society of Heating, Refrigerating and Air-Conditioning Engineers, Atlanta, GA.

ASHRAE (2007), www.ashrae.org, website accessed July 12, 2007.

ASHRAE (2008), *Advanced Energy Design Guide for Small Retail Buildings*, American Society of Heating, Refrigerating and Air-Conditioning Engineers, Atlanta, GA.

Cooper, C. D., and F. C. Alley (2002), *Air Pollution Control, A Design Approach*, Waveland Press, Inc., Prospect Heights, IL.

De Nevers, N. (2000), *Air Pollution Control Engineering*, 2d ed., McGraw-Hill, New York.

Distributed Energy (2007), www.distributedenergy.com, *Distributed Energy, The Journal for Onsite Power Solutions*, website accessed July 12, 2007.

EIA (2001), *Emissions of Greenhouse Gases in the United States 2001: Alternatives to Chlorofluorocarbons: Lowering Ozone Depletion Potentials vs. Raising Global Warming Potentials*, Energy Information Administration, U.S. Department of Energy, Washington, D.C., accessed webpage July 12, 2007, http://www.eia.doe.gov/oiaf/1605/gg02rpt/ozone.html.

EIA (2007), *Energy Market and Economic Impacts of a Proposal to Reduce Greenhouse Gas Intensity with a Cap and Trade System*, January 2007, Energy Information Administration, U.S. Department of Energy, Washington, D.C.

EPA (2007), http://www.epa.gov/solar/, U.S. Environmental Protection Agency Clean Energy Program, website accessed July 12, 2007.

Fay, J. A., and D. S. Golomb (2002), *Energy and the Environment*, Oxford University Press, New York.

Gonzalez, M. J., and J. G. Navarro (2006), "Assessment of the Decrease of CO_2 Emissions in the Construction Field through the Selection of Materials: Practical Case Study of Three Houses of Low Environmental Impact," *Building and Environment*, 41: 902–909, Elsevier, Amsterdam.

Green-e (2007), www.green-e.org, The Center for Resource Solutions (CRS), San Francisco, Calif., website accessed July 12, 2007.

Heinsohn, R. J., and R. L. Kabel (1999), *Sources and Control of Air Pollution*, Prentice-Hall, Upper Saddle River, N.J.

NBI (2009), *Advanced Buildings (TM) Core Performance (TM) Guide*, New Buildings Institute, White Salmon, WA, www.advancedbuildings.net, website accessed December 24, 2009.

NREL (2007), http://www.nrel.gov/, National Renewable Energy Laboratory, Denver, Colo., U.S. Department of Energy, webpage accessed July 12, 2007.

Refrigerants, Naturally! (2007), http://www.refrigerantsnaturally.com/, website accessed July 12, 2007.

Tri-State (2007), http://tristate.apogee.net/, Tri-State Generation and Transmission Association, Inc., Energy Library, Westminster, CO., accessed July 12, 2007.

USGBC (2003), *LEED-NC for New Construction, Reference Guide*, Version 2.1, 2d ed., May, U.S. Green Building Council, Washington, D.C.

USGBC (2005–2007), *LEED-NC for New Construction, Reference Guide*, Version 2.2, 1st ed., U.S. Green Building Council, Washington, D.C., October 2005 with errata posted through Spring 2007.

USGBC (2009), *LEED Reference Guide for Green Building Design and Construction*, 2009 Edition, U.S. Green Building Council, Washington, D.C., April 2009.
Varodompun, J., and M. Navvab (2007), "HVAC Ventilation Strategies: The Contribution for Thermal Comfort, Energy Efficiency, and Indoor Air Quality," *Journal of Green Building*, Spring, 2(2): 131–150.
Wark, K., C. F. Warner, and W. T. Davis (1998), *Air Pollution, Its Origin and Control*, Addison Wesley Longman, Menlo Park, CA.
WMO (1998), *The Scientific Assessment of Ozone Depletion, 1998*, World Meteorological Association's Global Ozone Research and Monitoring Project, Geneva, Switzerland.
WMO (2002), *The Scientific Assessment of Ozone Depletion, 2002*, World Meteorological Association's Global Ozone Research and Monitoring Project, Geneva, Switzerland.

Exercises

1. You are designing a new building in South Carolina where electricity costs about $0.085 per kWh and natural gas costs about $5.32 per MMBtu. The average of the four orientations of the base design uses about 12 kW of electricity and 130 MMBtu of natural gas each month for regulated uses.

A. Determine on an annual basis the BBP (baseline building performance).

B. You decide to implement some energy-saving programs for the regulated uses that reduce these electrical uses by 30 percent and natural gas uses by 25 percent. You are not going to use any renewable energy sources on-site. Determine the PDEM (proposed design energy model) cost and the PBP (proposed building performance).

C. Determine your AECS (annual energy cost savings) and the available points for EA credit 1 for LEED 2009.

2. You are designing a new building in a state where electricity costs about $0.095 per kWh and natural gas costs about $6.32 per MMBtu. The average of the four orientations of the base design uses about 14 kW of electricity and 140 MMBtu of natural gas each month for regulated uses.

A. Determine on an annual basis the BBP (baseline building performance).

B. You decide to implement some energy-saving programs for the regulated uses that reduce these electrical uses by 30 percent and natural gas uses by 25 percent. In addition, you would like to use on-site renewable sources of electricity with photovoltaic cells for 2 kW of the electricity. Determine the PDEM cost and the PBP.

C. Determine your AECS and the available points for EA credits 1 and 2 in LEED 2009.

3. Determine how much green power you need to purchase to get the LEED 2009 two points for credit EAc6 for the new building in Exercise 1.

4. Energy modelers have determined that the average BBP for your new project in St. Louis is about $1500 per month. Of this, 55 percent represents costs due to electrical usages, and the remainder of the costs is from natural gas usage. The PDEM gives a value of $1200 per month with 50 percent of the costs from electrical usages. All costs are based on the default energy costs of 2003. Determine the LEED 2009 points available for EAc1 and the amount of green power that must be purchased to obtain the two points for EAc6. Is this also eligible for RP points? Which ones? (See Chap. 7.)

5. Energy modelers have determined that the average BBP for your new project in Portland is about $1500 per month. Of this, 60 percent represents costs due to electrical usages, and the remainder is from natural gas. The PDEM gives a value of $1200 per month with 50 percent of the costs from electrical usages. All costs are based on the default energy costs of 2003. You are planning to install photovoltaic cells at the site that would generate on average 2 kW of power for on-site usage. Determine the LEED 2009 points available for EAc1 and the amount of green power that must be purchased to obtain the two points for EAc6. Is this also eligible for RP points? Which ones? (See Chap. 7.)

6. There are many options for bulbs in light fixtures. For instance, emergency exit lighting could possibly use incandescent, fluorescent, or LED (light-emitting diode) bulbs. In this exercise, assume that the three different types of lamps use 40, 10, and 5 W and have typical lamp lives of 0.7, 1.7, and 80 years, respectively.

- **A.** Calculate the energy usage for a single emergency exit lighting fixture with incandescent, compact fluorescent, and LED emergency bulb over 10 years.
- **B.** In your state, at current electrical rates, what are the electricity costs for 10 years?
- **C.** Estimate the additional bulb replacement costs as appropriate (both material and labor), and give the estimated 10-year life-cycle cost for these 3 types.

7. You are building a 10,000 ft^2 retail facility in Nebraska. You have opted to obtain some points for EAc1 by the prescriptive methods and will not do the whole building energy simulation. However, you would like to see what size solar power you would need to install on-site for EAc2. Based on LEED 2009, determine the minimum average power generation required from on-site solar power to obtain one point for EAc2. Determine the amount of green power you would need to purchase from off-site sources to obtain the EAc6 credit, with and without installation of the minimum on-site solar cell. Which, if any, RP points are available for these credits? (See Chap. 7.)

8. You are building a 4600 ft^2 pharmacy on Long Island. You will not do the whole building energy simulation due to the small size of the facility. However, there is a green power provider in the area that sells renewable energy certificates at 18.5 cents/kWh. Current electricity rates in the area are 17.3 cents/kWh. What is your annual cost premium to obtain EAc6, based on LEED 2009? Is this also eligible for an RP point? (See Chap. 7.)

9. You are designing a commercial strip with 10 storefronts and offices above. Each store uses a 5-ton packaged HVAC unit, and each office above needs a 2-ton split HVAC unit. The small refrigerators and water coolers in the store back rooms all use less than 0.5 lb of refrigerant and are exempt from this calculation. You have found manufacturers' specifications for the package units which list them as using 1.8 lb of refrigerant per ton of air conditioning for each unit and HFC-410A for the refrigerant. The cut sheet for the split units lists 3.1 lb of HCFC-22 per ton of air conditioning.

Calculate the equivalent annual life-cycle-based global warming potential ($LCGWP_i$) and the equivalent annual life-cycle-based ozone depleting potential ($LCODP_i$) for each unit type. Determine if LEED 2009 EAc4 is met.

10. You are designing a commercial office tower with 10 offices or stores on the first floor. Each of the first-floor spaces needs a 2-ton split HVAC unit listed at 3.1 lb of HCFC-22 per ton of air conditioning. The upper floors are all used for an engineering consulting firm, and they are cooled by a single 400-ton_{ac} centrifugal chiller using HCFC-123, which lists 1.8 lb of refrigerant per ton of air conditioning.

Calculate the equivalent annual life-cycle-based global warming potential ($LCGWP_i$) and the equivalent annual life-cycle-based ozone depleting potential ($LCODP_i$) for each unit type. Determine if LEED 2009 EAc4 is met.

11. For a typical 10-year life HVAC&R unit, calculate the maximum allowable equipment refrigerant charge (Rc_i) per ton_{ac} that a single unit may have, based on the default values for leakages and losses in order to be compliant with LEED 2009 EAc4 for the following refrigerants:

- **A.** HCFC-22
- **B.** HCFC-123
- **C.** HFC-134a

D. HFC-245fa
E. HFC-407c
F. HFC-410a

12. For a typical 15-year life HVAC&R unit, calculate the maximum allowable equipment refrigerant charge (Rc_i) per ton_{ac} that a single unit may have, based on the default values for leakages and losses in order to be compliant with LEED 2009 EAc4 for the following refrigerants:
A. HCFC-22
B. HCFC-123
C. HFC-134a
D. HFC-245fa
E. HFC-407c
F. HFC-410a

13. For a typical 20-year life HVAC&R unit, calculate the maximum allowable equipment refrigerant charge (Rc_i) per ton_{ac} that a single unit may have, based on the default values for leakages and losses in order to be compliant with LEED 2009 EAc4 for the following refrigerants:
A. HCFC-22
B. HCFC-123
C. HFC-134a
D. HFC-245fa
E. HFC-407c
F. HFC-410a

14. For a typical 23-year life HVAC&R unit, calculate the maximum allowable equipment refrigerant charge (Rc_i) per ton_{ac} that a single unit may have, based on the default values for leakages and losses in order to be compliant with LEED 2009 EA credit 4 for the following refrigerants:
A. HCFC-22
B. HCFC-123
C. HFC-134a
D. HFC-245fa
E. HFC-407c
F. HFC-410a

15 Simplify Eqs. (4.4.1) and (4.4.2) for a unit with a 15-year life using the default values of leakage and loss.

$$LCODP_i = [Lr_i + (Mr_i/Life_i)](ODP_r \times Rc_i) \quad (4.4.1)$$

$$LCGWP_i = [Lr_i + (Mr_i/Life_i)](GWP_r \times Rc_i) \quad (4.4.2)$$

16 Simplify Eqs. (4.4.1) and (4.4.2) for a unit with a 20-year life using the default values of leakage and loss.

17. Simplify Eqs. (4.4.1) and (4.4.2) for a unit with a 23-year life using the default values of leakage and loss.

18. Derive Eqs. (4.7.2) and (4.7.3) from the information given in the text and verify the point total.

19. Which, if any, of the EA credits are available for Regional Priority points in the zip code of your work or school? (See Chap. 7.)

20. Which, if any, of the EA credits are available for Regional Priority points in the following major cities? (See Chap. 7.)

A. New York City
B. Houston, TX
C. Los Angeles, CA
D. Portland, OR
E. Miami, FL
F. Kansas City, MO

21. NO_x are a collection of nitrogen oxide air pollutants produced from combustion, mainly from transportation (mobile) and energy generation (stationary) sources. In a city, assume that 60 percent of the NO_x emissions are from mobile sources and 40 percent are from stationary fossil fuel power sources. If the background NO_x concentration coming into this city is 30 $\mu g/m^3$ from agricultural and other upwind sources, and this together with emissions within the city produces an average "Box Model" concentration of 110 $\mu g/m^3$, then what would the new average concentration be estimated at if the stationary source emissions were reduced by 35 percent due to LEED practices?

22. You wish to use the prescriptive method for EAc1. Look for the "ASHRAE Advanced Energy Design Guides" on the Internet and download one of the guides. Prepare a minimum 500 word summary of this guide for the region where your home town is (name your home town and region). If your last name starts with A–F, summarize the guide for Small Warehouses and Self-Storage Buildings, with G–L, summarize the guide for Small Office Buildings, with M–Q, summarize the guide for K–12 School Buildings, or with R–Z, summarize the guide for Small Retail Buildings.

CHAPTER 5

LEED Materials and Resources

The Materials and Resources category of the U.S. Green Building Council (USGBC) rating system for new construction version 2.2 (LEED-NC 2.2) and version 2009 (LEED-NC 2009) consist of one prerequisite subcategory and seven credit subcategories which together may earn a possible 13 points (credits) maximum for version 2.2 and 14 possible points for version 2009, not including EP points. The notation format for the prerequisites and credits is, for example, MRp1 for Materials and Resources prerequisite 1 and MRc1.1 for Materials and Resources credit 1.1. Also in this chapter, to facilitate easier cross-referencing between this text and the USGBC rating system, the second digit in a section heading, equation, table, or figure number represents the credit subcategory number for sections that deal directly with a USGBC LEED-NC 2.2 or LEED-NC 2009 credit subcategory.

The Materials and Resources (MR) portion deals with issues that reduce the use of new materials and resources, encourages the use of materials and resources that have a smaller impact on the environment, and promotes the reuse or recycling of materials so that more virgin materials and resources are not used on LEED-certified projects. The life cycles of many products and materials are taken into account to reduce the impact on the environment of their use. This may include transportation impacts, harvesting impacts, manufacturing impacts, and the benefit of using recycled materials in the production of the product.

The one prerequisite subcategory is MR Prerequisite 1: *Storage and Collection of Recyclables* (MRp1). As is typical for prerequisites, there are no points associated with MRp1.

The points for the credits in the Materials and Resources category are not necessarily individually related to a specific credit under a subcategory in LEED 2009. The first subcategory (Building Reuse) has two sections for which the first section has multiple points and the second section has a single point. There are multiple points for the second through fifth credit subcategories. The sixth and seventh credit subcategories each are worth one point in LEED 2009. Table 5.0.0 is a summary of the prerequisites and credits in both versions (2.2 and 2009) with a comparison of the points available and a summary of many of the major differences. Note that if the credit subcategory in LEED 2.2 had separate credit numbers for the same general requirements met to a greater degree (e.g., percent conformance), LEED 2009 now only uses one number, with a range of points. This conversion also resulted in MRc1.3 for LEED 2.2 being renumbered to MRc1.2 for LEED 2009.

Subcategory 2009	Subcategory 2.2	Pts 2009	Pts 2.2	EP Pts 2009	EP Pts 2.2	Major Comments
MRp1: Storage and Collection of Recyclables	MRp1: Storage and Collection of Recyclables	na	na	na	na	–
MRc1.1: Building Reuse Maintain Existing Walls, Floors, and Roof	MRc1.1: Building Reuse, Maintain 75% of Existing Walls, Floors, and Roof	1–3	1	–	–	• LEED 2009 gives 1 point for 55%, 2 points for 75%, and 3 points for 95%. LEED 2.2 allows 1 point for 75% and 2 points for 95%. • In LEED 2009, only use one side of existing interior structural walls in the calculations. Both sides were included in LEED 2.2.*
	MRc1.2: Building Reuse, Maintain 95% of Existing Walls, Floors, and Roof		1			
MRc1.2: Building Reuse Maintain Existing Interior Non-structural Elements	MRc1.3: Building Reuse Maintain 50% of Interior Nonstructural Elements	1	1	–	–	• LEED 2009 includes the interior finishes on exterior structural and party walls in the calculations.*
MRc2: Construction Waste Management	MRc2.1: Construction Waste Management Divert 50% from Disposal	1–2	1	1	1	• As in LEED 2.2, LEED 2009 allows 1 point for 50%, 2 points for 75%, and a third EP point for 95%.*
	MRc2.2: Construction Waste Management Divert 75% from Disposal		1			
MRc3: Materials Reuse	MRc3.1: Materials Reuse 5%	1–2	1	1	1	• As in LEED 2.2, LEED 2009 allows 1 point for 5%, 2 points for 10%, and a third EP point for 15%.*
	MRc3.2: Materials Reuse 10%		1			

MRc4: Recycled Content	MRc4: Recycled Content 10% (postconsumer + ½ preconsumer)	1–2	1	1	1	• As in LEED 2.2, LEED 2009 allows 1 point for 10%, 2 points for 20%, and a third EP point for 30%. • LEED 2009 warns about possible problematic air emissions in recycled materials. Include in IEQc4 for VOCs if applicable.
	MRc4: Recycled Content 20% (postconsumer + ½ preconsumer)		1			
MRc5: Regional Materials	MRc5: Regional Materials, 10% Extracted, Processed, and Manufactured Regionally	1–2	1	1	1	• As in LEED 2.2, LEED 2009 allows 1 point for 10% and 2 points for 20%. However, in LEED 2009 a third EP point is allowed for only 30%, whereas this third point was credited to 40% in LEED 2.2.
	MRc5: Regional Materials, 20% Extracted, Processed, and Manufactured Regionally		1			
MRc6: Rapidly Renewable Materials	MRc6: Rapidly Renewable Materials (>2.5%)	1	1	1	1	• As in LEED 2.2, LEED 2009 allows 1 point for 2.5% and a second EP point for 5%.[†]
MRc7: Certified Wood	MRc7: Certified Wood (>50%)	1	1	1	1	• As in LEED 2.2, LEED 2009 allows 1 point for 50% and a second EP point for 95%.
	Total	**14**	**13**	**6**	**6**	

*LEED 2009 requires: "If an existing building is found to contain contaminated substances, such as lead or asbestos, these materials should be remediated as required by the EPA."

†LEED 2009 lists additional examples of rapidly renewable materials: sunflower seed board panels, bio based paints, coir and jute geotextiles, soy-based insulation and form, release agents, and straw bales.

Table 5.0.0 MR Summary Table of Major Differences between Versions 2.2 and 2009 and Associated Points (Pts)

The prerequisites and credits in the Materials and Resources category can be grouped into the following three overall sets of goals:

- MRp1 and the first four Materials and Resources credit subcategories (1 through 4) all encourage the continued use, reuse, or recycling of materials or products to decrease the use of more raw materials and decrease the impact to the waste disposal stream.
- The two credits in the Materials and Resources subcategory 5 encourage the use of materials or products that do not require long transportation distances to reach the site. This reduces the negative environmental impacts of transportation and increases sustainability in the area.
- The Materials and Resources subcategories 6 and 7 encourage stewardship in the use of products made from renewable materials, particularly promoting rapidly renewable materials in MRc6.

The one prerequisite and the seven credit subcategories (including the two sections in the first subcategory) in the Materials and Resources category for LEED 2009 are as follows:

- **MR Prerequisite 1** (MRp1): *Storage and Collection of Recyclables*
- **MR Credit 1.1** (MRc1.1): *Building Reuse—Maintain Existing Walls, Floors, and Roof*
- **MR Credit 1.2** (MRc1.2): *Building Reuse—Maintain Existing Interior Non-Structural Elements*
- **MR Credit 2** (MRc2): *Construction Waste Management*
- **MR Credit 3** (MRc3): *Materials Reuse*
- **MR Credit 4** (MRc4): *Recycled Content* (EB)
- **MR Credit 5** (MRc5): *Regional Materials* (EB)
- **MR Credit 6** (MRc6): *Rapidly Renewable Materials* (EB)
- **MR Credit 7** (MRc7): *Certified Wood* (EB)

LEED-NC 2.2 and 2009 provide tables which summarize some of these prerequisite and credit characteristics, including the main phase of the project for submittal document preparation (Design or Construction). Even though the entire LEED process encompasses all phases of a project, the prerequisites or credits which are noted as Construction Submittal credits may have major actions going on through the construction phase, which, unless documented through that time, may not be able to be verified from preconstruction documents or the built project. Sometimes these credits have activities associated with them that cannot or need not be performed until or subsequent to the construction phase. All the Materials and Resources credits have been included in the construction submittal column and are thus noted in Table 5.0.1.

Four of the credit subcategories listed in Table 5.0.1 have the icon EB noted after their title (Recycled Content, Regional Materials, Rapidly Renewable Materials, and Certified Wood). This icon is a tool to help those who wish to proceed with the continuing LEED-EB certification (Existing Building) after certification of LEED-NC is obtained. Those credits noted with this icon are usually significantly more cost-effective to

Prerequisite or Credit	Construction Submittal	EB Icon	Regional Variations May Be Applicable in LEED 2009	Location of Site Boundary and Site Area Important in Calculations
MRp1: Storage and Collection of Recyclables		n.a.	Yes	No
MRc1: Building Reuse—Maintain Existing Walls, Floors, and Roof	*	No	Yes	No
MRc1.2: Building Reuse—Maintain Existing Interior Non-structural Elements	*	No	Yes	No
MRc2.1: Construction Waste Management	*	No	Yes	Yes
MRc3: Materials Reuse	*	No	Yes	No
MRc4: Recycled Content (post consumer + ½ pre consumer)	*	Yes	Yes	No
MRc5: Regional Materials—Extracted, Processed, and Manufactured Regionally	*	Yes	Yes	No
MRc6: Rapidly Renewable Materials	*	Yes	Yes	No
MRc7: Certified Wood	*	Yes	Yes	No

TABLE 5.0.1 MR EB Icon, Regional Variation Applicability, and Miscellaneous LEED 2009

implement during the construction of the building than later during its operation. They are also shown in Table 5.0.2. There are many prerequisites or credits which might have variations applicable to the design or implementation of the project related to regional conditions in the United States. All the prerequisites and credits for which regional variations might be applicable are noted in the LEED 2009 Reference Guide and are so listed in Table 5.0.1 for the Material and Resources category.

There are EP points under the Innovation and Design category available for exceeding each of the respective credit criteria to a minimum level for some of the credits. All the credit subcategories other than Building Reuse (MRc1.1 and 1.2) have an EP point available. The available EP points for both versions 2.2 and 2009 are listed in Table 5.0.1. Note that only a maximum total of four EP points are available in total for a project in LEED 2.2 and a maximum of three in LEED 2009 and they may be from any of the noted EP options in any of the SS, WE, EA, MR, or IEQ LEED categories (see Chap. 7).

One of the variables that must be determined for each project and that must be kept consistent throughout the credits in a LEED submittal is the size and location of the project site. The location of the site boundary is particularly important to the construction waste management credit (MRc2), since existing concrete, masonry, or asphalt from pavement on a site which is crushed and reused on-site or elsewhere should be included in the weight or volume of the materials diverted from disposal. There may be exceptions to this for special circumstances such as campus settings. Therefore, there might be a need for narratives describing any such variations and how the project still meets the intent of the credit. The importance of the LEED boundary to MRc2 is also noted in Table 5.0.1.

Material and Resources Credit Calculation Summaries

There are so many different types and uses of materials that go into a building, and they have vastly different values based on weight, cost, or application. Therefore, to determine percentages of material usages, it is important to define which materials are included in the calculations and what units the calculations are based on. The LEED-NC 2.2 and 2009 Reference Guides provide tables of MR credit metrics which help in organizing and standardizing the calculations and credit criteria. The process for classification can be summarized in four steps:

1. Segregation of materials into material uses
2. Segregation of the materials for the MR credits into applicable material types or items
3. Determination of the units
4. Setting of the bases (denominator) for each calculation

The various calculations in the MR category are based on certain groups of materials. This has not really changed from the 2.2 to the 2009 version. However, the categories of materials as tracked on a project that are used in these calculations are segregated as based on the Construction Specifications Institute's MasterFormat Division Listing and this listing has been revised in the interim. LEED 2.2 is based on the CSI 1995 MasterFormat and LEED 2009 is based on the CSI 2004 MasterFormat. The main divisions in these two versions of the MasterFormat are summarized in Table 5.0.2. Table 5.0.3 is an overview on how the materials are grouped and utilized for the LEED 2009 calculations based on the CSI 2004 MasterFormat.

Note that items used for mechanical, electrical, and plumbing (MEP) purposes are not included in any of the calculations in the Materials and Resources category except for construction waste management. The calculations for the items in subcategories 3 through 7 are based on value or cost. MEP items tend to be costly per unit weight compared to many of the other bulk building materials. The intent of these subcategories, to the greatest extent possible, is to be able to recycle or reuse or purchase local materials. Therefore, it is assumed that having the high-dollar MEP items in these calculations might bias the equations, without having the bulk materials included.

Note also that furniture and furnishings are optional, but only if fully included consistently in subcategories 3 through 7. The choice is up to the project team. There are many manufacturers of furnishings and furniture items which are providing product options that can meet recycle content, certified wood, and regional material needs among others. Figure 5.0.1a and b depicts many furniture options provided by some manufacturers in the United States.

CSI 1995	CSI 2004	Description 1995	Description 2004
1	1, 2	General Requirements	General Requirements, Existing Conditions
2	31, 32, 33, 35	Site Construction	Earthwork, Exterior Improvements, Utilities, Waterway, and Marine Construction
3	3	Concrete	Concrete
4	4	Masonry	Masonry
5	5	Metals	Metals
6	6	Wood and Plastics	Wood, Plastics, and Composites
7	7	Thermal and Moisture Protection	Thermal and Moisture Protection
8	8	Doors and Windows	Openings
9	9	Finishes	Finishes
10	10	Specialties	Specialties
11	11, 42, 43, 44, 45	Equipment	Equipment and Various Other Equipment Categories (See CSI for Details)
12	12	Furnishings	Furnishings
13	13, 25	Special Construction	Special Construction, Integrated Automation
14	14, 34, 41	Conveying Systems	Conveying Systems, Transportation, Material Processing, and Handling Equipment
15	21, 22, 23	Mechanical	Fire Suppression, Plumbing, HVAC
16	26, 27, 28, 48	Electrical	Electrical, Communications, Electronic Safety and Security, Electrical Power Generation
–	40	–	Process Integration

Table 5.0.2 Construction Specification Institute CSI MasterFormat 1995 and 2004 Divisions

The approximate classifications used in LEED-NC 2.2 for steps 2 through 4 are summarized in Table 5.0.4. These will also be further explained under each individual subcategory.

The last summary detail in the credit calculations for this category is to understand which items can be included in multiple credits and which cannot. This is summarized in Table 5.0.5. Note that some individual items may be made of several types of materials. Values and allocations for these are usually by weight percent of the item.

Credits	CSI Divisions Used for Credit Calculations			
	3-10, 31, 32* (Formerly)	12 (Furniture and Furnishings)	21, 22, 23 (Formerly: Mechanical and Plumbing)	26, 27, 28, 29 (Formerly 16: Electrical)
MRc1.1 and 1.2: Building Reuse	Yes	n.a.	n.a.	n.a.
MRc2: Construction Waste Management	Yes	Yes	Yes	Yes
MRc3: Materials Reuse	Yes	Items may be included if consistent for credits 3–7	n.a.	n.a.
MRc4: Recycled Content	Yes		n.a.	n.a.
MRc5: Regional Materials	Yes		n.a.	n.a.
MRc6: Rapidly Renewable Materials	Yes		n.a.	n.a.
MRc7: Certified Wood	Yes		n.a.	n.a.

(*Specifically 31.60.00 Foundations, 32.10.00 Paving, 32.30.00 Site Improvements, and 32.90.00 Plantings. Other construction and demolition (C&D) waste excluding hazardous materials, land clearing debris, and soils may be included in MRc2. MRc1.1 includes CSI 31)

Table 5.0.3 CSI Division Uses Applicable for Materials and Resources Calculations for Each MR Credit Subcategory in LEED 2009

Details about the specific items included in each subcategory and many of the exceptions can be found in the sections for each individual credit.

Materials and Resources Prerequisites

MR Prerequisite 1: *Storage and Collection of Recyclables*

This prerequisite covers what we commonly consider to be building user consumption recycling. It requires that the common practices of recycling items such as drink bottles and office paper be incorporated into the design and operations of the facility. It has two components. The first is to make sure that there are easily accessible and marked recycling areas in the building design so that all the occupants can recycle these items readily. Areas should be designated for easy user access in commercial and institutional facilities and, as appropriate, for central collection, especially for residential complexes. The second is to make sure that recycling is part of the operations and is actually implemented.

USGBC Rating System

LEED-NC 2.2 lists the Intent, Requirements, and Potential Technologies and Strategies for this prerequisite as follows (LEED 2009 is essentially the same):

(a)

(b)

Figure 5.0.1 Examples of furniture that may optionally be included in various MR credits. The furniture is from (a) Steelcase, Inc., and (b) Haworth, Inc. The carpets are from Milliken & Company and are specially adapted for recycling. (*Photographs taken at the Strom Thurmond Building on Fort Jackson, Columbia, S.C., June 2007.*)

Credits	Items	Unit of Items	Basis	Unit of Basis	Default Unit of Basis
MRc1.1: Building Reuse: Existing Walls, Floors, and Roof	Existing, to remain, structural walls and decks (LEED 2009 only counts one side of interior existing structural walls.)	Area (exclude doors, windows unsound or hazardous materials)	All existing structural walls and decks (LEED 2009 only counts one side of interior existing structural walls.)	Area (exclude doors, windows unsound or hazardous materials)	n.a.
MRc1.2: Building Reuse—Interior Non-structural Elements	Existing, to be reused, nonstructural interior walls, doors, windows, flooring, and ceiling materials	Visible area of all these surfaces, but only count one side of interior doors and windows, and include interior side of exterior walls	Final design nonstructural interior walls, doors, windows, flooring and ceiling materials	Visible area of all these surfaces, but only count one side of interior doors and windows, and include interior side of exterior walls	n.a.
MRc2: Construction Waste Management	Diverted C&D debris*	Weight or volume	All C&D debris*	Weight or volume	n.a.
MRc3: Materials Reuse	Installed reused or salvaged items†	Greater $ value of new or salvage cost of these items	All installed materials in applicable CSI uses†	$ Value of all installed applicable CSI materials	45% of total construction costs ($)†,§
MRc4: Recycled Content	Installed items all or partially made from recycled materials†,‡	$ Value of recycled portions‡ of these materials (based on weight %)	All installed materials in applicable CSI uses†	$ Value of all installed applicable CSI materials	45% of total construction costs ($)†,§
MRc5: Regional Materials	Installed materials extracted, processed, and manufactured regionally†	$ Value of regional materials (based on weight %)	All installed materials in applicable CSI uses†	$ Value of all installed applicable CSI materials	45% of total construction costs ($)†,§

MRc6: Rapidly Renewable Materials (<10-Year Growth)	Installed rapidly renewable materials[†]	$ Value of rapidly renewable materials	All installed materials in applicable CSI uses[†]	$ Value of all installed applicable CSI materials	45% of total construction costs ($)[†,§]
MRc7: certified wood	Installed items made from FSC new wood[†,§]	$ Value of FSC new wood portion of items	All installed items made from new wood[†,§]	$ Value of new wood portion of all installed items	n.a.

[*]May include weight or volume of existing structural items which are remaining if MRc1.1 is not applied for and may include weight or volume of existing nonstructural items which are remaining if MRc1.2 is not applied for. May also include weight or volume of other items which are salvaged from the subject site and reinstalled in the project if MRc3 is not applied for. Land clearing debris, hazardous waste, and soils are not included.

[†]Usually only items in CSI 3–10, 31.60, 32.10, 32.30, and 32.90 are included; however, furniture and furnishing items can be included if these items are used consistently throughout credit subcategories 3 through 7. In this case, the default 45% should not be used unless these additional items are added in.

[‡]These are separated into postconsumer and preconsumer recycled portions, and the preconsumer portions are only given one-half their value.

[§]When using total construction costs, use only hard construction costs (i.e., not design, permitting costs, etc.).

TABLE 5.0.4 Material Types and Unit Bases for Materials and Resources Calculations

	LEED 2009 Credits for Which These Materials Can Also Be Included as Applicable[*]							
Materials Included for Credits	**MRc1.1**	**MRc1.2**	**MRc2**	**MRc3**	**MRc4**	**MRc5**	**MRc6**	**MRc7**
MRc1.1: Building Reuse: Existing Walls, Floors, and Roof	—	No	No[†]	No	No	No	No	No
MRc1.2: Building Reuse: Interior Nonstructural Elements	No	—	No[‡]	No	No	No	No	No
MRc2: Construction Waste Management	No[†]	No[‡]	—	No[§]	No	No	No	No
MRc3: Materials Reuse	No	No	No[§]	—	No	Yes	No	No
MRc4: Recycled Content	No	No	No	No	—	Yes	No	No
MRc5: Regional Materials	No	No	No	Yes	Yes	—	Yes	Yes
MRc6: Rapidly Renewable Materials	No	No	No	No	No	Yes	—	No
MRc7: Certified Wood	No	No	No	No	No	Yes	No	—

[*]Individual items that are made of materials applicable to several of the categories should be allocated to the different categories on a weight percent basis.

[†]MRc2 may include weight of existing structural items which are to remain if MRc1.1 is not applied for.

[‡]MRc2 may include weight of existing nonstructural shell items which are to remain if MRc1.2 is not applied for.

[§]MRc2 may also include weight of other building materials not addressed in MRc1.1 which are to remain if MRc3 is not applied for.

Table 5.0.5 Materials Which May Be Included in Multiple MR Credit Subcategories

Intent

Facilitate the reduction of waste generated by building occupants that is hauled to and disposed of in landfills.

Requirements

Provide an easily accessible area that serves the entire building and is dedicated to the collection and storage of materials for recycling, including (at a minimum) paper, corrugated cardboard, glass, plastics and metals.

Potential Technologies and Strategies

Coordinate the size and functionality of the recycling areas with the anticipated collection services for glass, plastic, office paper, newspaper, cardboard and organic wastes to maximize the effectiveness of the dedicated areas. Consider employing cardboard balers, aluminum can crushers, recycling chutes and collection bins at individual workstations to further enhance the recycling program.

Calculations and Considerations (LEED 2009 MRp1)

There are no calculations required for the submittal, but floor plans and recycling schedules/contracts should be provided to verify compliance. The main consideration in the design of

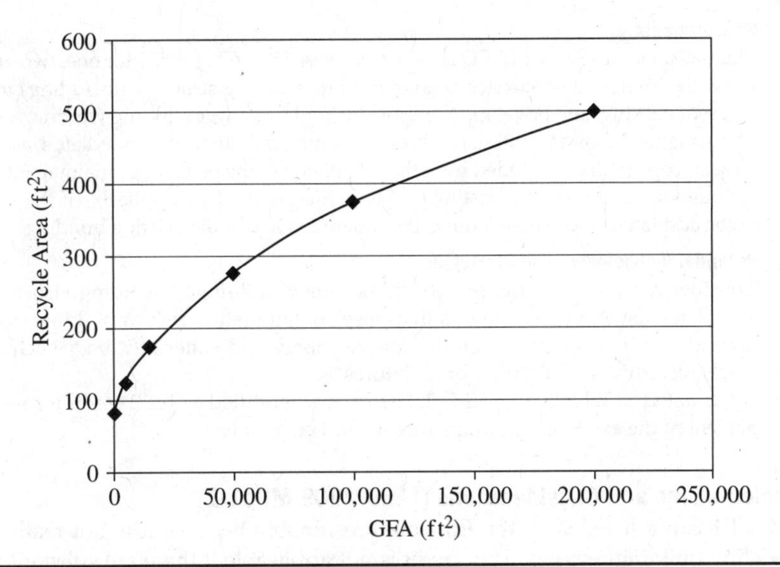

FIGURE 5.0.2 LEED-NC 2.2 commercial recycling minimum area guidelines.

a facility is to provide adequate and appropriate space for the recycling activities. The LEED-NC 2.2 Reference Guide does provide a guideline for the amounts of gross floor area (GFA) that might typically service different commercial building sizes. These are reiterated and smoothed into a graph in Fig. 5.0.2.

5.1 MR Credit Subcategory 1: *Building Reuse*

MR Credit 1.1: *Building Reuse—Maintain Existing Walls, Floors, and Roof (LEED 2009)*

In LEED 2.2, this was subdivided into MRc1.1 and MRc1.2, with MR credit 1.1 worth one point for 75 percent compliance, and MR credit 1.2 worth one point in addition to MR credit 1.1 for 95 percent compliance. LEED 2009 has put these into the one notation of MRc1.1 and has varying point values for compliance. One point for 55 percent, two for 75 percent, and three for 95 percent. Refer to Table 5.0.0 for a summary of these changes. Both versions encourage the reuse of structural materials of existing structures if building on a site with an existing facility.

USGBC Rating System MR Credit 1.1

LEED-NC 2.2 lists the Intent, Requirements, and Potential Technologies and Strategies for this credit as follows, with notations for the LEED 2009 changes included:

Intent

Extend the life cycle of existing building stock, conserve resources, retain cultural resources, reduce waste and reduce environmental impacts of new buildings as they relate to materials manufacturing and transport.

Requirements

Maintain at least 75% (for LEED 2009 this is either 55%, 75% or 95% for one, two or three points respectively) (based on surface area) of existing building structure (including structural floor and roof decking) and envelope (exterior skin and framing, excluding window assemblies and non-structural roofing material). Hazardous materials that are remediated as a part of the project scope shall be excluded from the calculation of the percentage maintained. If the project includes an addition to an existing building, this credit is not applicable if the square footage of the addition is more than 2 times the square footage of the existing building.

Potential Technologies and Strategies

Consider reuse of existing, previously occupied buildings, including structure, envelope and elements. Remove elements that pose contamination risk to building occupants and upgrade components that would improve energy and water efficiency such as windows, mechanical systems and plumbing fixtures.

(As noted in Table 5.0.0, LEED 2.2 has a now modified MRc1.2 credit for maintaining 95 percent of the existing building structure and envelope.)

Calculations and Considerations (LEED 2009 MRc1.1)

MRc1.1 is not listed with the EB icon, presumably because it is not really applicable for existing building reviews. This credit is not applicable if there are substantial additions to the building, totaling more than double the existing gross floor area of the project. If GFA_{PEX} is defined as being the gross floor area of the existing building portion of the project which is to remain and be added on to, and GFA_P is the proposed total gross floor area of the project, both existing and new, then MRc1.1 cannot be obtained if

$$GFA_P > 3 \times GFA_{PEX} \tag{5.1.1}$$

It is important to note that the calculations are based on the area of the structural walls or deckings (floor or roof) and the cladding less the window and door areas. This is simplified from the LEED-NC 2.1 calculations which had some volume and some area calculations based on the volume of many structural units and the area of the cladding items. A simplified summary of how to itemize each of the components is as follows:

- *Deckings:* Itemize separately the square footage of structural roof decking, interior floor decking, and the foundation with decking or slab on grade. Exclude any areas that are considered to be structurally unsound or hazardous materials. Exclude nonstructural roof surfaces such as asphalt shingle. The total summation of these can be referred to as $STRDECK_T$. The summation of the decking areas that are to be retained can be referred to as $STRDECK_{RE}$.
- *Exterior walls:* Itemize separately each exterior wall area excluding windows and doors. Exclude any areas that are considered to be structurally unsound or hazardous materials. The total summation of these can be referred to as $STREXWALL_T$. The summation of the exterior wall areas that are to be retained can be referred to as $STREXWALL_{RE}$.
- *Interior structural walls:* Itemize separately each interior *structural* wall area excluding openings, and count only one side of the wall elements for LEED 2009 (both sides were previously counted in LEED 2.2). (It is assumed that the interior sides of exterior walls are already included in the exterior wall variables.) Exclude any areas that are considered to be structurally unsound or hazardous materials. Also, do not include any nonstructural interior wall sections. The total

summation of these can be referred to as $STRINWALL_T$. The summation of the interior structural wall areas that are to be retained can be referred to as $STRINWALL_{RE}$.

In LEED 2009 one point is earned for MRc1.1 if the following is true:

$$(STRDECK_{RE} + STREXWALL_{RE} + STRINWALL_{RE}) \geq 0.55(STRDECK_T + STREXWALL_T + STRINWALL_T) \tag{5.1.2}$$

An additional point is earned for MRc1.1 if the following is true:

$$(STRDECK_{RE} + STREXWALL_{RE} + STRINWALL_{RE}) \geq 0.75(STRDECK_T + STREXWALL_T + STRINWALL_T) \tag{5.1.3}$$

And a second additional point is earned for MRc1.1 if the following is true:

$$(STRDECK_{RE} + STREXWALL_{RE} + STRINWALL_{RE}) \geq 0.95(STRDECK_T + STREXWALL_T + STRINWALL_T) \tag{5.1.4}$$

Special Circumstances and Exemplary Performance (LEED 2009 MRc1.1)

The calculations for LEED 2009 MR credit 1.1 are summarized in Table 5.1.1. They will be further explained in later sections as they relate to the other credits. Note that MRc2 (construction waste management) may include the weight or volume of existing structural items which are to remain if and only if MRc1.1 is not applied for. There is no exemplary point available as 95 percent is already a very high achievement.

MR Credit 1.2: *Building Reuse—Maintain Interior Nonstructural Elements (LEED 2009)*

LEED 2009 MR credit 1.2 is worth one point and is not dependent on applying for MR credit 1.1. (In LEED 2.2 it was previously referred to as MRc1.3.) This credit encourages the reuse of interior nonstructural materials of existing structures if building on a site with an existing facility.

	Not Applicable If	Worth	Criterion
MRc1.1	$GFA_P > 3 \times GFA_{PEX}$	1 Point and 2 Points and 3 Points	$(STRDECK_{RE} + STREXWALL_{RE} + STRINWALL_{RE}) \geq 0.55(STRDECK_T + STREXWALL_T + STRINWALL_T)$ and $(STRDECK_{RE} + STREXWALL_{RE} + STRINWALL_{RE}) \geq 0.75(STRDECK_T + STREXWALL_T + STRINWALL_T)$ and $(STRDECK_{RE} + STREXWALL_{RE} + STRINWALL_{RE}) \geq 0.95(STRDECK_T + STREXWALL_T + STRINWALL_T)$
MRc1.2	$GFA_P > 3 \times GFA_{PEX}$	1 Point	$INTFIN_{RE} \geq 0.5 \times INTFIN_T$

Table 5.1.1 Building Reuse Equation Summary for LEED 2009

USGBC Rating System

LEED-NC 2.2 lists the Intent, Requirements, and Potential Technologies and Strategies for this credit as follows (LEED 2009 is essentially the same):

Intent

Extend the life cycle of existing building stock, conserve resources, retain cultural resources, reduce waste and reduce environmental impacts of new buildings as they relate to materials manufacturing and transport.

Requirements

Use existing interior non-structural elements (interior walls, doors, floor coverings and ceiling systems) in at least 50% (by area) of the completed building (including additions). If the project includes an addition to an existing building, this credit is not applicable if the square footage of the addition is more than 2 times the square footage of the existing building.

Potential Technologies and Strategies

Consider reuse of existing buildings, including structure, envelope and interior non-structural elements. Remove elements that pose contamination risk to building occupants and upgrade components that would improve energy and water efficiency, such as mechanical systems and plumbing fixtures. Quantify the extent of building reuse.

Calculations and Considerations (LEED 2009 MRc1.2)

This credit is not listed with the EB icon, presumably because it is not really applicable for existing building reviews. Just as for MRc1.1, this credit is not applicable if there are substantial additions to the building, totaling more than double the existing GFA. If GFA_{PEX} is defined as being the gross floor area of the existing building portion of the project which is to be retained and added on to, and GFA_p is the proposed total gross floor area of the project including both the retained and the new areas, then MRc1.2 cannot be applied for if

$$GFA_p > 3 \times GFA_{PEX} \tag{5.1.1}$$

This credit is intended to reuse as much as possible of the interior finished (nonstructural) surfaces in the final structure. First, all interior nonstructural surfaces in the *proposed design* (including the additions if applicable) shall be totaled. Since it is usually based on the proposed design and the finished product, there is no need to exclude hazardous materials, as these should not be included in any newly finished or renovated building in the first place. These interior nonstructural surfaces include all finished interior ceilings, flooring surfaces, doors, windows, casework, demountable partitions (count both sides), walls (count both sides for interior walls for which the finishes are remaining on both sides, but only one side for the interior sides of exterior walls), and visible casework surfaces. It is easy to understand if one pictures the finished interior surfaces that can be seen in all the rooms, but with the interior doors and windows and sidelights counting only once. (LEED 2.2 did not include any exterior doors and windows, even their interior side, but this has been modified in LEED 2009 as previously noted.) Let the total of all these areas be represented by $INTFIN_T$. Let $INTFIN_{RE}$ be the total of all these finished interior surfaces that have been reused from the existing building, even those which may have been moved. Then one point is earned for MRc1.2 if

$$INTFIN_{RE} \geq 0.5 \times INTFIN_T \tag{5.1.2}$$

Special Circumstances and Exemplary Performance (LEED 2009 MRc1.2)

The calculations are also summarized in Table 5.1.1 and will be further explained in later sections as they relate to the other credits. Note that MRc2 (construction waste management) may include the weight or volume of existing nonstructural interior items which are to remain if and only if MRc1.2 is not applied for. Also, MRc1.2 is not dependent on obtaining MRc1.1.

There is no EP point related to MRc1.2. It is the author's opinion that this is so because it is not very reasonable to encourage more than a 50 percent reuse of interior surfaces as many are worn over time or may have been made from materials not considered to be as sustainable as products encouraged in the USGBC rating system.

5.2 MR Credit Subcategory 2: *Construction Waste Management*

The construction waste management subcategory covers what is commonly called *construction and demolition* (C&D) debris. In LEED 2.2, one point is awarded for MRc2.1 (50% diversion), and an additional one point can be awarded for MRc2.2 (75% diversion) in addition to the one for MR credit 2.1. LEED 2009 has consolidated both into the single MRc2 subcategory with similar varying points depending on compliance, one point for 50 percent and two points for 75 percent.

MR Credit 2: *Construction Waste Management—Divert from Disposal (LEED 2009)*

USGBC Rating System MR Credit 2

LEED-NC 2.2 lists the Intent, Requirements, and Potential Technologies and Strategies for this credit as follows, with notations for the LEED 2009 changes included:

Intent

Divert construction and demolition debris from disposal in landfills and incinerators. Redirect recyclable recovered resources back to the manufacturing process. Redirect reusable materials to appropriate sites.

Requirements

Recycle and/or salvage at least 50% (for one point and 75% for two points in LEED 2009) of non-hazardous construction and demolition debris. Develop and implement a construction waste management plan that, at a minimum, identifies the materials to be diverted from disposal and whether the materials will be sorted on-site or commingled.

Excavated soil and land-clearing debris do not contribute to this credit. Calculations can be done by weight or volume, but must be consistent throughout. (As noted in Table 5.0.0, LEED 2.2 has a now modified MRc2.2 credit for diverting 95 percent.)

Potential Technologies and Strategies

Establish goals for diversion from disposal in landfills and incinerators and adopt a construction waste management plan to achieve these goals. Consider recycling cardboard, metal, brick, acoustical tile, concrete, plastic, clean wood, glass, gypsum wallboard, carpet and insulation. Designate a specific area(s) on the construction site for segregated or commingled collection of recyclable materials, and track recycling efforts throughout the construction process. Identify construction haulers and recyclers to handle the designated materials. Note that diversion may include donation of materials to charitable organizations and salvage of materials on-site.

Calculations and Considerations (LEED 2009 MRc2)

The intention of the two MRc2 points and the EP point for diversion of 50, 75, or 95 percent, respectively, of the construction and demolition (C&D) waste stream from landfills is to find reuses for as many materials as possible from either demolition activities or construction activities. Note that there are two different sources of these waste materials. One source is from the existing manmade buildings or other structural items on the site which are to be either reused or demolished, and one source is from the construction process. If a project has no demolition on-site since the land was originally not built upon, then the source is solely from the construction phase.

It is very important to establish a construction waste management plan that involves the construction contractor to adequately verify that these goals have been met. Good waste management plans will include proper designations of areas on-site for the recycling activities, plans to train the workers in the recycling procedures, some feedback or reporting system to oversee the activities, and proper labeling of all recycling facilities, particularly in the languages of the workers. Care should be taken to avoid tight or not readily accessible areas where either disposal or pickup will be difficult. Care should also be taken to avoid problems with stormwater in and around the recycling facilities.

A significant amount of documentation is needed to verify the items throughout the demolition and construction phases. These might include tipping fee receipts from the recyclers or landfills and accounting of all the dumpster and waste removal procedures on-site. For instance, dumpsters come in different sizes and may be removed when full or partially full, depending on the pickup times. It is difficult to determine the volume of a waste in a container if it is not documented as to its fullness at the time of pickup. In some cases the weights will be given regardless from the tipping accounts, but some debris items are valuable (such as metals) and recyclers may contract to haul them free of charge, in which case the general contractor may not receive the necessary records unless they are specifically requested in the contracts.

Good waste management plans will have the contracts and oversight in place necessary to have dumpsters usually as full as possible when picked up, since this will typically reduce overall hauling costs and environmental impacts of the transport phase. Commonly used dumpsters in the United States range from a small size of 6 yards to a large size of 30 yards or more. A yard in this case actually refers to a cubic yard. Dumpster sizes are usually chosen based on the expected amount of different types of wastes generated. Some wastes may also never be put into dumpsters, but instead either reused on-site or loaded into dump trucks or other hauling vehicles and then transported off-site. Typical examples of this are recycled asphalt or concrete pavements which are usually first crushed and then piled on-site before they are reused.

The calculations are fairly simple if there is adequate control, recording, and documentation resulting from the waste management plan. What is important is to make sure that the items included in these calculations have been adequately documented.

The items used in the calculations for MRc2 are usually defined by the following criteria:

- In all cases, hazardous materials, land-clearing debris, and soils removed from the site are not included in the calculations.
- All the demolition wastes that go into the dumpsters and are hauled off-site are included.
- All the construction wastes that go into the dumpsters and are hauled off-site are included.

- Any pavements that are crushed and reused on-site are included.
- Existing building items which are to remain as addressed in LEED 2009 MRc1.1 (building reuse—structural) can also be included but only if MRc1.1 is not applied for.
- Existing building items which are to remain as addressed in LEED 2009 MRc1.2 (building reuse—nonstructural) can also be included but only if LEED 2009 MRc1.2 is not applied for.
- Other building materials salvaged and reused on-site can contribute to this credit if they are not included in the MR credit 3 calculations.

Using these criteria, the following variables can be defined:

DeBLDGREUSE	The items reused in the building as defined in LEED 2009 MRc1.1 if MRc.1.1 is *not* applied for, and the items reused in the building as defined in LEED 2009 MRc1.2 if MRc.1.2 is *not* applied for and any other building materials salvaged and reused on-site if they are not included in MRc3 calculations
DeCARD	Cardboard recycled off-site
DeDIVERT	C&D debris diverted
DeGYPSUM	Gypsum-type wallboard recycled off-site
DeLANDFILL	All C&D debris excluding hazardous wastes, land-clearing debris, and soils that are disposed of off-site in a landfill or other waste facility without recycling or reuse
DeMETAL	Metals recycled off-site
DeMISC	Miscellaneous items recycled off-site
DePAVEREUSE	Existing pavement debris reused on-site
DeRUBBLE	Any rubble (bricks, concrete, asphalt, masonry, etc.) recycled off-site
DeWOOD	Any wood recycled off-site

The units on these variables can be either volume-based or weight-based, but must be consistent throughout. There has been very limited research on the conversion factors needed to bring some volume measurements into weight measurements (or vice versa). The LEED-NC 2.2 and LEED 2009 Reference Guides specify that either the solid waste conversion factors listed in the LEED 2.2 or LEED 2009 Reference Guides Table 2 under MRc2 can be used, or another justifiable set of conversion factors. The default LEED conversion factors are represented in Table 5.2.1. Typically, these are assumed to be waste management dumpster volumes. In addition, another set of many of the conversion factors was developed based on construction debris (no demolition debris was included in these factors) from a LEED certified project in Columbia, S.C., in 2005. These conversion factors were derived from the weights and volumes of the dumpsters and were referred to as "dumpster densities" in that report. They are also given in Table 5.2.1 as are other conversion factors found in a compilation of studies on recycling operations elsewhere, but which were usually from receiving facilities and not necessarily LEED-related. Table 5.2.1 is taken from L. Haselbach and S. Bruner, "Determining Construction Debris Recycling Dumpster Densities," *Journal of Green Building*, Summer 2006, 1(3).

Solid Waste Material	AISC (1980) Source Material Density (lb/CY)	LEED-NC 2.2 and LEED 2009 Conversion Factors (lb/CY)	Cal-Recovery Bulk Density (lb/CY)	Columbia, S.C., Study Dumpster Avg. Density (lb/CY)	Columbia, S.C., Study Density Statistics (lb/CY)	No. Data Points in Study
General trash (mixed waste)	n.a.	350	n.a.	600	Min. 280 Max. 1540 SD 290	34
Masonry (rubble)	3000–4000	1400	1610 (brick) 1855 (conc.)	1680	Min. 1220 Max. 2620 SD 320	15
Wood	680–1100	300	330	280	Min. 140 Max. 480 SD 110	6
Steel (metals)	13,200	1000	440	320	Min. 260 Max. 380 SD 50	7
Sheetrock (gypsum wallboard)	1300–1950	500	390	440	Min. 280 Max. 620 SD 170	4
Cardboard	—	100	—	Not studied	Not studied	—

Source: Courtesy *Journal of Green Building* Vol. 1, No. 3, College Publishing.

TABLE 5.2.1 Calculated Average, Maximum, Minimum, and Standard Deviation (SD) of Solid Waste Conversion Factors from the Columbia, S.C., Study by Haselbach and Bruner as Compared to LEED-NC 2.2, Cal-Recovery Values and Density of the Solid Source Material

Note that the density of the solid source material as given in Table 5.2.1 is always much greater than the conversion factor. This is so because there are many void spaces in the dumpsters and hauling trucks since the contents are not compacted to the density of the material. The source material densities may be better used in converting volumes to weight of reused building reuse items. It is recommended that the Columbia study value for steel and metal be used for cases where there is only construction debris. It was noted in that study that recycled metals in the dumpsters for construction-only phases were rarely very large, dense pieces such as structural steel, but were rather usually items such as pieces of ductwork, piping, and flashing.

The equations needed to determine if the construction waste management credits in the construction waste management subcategory are earned are as follows:

$$\text{DeDIVERT} = \text{DeBLDGREUSE} + \text{DeCARD} + \text{DeGYPSUM} + \text{DeMETAL} + \text{DeMISC} + \text{DePAVEREUSE} + \text{DeRUBBLE} + \text{DeWOOD} \quad (5.2.1)$$

One point for LEED 2009 MRc2 is earned if

$$\text{DeDIVERT} \geq 0.50(\text{DeLANDFILL} + \text{DeDIVERT}) \quad (5.2.2)$$

One additional point for LEED 2009 MRc2 is earned if

$$\text{DeDIVERT} \geq 0.75(\text{DeLANDFILL} + \text{DeDIVERT}) \quad (5.2.3)$$

Figure 5.2.1 shows how one type of material can get recycled and save energy and transportation costs for many different groups. It is a pile of crushed concrete from the demolition of a dormitory at the University of South Carolina. It was hauled only a few miles away to an asphalt plant where it is stockpiled for reuse in many ways.

Figure 5.2.1 Stockpile of crushed concrete at the Sloan Construction asphalt plant in Columbia, S.C., ready for reuse. The concrete comes from the demolition of dormitories at the University of South Carolina, only a few miles away. (*Photograph taken August 2007.*)

Special Circumstances and Exemplary Performance (LEED 2009 MRc2)

An EP point can be awarded in the construction waste management subcategory if the diversion rate reaches at least 95 percent. One additional point for EP is earned if

$$\text{DeDIVERT} \geq 0.95(\text{DeLANDFILL} + \text{DeDIVERT}) \qquad (5.2.4)$$

5.3 MR Credit Subcategory 3: *Materials Reuse*

MR Credit 3: *Materials Reuse (LEED 2009)*

In LEED 2.2, MRc3.1 is worth one point for 5 percent reuse, and MRc3.2 for 10 percent reuse is worth one point in addition to the point for MRc3.1. LEED 2009 has consolidated both into the single MRc3 subcategory with similar varying points depending on compliance, one point for 5 percent and two points for 10 percent. The intention of this subcategory is to promote the reuse of salvaged goods in the marketplace.

USGBC Rating System MR Credit 3

LEED-NC 2.2 lists the Intent, Requirements, and Potential Technologies and Strategies for this credit as follows, with notations for LEED 2009 changes included:

Intent

Reuse building materials and products in order to reduce demand for virgin materials and to reduce waste, thereby reducing impacts associated with the extraction and processing of virgin resources.

Requirements

Use salvaged, refurbished or reused materials such that the sum of these materials constitutes at least 5% (for one point and 10% for two points in LEED 2009), based on cost, of the total value of materials on the project. Mechanical, electrical and plumbing components and specialty items such as elevators and equipment shall not be included in this calculation. Only include materials permanently installed in the project. Furniture may be included, providing it is included consistently in MR Credits 3–7. (As noted in Table 5.0.0, LEED 2.2 has a now modified MRc3.2 credit for accomplishing a 10% materials reuse rate.)

Potential Technologies and Strategies

Identify opportunities to incorporate salvaged materials into building design and research potential material suppliers. Consider salvaged materials such as beams and posts, flooring, paneling, doors and frames, cabinetry and furniture, brick and decorative items.

Calculations and Considerations (LEED 2009 MRc3)

The items that can be included in the Materials Reuse category can come from many sources. Usually, only permanent items in the project from the 2004 CSI categories 3 through 10, 31.60, 32.10, 32.30, and 32.90 are included, but furniture and furnishings can be included if they are consistently used throughout credit subcategories 3 through 7 as noted in Table 5.0.3. If they are included, then they should also be added into the denominators in the calculations. LEED 2009 requires that if salvaged furniture is taken from the occupant's previous facility, then they must have been acquired at least 2 years prior to the initiation of the current project.

In general, there are two different classes of what may be referred to as salvaged items. First, there are the items that are reused or salvaged from an existing facility on-site, and then there are items salvaged from elsewhere.

Most of these items from the site are either already included in the first credit subcategory for Building Reuse, or if MRc1.1 is not applied for, then are included as being diverted per the second credit subcategory (construction waste management) and cannot be included in the Materials Reuse (salvage) category. However, some additional items can be included. For instance, exterior doors and windows and associated hardware are items which are not always appropriately included in building reuse. If they are appropriately refurbished or refinished and reused somewhere in the project, then they can be included, even if used for a different purpose.

Materials contributing toward achievement of MRc3 cannot be applied to MR credits 1, 2, 4, 6, or 7; but if MRc3 is not being attempted, applicable materials can be applied to another LEED credit if eligible.

The second group includes items that have been used in another project somewhere and have been salvaged, many times with refurbishing or refinishing, and reused in the project. Some areas of the country have a large market for salvaged items. Some items commonly salvaged include wood flooring and beams, stone, pavers, ornate ironwork, and bricks. Figures 5.3.1 and 5.3.2 show some features made from salvaged wood at a LEED-NC-certified building in Columbia, S.C.

FIGURE 5.3.1 Salvaged wood staircase at Cox and Dinkins, Engineers and Surveyors, Columbia, S.C. (*Photograph taken June 2007.*)

Figure 5.3.2 Salvaged wood alcove area at Cox and Dinkins, Engineers and Surveyors, Columbia, S.C. (*Photograph taken June 2007.*)

It is impossible to compare different types of construction materials unless there is a basis by which they can be compared. In the energy category, when there are different energy sources and types, this common rating system is cost or value. Again, in the remaining Materials and Resources categories, the calculations are based on cost or value. This is summarized in Table 5.0.5.

To promote as many salvaging activities as possible, and at the same time promote as economical a project as possible, the cost values of the salvaged items used in the calculations represent either the cost of salvaging (either the costs to refurbish the on-site items or the cost to purchase the off-site items) or the market value of the equivalent new item, whichever is greater.

Let the following variables be defined:

$MATL\$_T$ Total cost of all project materials in CSI categories 3 through 10, 31.60, 32.10, 32.30, and 32.90 plus any furniture or furnishing costs included consistently in MR subcategories 3 through 7. LEED 2009 specifically includes taxes, transportation, and delivery fees to the site as part of the material acquisition costs. (Total default costs if not itemized can be calculated as 45 percent of total construction costs in the same 2004 CSI categories. Use only hard construction costs. Add in furniture or furnishing costs if included consistently in MR subcategories 3 through 7. Salvaged items are valued at the greater of market value or actual salvage cost.)

$MATL\$_{SAL}$ Total cost of salvaged materials plus any salvaged furniture or furnishing costs if included consistently in MR subcategories 3 through 7 (Salvaged items are valued at the greater of market value or actual cost.)

In LEED 2009, one point is earned for MRc3 if the following is true:

$$100(\text{MATL\$}_{\text{SAL}}/\text{MATL\$}_T) \geq 5 \quad (5.3.1)$$

An additional point is earned for MRc3 if the following is true:

$$100(\text{MATL\$}_{\text{SAL}}/\text{MATL\$}_T) \geq 10 \quad (5.3.2)$$

Special Circumstances and Exemplary Performance (LEED 2009 MRc3)

An EP point can be earned in the MR subcategory for increasing the performance by 50 percent more than required for the second point, that is, for using salvaged or reused items which are valued at 15 percent or more of the applicable project material costs. Therefore, an EP point can be earned for material reuse if the following is true:

$$100(\text{MATL\$}_{\text{SAL}}/\text{MATL\$}_T) \geq 15 \quad (5.3.3)$$

Since it is usually not certain whether salvaged materials can be appropriately refurbished or reused for many applications in the design phase, this subcategory is listed as a construction phase submittal. The contractor should keep records of salvaging activities and costs.

5.4 MR Credit Subcategory 4: *Recycled Content*

The intent of this credit subcategory is to encourage the use of items made from recycled materials to lessen the resource and environmental impacts of making many products from raw materials and to lessen the solid waste load in the country. In LEED 2.2, MRc4.1 is worth one point for 10 percent, and MRc4.2 is worth one point for achieving 20 percent in addition to the point for MRc4.1. LEED 2009 has consolidated both into the single MRc4 subcategory with similar varying points depending on compliance, one point for 10 percent and two points for 20 percent.

MR Credit 4: *Recycled Content (postconsumer + ½ preconsumer)*

USGBC Rating System MR Credit 4

LEED-NC 2.2 lists the Intent, Requirements, and Potential Technologies and Strategies for this credit as follows, with notations for LEED 2009 changes included:

Intent

Increase demand for building products that incorporate recycled content materials, thereby reducing impacts resulting from extraction and processing of virgin materials.

Requirements

Use materials with recycled content such that the sum of post-consumer recycled content plus one-half of the pre-consumer content constitutes at least 10% (for one point and 20% for two points in LEED 2009) (based on cost) of the total value of the materials in the project. The recycled content value of a material assembly shall be determined by weight. The recycled fraction of the assembly is then multiplied by the cost of assembly to determine the recycled content value. Mechanical, electrical and plumbing components and specialty items such as elevators shall not be included in this calculation. Only include materials permanently installed in the project. Furniture may be included, providing it is included consistently in MR Credits 3–7. Recycled content shall be defined in accordance with the

International Organization of Standards document, *ISO 14021—Environmental labels and declarations—Self-declared environmental claims (Type II environmental labeling)*. Post-consumer material is defined as waste material generated by households or by commercial, industrial and institutional facilities in their role as end-users of the product, which can no longer be used for its intended purpose. Pre-consumer material is defined as material diverted from the waste stream during the manufacturing process. Excluded is reutilization of materials such as rework, regrind or scrap generated in a process and capable of being reclaimed within the same process that generated it. (As noted in Table 5.0.0, LEED 2.2 has a now modified MRc4.2 credit for accomplishing a 20% recycle rate.)

Potential Technologies and Strategies

Establish a project goal for recycled content materials and identify material suppliers that can achieve this goal. During construction, ensure that the specified recycled content materials are installed. Consider a range of environmental, economic and performance attributes when selecting products and materials.

Calculations and Considerations (LEED 2009 MRc4)

The intention of these credits is to use items that are made from materials which have been made from recycled materials to reduce the use and costs (economic, social, and environmental) of using virgin material sources. In this case recycled differs from salvaged in the sense that the salvaged items as addressed in the materials reuse subcategory (MRc3) are recycled in their final functional form and recycled items in the recycled content subcategory (MRc4) are made in part from materials recycled either after final use (postconsumer) or from by-products of industrial processes (preconsumer). Materials that are preconsumer but that could be put back into the industrial stream from whence they came are not eligible. There is greater value given for the items recycled postconsumer, since they have already gone through two types of material streams: manufacturing and final product use.

As in many of the MR credit subcategories, these credits are based on material cost and values. There are so many different types of materials and material assemblies (items) installed in the built environment, and cost (or value) is really the only common element that can be used for comparison in typical construction analyses. The credit criteria are based on the percentage of total material costs that come from preconsumer and postconsumer recycled materials. The values of the preconsumer and postconsumer recycled portions of the materials can be defined as follows:

$MATL\$_{RECPOC}$ Total value of postconsumer recycled portion of materials (CSI 3 through 10, 31.60, 32.10, 32.30, and 32.90) plus any value of postconsumer recycled portions of furniture or furnishings if included consistently in MR subcategories 3 through 7. Value determinations are based on postconsumer recycled weight percent of the value of each of the individual items.

$MATL\$_{RECPRC}$ Total value of pre-consumer recycled portion of materials (CSI 3 through 10, 31.60, 32.10, 32.30, and 32.90) plus any value of preconsumer recycled portions of furniture or furnishings if included consistently in MR subcategories 3 through 7. Value determinations are based on preconsumer recycled weight percent of the value of each of the individual items.

$MATL\$_{T}$ See definition under MRc3.

Given these, in LEED 2009, one point is earned for MRc4 if the following is true:

$$100[\text{MATL\$}_{RECPOC} + (0.5 \times \text{MATL\$}_{RECPRC})]/\text{MATL\$}_T \geq 10 \qquad (5.4.1)$$

An additional point is earned for MRc4 if the following is true:

$$100[\text{MATL\$}_{RECPOC} + (0.5 \times \text{MATL\$}_{RECPRC})]/\text{MATL\$}_T \geq 20 \qquad (5.4.2)$$

Most construction items are not 100 percent made from recycled materials. Therefore, the recycled value is only a portion of the value of the item, whether these items are one piece (such as a steel beam) or an assembly (such as a prebuilt window unit). The value determination is usually made on a mass percent basis. So for any item the following can be defined:

$\text{MASS}_{RECPOCi}$	Mass (weight) of the postconsumer recycled portion of unit *i*
$\text{MASS}_{RECPRCi}$	Mass (weight) of the preconsumer recycled portion of unit *i*
MASS_{UNITi}	Total mass (weight) of unit *i*
$\text{MATL\$}_{UNITi}$	Total cost of unit *i*
$\text{MATL\$}_{RECPOCi}$	Value of postconsumer recycled portions of unit *i*
$\text{MATL\$}_{RECPRCi}$	Value of preconsumer recycled portions of unit *i*

The value of the preconsumer and postconsumer recycled portions for each individual unit *i* can be determined from the following equations:

$$\text{MATL\$}_{RECPOCi} = \text{MATL\$}_{UNITi}\,(\text{MASS}_{RECPOCi})/\text{MASS}_{UNITi} \qquad (5.4.3)$$

$$\text{MATL\$}_{RECPRCi} = \text{MATL\$}_{UNITi}\,(\text{MASS}_{RECPRCi})/\text{MASS}_{UNITi} \qquad (5.4.4)$$

And, in similar manner, the total values of the recycled portions needed in Eqs. (5.4.1) and (5.4.2) can be determined from the summation of the individual unit contributions as follows:

$$\text{MATL\$}_{RECPOC} = \Sigma\, \text{MATL\$}_{RECPOCi} \quad \text{over all units } i \qquad (5.4.5)$$

$$\text{MATL\$}_{RECPRC} = \Sigma\, \text{MATL\$}_{RECPRCi} \quad \text{over all units } i \qquad (5.4.6)$$

Special Circumstances and Exemplary Performance (LEED 2009 MRc4)

There are some cases in which the value of the recycled content in some materials is disproportionately high compared to the costs of the other material contents. In these cases, it may be appropriate to use an alternative to the listed weight-based calculation, but these should be justified. An already accepted alternative is in the case of Portland cement–based concrete. Many concretes are made with a combination of Portland cement and fly ash or other supplementary cementitious materials (SCMs), usually up to 25 percent. Costs for cement and other cementitious materials are generally a large percent of the cost of concrete, and therefore it is appropriate to calculate the value of the recycled cementitious material content in concrete as the weight percent of only the cementitious materials and multiply by the cost of the cementitious materials.

As related to the recycled content credit subcategory, this use of fly ash is promoted as an environmental benefit as it recycles waste from another process which might otherwise be landfilled, and it also promotes a reduction in the use of energy and the production of carbon dioxide from the cement-making process.

There is also a default postconsumer recycle content value for steel of 25 percent which can be used regardless of documentation as steel is usually made from at least 25 percent postconsumer recycled steel. In cases where the steel recycled content is greater than 25 percent, documentation should be provided.

In LEED 2009, an additional EP point can be earned for recycled content if the following is true:

$$100[\mathrm{MATL\$}_{RECPOC} + (0.5 \times \mathrm{MATL\$}_{RECPRC})]/\mathrm{MATL\$}_T \geq 30 \qquad (5.4.7)$$

Since it is usually not certain during the design phase of the exact amounts of recycled content that might be purchased for each item, this subcategory is listed as a construction phase submittal. The contractor should keep records of recycled content and costs. LEED 2009 cautions about indoor air emissions from certain recycled materials and if they are used, then their emissions should be included in the IEQc4 if applicable. The equations for this credit subcategory are summarized in Tables 5.4.1 and 5.4.2.

Some examples of carpet systems made with materials which can be included in the recycle content subcategory can be seen in Fig. 5.4.1.

	Postconsumer Recycled Content Value of Unit *i* ($\mathrm{MATL\$}_{RECPOCi}$)	**Total Postconsumer Recycled Content Value ($\mathrm{MATL\$}_{RECPOC}$)**	**Preconsumer Recycled Content Value of Unit *i* ($\mathrm{MATL\$}_{RECPRCi}$)**	**Total Preconsumer Recycled Content Value ($\mathrm{MATL\$}_{RECPRC}$)**
Equations	$\mathrm{MATL\$}_{UNITi} \times (\mathrm{MASS}_{RECPOCi})/\mathrm{MASS}_{UNITi}$	$\Sigma\mathrm{MATL\$}_{RECPOCi}$	$\mathrm{MATL\$}_{UNITi} \times (\mathrm{MASS}_{RECPRCi})/\mathrm{MASS}_{UNITi}$	$\Sigma\mathrm{MATL\$}_{RECPRCi}$
Some exceptions	Value of recycled steel content is at a minimum of 25% or greater with documentation		Recycled supplementary cementitious materials used in concrete are valued per weight percent of cementitious materials	

TABLE 5.4.1 Summary of Recycled Content Material Value Equations

	Worth	Criterion
MRc4 (MRc4.1 in LEED 2.2)	1 point	$100[MATL\$_{RECPOC} + (0.5 \times MATL\$_{RECPRC})]/MATL\$_T \geq 10$
MRc4 (MRc4.2 in LEED 2.2)	1 additional point	$100[MATL\$_{RECPOC} + (0.5 \times MATL\$_{RECPRC})]/MATL\$_T \geq 20$
Exemplary performance	1 point additional to MRc4	$100[MATL\$_{RECPOC} + (0.5 \times MATL\$_{RECPRC})]/MATL\$_T \geq 30$

TABLE 5.4.2 Recycled Content Compliance Equation Summary LEED 2009

FIGURE 5.4.1 Carpets which may count toward the recycle content credit as provided by Interface, Inc., and Lees Carpets. *(Photograph taken at the Sustainable Interiors ribbon cutting at the Strom Thurmond Building on Fort Jackson, Columbia, S.C., June 2007.)*

5.5 MR Credit Subcategory 5: *Regional Materials*

MR Credit 5: *Regional Materials—Extracted, Processed, and Manufactured Regionally*

The intent of this credit subcategory is to encourage the use of regional materials to lessen the environmental impacts of transportation and to make regions more self-sufficient. In LEED 2.2, MRc5.1 is worth one point, and MRc5.2 is worth one point in addition to the point for MRc5.1. LEED 2009 has consolidated both into the single MRc5 subcategory with similar varying points depending on compliance, one point for 10 percent and two points for 20 percent.

USGBC Rating System MR Credit 5

LEED-NC 2.2 lists the Intent, Requirements, and Potential Technologies and Strategies for this credit as follows, with notations for LEED 2009 changes included:

Intent

Increase demand for building materials and products that are extracted and manufactured within the region, thereby supporting the use of indigenous resources and reducing the environmental impacts resulting from transportation.

Requirements

Use building materials or products that have been extracted, harvested or recovered, as well as manufactured, within 500 miles of the project site for a minimum of 10% (for one point and 20% for two points in LEED 2009) (based on cost) of the total materials value. If only a fraction of a product or material is extracted/harvested/recovered and manufactured locally, then only that percentage (by weight) shall contribute to the regional value. Mechanical, electrical and plumbing components and specialty items such as elevators and equipment shall not be included in this calculation. Only include materials permanently installed in the project. Furniture may be included, providing it is included consistently in MR Credits 3–7. (As noted in Table 5.0.0, LEED 2.2 has a now modified MRc5.2 credit for accomplishing a 20% regional material rate.)

Potential Technologies and Strategies

Establish a project goal for locally sourced materials, and identify materials and material suppliers that can achieve this goal. During construction, ensure that the specified local materials are installed and quantify the total percentage of local materials installed. Consider a range of environmental, economic and performance attributes when selecting products and materials.

Calculations and Considerations (LEED 2009 MRc5)

As in many of the other MR credit subcategories, these credits are based on material cost and values. There are so many different types of materials and material assemblies (items) installed in the built environment, and cost (or value) is really the only common element that can be used for comparison in typical construction analyses. The credit criteria are based on the percentage of the total material costs that represents items or portions of items which have been extracted, processed, *and* manufactured within a 500-mi radius of the project site. For salvaged or reused items, the point of extraction is considered to be the location that it is being salvaged from, and the manufacturing point is the location of the salvage dealer.

Given the following definition:

$MATL\$_{REG}$ Total value of regionally extracted, processed, and manufactured portions of materials (CSI 3 through 10, 31.60, 32.10, 32.30, and 32.90) plus any similar portions of furniture or furnishings if included consistently in MR subcategories 3 through 7. Value determinations are based on the weight percent of the regional portion of each of the individual item values.

$MATL\$_T$ See definition under MRc3.

In LEED 2009, one point is earned for MRc5 if the following is true:

$$100(MATL\$_{REG})/MATL\$_T \geq 10 \tag{5.5.1}$$

An additional point is earned for MRc5 if the following is true:

$$100(MATL\$_{REG})/MATL\$_T \geq 20 \tag{5.5.2}$$

Most construction items are not 100 percent made from regional materials. Therefore, the regional value is only a portion of the value of the item, whether these items are one piece (such as a steel beam) or an assembly (such as a prebuilt window unit). The value determination is made on a mass percent basis. So for any item the following can be defined:

$MASS_{REGi}$ Mass (weight) of the regional portion of unit i

$MASS_{UNITi}$ Total mass (weight) of unit i

$MATL\$_{UNITi}$ Total cost of unit i

$MATL\$_{REGi}$ Value of the regional portions of unit i

The value of the regional portions for each individual unit i can be determined from the following:

$$MATL\$_{REGi} = MATL\$_{UNITi}\,(MASS_{REGi})/MASS_{UNITi} \tag{5.5.3}$$

And, in a similar manner, the total values of the regional portions needed in Eqs. (5.5.1) and (5.5.2) can be determined from the summation of the individual unit contributions as follows:

$$MATL\$_{REG} = \Sigma\, MATL\$_{REGi} \quad \text{over all units } i \tag{5.5.4}$$

Special Circumstances and Exemplary Performance (LEED 2009 MRc5)

Since it is usually not certain during the design phase of the exact amounts of regional items or content that might be purchased, this subcategory is listed as a construction phase submittal. The contractor should keep records of the locations of extraction, processing, and manufacturing.

Any materials used for obtaining the points in this category can also be used for MR subcategories 3 (Materials Reuse), 4 (Recycled Content), 6 (Rapidly Renewable Materials), and 7 (Certified Wood).

An additional EP point can be earned for regional content if the following is true:

$$100(MATL\$_{REG})/MATL\$_{T} \geq 40 \tag{5.5.5}$$

The equations for this credit subcategory are summarized in Tables 5.5.1 and 5.5.2.

	Regional Content Value of Unit i ($MATL\$_{REGi}$)	Total Regional Content Value ($MATL\$_{REG}$)
Equations	$MATL\$_{UNITi}(MASS_{REGi})/MASS_{UNITi}$	$\Sigma MATL\$_{REGi}$ over all units i

TABLE 5.5.1 Summary of Regional Materials Value Equations

	Worth	Criterion
MRc5 (MRc5.1 in LEED 2.2)	1 point	$100(MATL\$_{REG})/MATL\$_{T} \geq 10$
MRc5 (MRc5.2 in LEED 2.2)	1 additional point	$100(MATL\$_{REG})/MATL\$_{T} \geq 20$
Exemplary performance	1 point in addition to MRc5	$100(MATL\$_{REG})/MATL\$_{T} \geq 30$*

*LEED 2.2 had 40 percent compliance for the EP point.

TABLE 5.5.2 Regional Materials Compliance Equation Summary LEED 2009

5.6 MR Credit 6: *Rapidly Renewable Materials*

The intent of MR credit 6 is to encourage the use of rapidly renewable materials, that is, materials that are made from living products that have a short growth span. One point is available for MRc6 in both versions 2.2 and 2009.

USGBC Rating System

LEED-NC 2.2 lists the Intent, Requirements, and Potential Technologies and Strategies for this credit as follows (LEED 2009 is essentially the same):

> **Intent**
>
> Reduce the use and depletion of finite raw materials and long-cycle renewable materials by replacing them with rapidly renewable materials.
>
> **Requirements**
>
> Use rapidly renewable building materials and products (made from plants that are typically harvested within a ten-year cycle or shorter) for 2.5% of the total value of all building materials and products used in the project, based on cost.
>
> **Potential Technologies and Strategies**
>
> Establish a project goal for rapidly renewable materials and identify products and suppliers that can support achievement of this goal. Consider materials such as bamboo, wool, cotton insulation, agrifiber, linoleum, wheatboard, strawboard and cork. During construction, ensure that the specified renewable materials are installed.

Calculations and Considerations (LEED 2009 MRc6)

Rapidly renewable materials are considered to be sustainable since they tend to use less land and have a faster economic cycle than materials with a growing span longer than 10 years. They are not limited to plant products, but may also be animal based. Some common examples of products made wholly or partially from rapidly renewable materials are sunflower seed board panels, bio-based paints, coir and jute geotextiles, soy-based insulation and form-release agents, straw bales, linoleum, bamboo flooring, wheatboard, cotton and wool products, and cork.

The determinations are based on the following definitions and equations.

$MATL\$_{RRM}$ Total value of rapidly renewable materials (CSI 3 through 10 , 31.60, 32.10, 32.30, and 32.90) plus any similar portions of furniture or furnishings made from rapidly renewable materials if furniture and furnishings are included consistently in MR subcategories 3 through 7. Value determinations are based on the weight percent of the rapidly renewable portion of each of the individual item values.

$MATL\$_T$ See definition under MRc3.

Then one point is earned for MRc6 if the following is true:

$$100(MATL\$_{RRM})/MATL\$_T \geq 2.5 \tag{5.6.1}$$

Most applicable construction items are not 100 percent made from rapidly renewable materials, but rather are composites such as bamboo flooring or linoleum. Therefore, the rapidly renewable value is only a portion of the value of the item, whether these items are one piece (such as linoleum flooring) or an assembly (such as wheatboard casework). The value determination is made on a mass percent basis. So for any item the following can be defined:

$MASS_{RRMi}$	Mass (weight) of the rapidly renewable portion of unit i
$MASS_{UNITi}$	Total mass (weight) of unit i
$MATL\$_{UNITi}$	Total cost of unit i
$MATL\$_{RRMi}$	Value of the rapidly renewable portions of unit i

The value of the rapidly renewable portions for each individual unit i can be determined from the following:

$$MATL\$_{RRMi} = MATL\$_{UNITi}\,(MASS_{RRMi})/MASS_{UNITi} \qquad (5.6.2)$$

And in similar manner, the total value of the rapidly renewable portion needed in Eq. (5.6.1) can be determined from the summation of the individual unit contributions as follows:

$$MATL\$_{RRM} = \Sigma MATL\$_{RRMi} \quad \text{over all units } i \qquad (5.6.3)$$

Special Circumstances and Exemplary Performance (LEED 2009 MRc6)

There may be some concern that many of these rapidly renewable products are not available regionally or are exotic. For instance, bamboo is not commonly associated with forests in the United States. However, many of the rapidly renewable products can be grown and produced within various regions of the United States. Figure 5.6.1 shows a wild growing bamboo stand in an uncultivated area around Lake Murray in South Carolina.

Figure 5.6.1 Bamboo growing wild near Lake Murray, S.C. (*Photograph taken July 2007.*)

	Rapidly Renewable Content Value of Unit *i* ($MATL\$_{RRMi}$)	Total Rapidly Renewable Content Value ($MATL\$_{RRM}$)
Equations	$MATL\$_{UNITi}(MASS_{RRMi})/MASS_{UNITi}$	$\Sigma MATL\$_{RRMi}$ over all units *i*

Table 5.6.1 Summary of Rapidly Renewable Materials Value Equations

	Worth	Criterion
MRc6	1 point	$100(MATL\$_{RRM})/MATL\$_T \geq 2.5$
Exemplary performance	1 point in addition to MRc6	$100(MATL\$_{RRM})/MATL\$_T \geq 5$

Table 5.6.2 Rapidly Renewable Materials Compliance Equation Summary LEED 2009

In both versions 2.2 and 2009, an additional EP point can be earned for rapidly renewable content if the following is true:

$$100(MATL\$_{RRM})/MATL\$_T \geq 5 \qquad (5.6.4)$$

The equations for this credit subcategory are summarized in Tables 5.6.1 and 5.6.2.

5.7 MR Credit 7: *Certified Wood*

The intention of this credit is to use wood products which are made from wood that has been grown and processed in a manner that is environmentally friendly. One point is available for MRc7 in both LEED 2.2 and LEED 2009.

USGBC Rating System

LEED-NC 2.2 lists the Intent, Requirements, and Potential Technologies and Strategies for this credit as follows (LEED 2009 is essentially the same):

Intent

Encourage environmentally responsible forest management.

Requirements

Use a minimum of 50% of wood-based materials and products, which are certified in accordance with the Forest Stewardship Council's (FSC) Principles and Criteria, for wood building components. These components include, but are not limited to, structural framing and general dimensional framing, flooring, sub-flooring, wood doors and finishes.

Include materials permanently installed in the project. Furniture may be included, providing it is included consistently in MR Credits 3–7. Wood products purchased for temporary use on the project (e.g., formwork, bracing, scaffolding, sidewalk protection, and guard rails) may be included in the calculation at the project team's discretion. If any such materials are included, all such materials must be included in the calculation. If such materials are purchased for use on multiple projects, the applicant may include these materials for only one project, at its discretion.

Potential Technologies and Strategies

Establish a project goal for FSC-certified wood products and identify suppliers that can achieve this goal. During construction, ensure that the FSC-certified wood products are installed and quantify the total percentage of FSC-certified wood products installed.

Calculations and Considerations (LEED 2009 MRc7)

This credit encourages the use of Forest Stewardship Council (FSC) labeled wood products. FSC products will be labeled as either FSC Pure or FSC Mixed, and both types of labels ensure that the wood portions of the item are a specific percent FSC certified. FSC-certified products will have a Chain-of-Custody (CoC) certification. CoC numbers for all items are required in the submittals.

Note that this credit is based only on new wood, in both the numerator and the denominator. Any reused or salvaged wood and wood products are included in other credits. Also, even though the equations developed in this chapter to determine the values of the FSC new wood and all the new wood on the project are based on mass (weight) percentages of the items, there is the option for any item to be instead based on volume percent or value percent if these are easier or more readily available ways to determine the percent value. For instance, a cost percent value may be easier to do with veneers. Some of the variables used in the equations are as follows:

$WOOD\$_T$	Total value of *new* wood portions of all items installed in the project
$WOOD\$_{FSC}$	Total value of FSC *new* wood portions of all items installed in the project
$MASS_{FSCi}$	Mass (weight) of the FSC *new* wood portion of unit *i* (Volume or value can be substituted for mass as appropriate for an item.)
$MASS_{WOODi}$	Total mass (weight) of all *new* wood in unit *i* (Volume or value can be substituted for mass as appropriate for an item.)
$MASS_{UNITi}$	Total mass (weight) of unit *i* (Volume or value can be substituted for mass as appropriate for an item.)
$MATL\$_{UNITi}$	Total cost of unit *i*
$MATL\$_{FSCi}$	Value of the FSC *new* wood portions of unit *i*
$MATL\$_{WOODi}$	Value of all the *new* wood portions of unit *i*

For both versions 2.2 and 2009, one point is earned for MRc7 if the following is true:

$$100(WOOD\ \$_{FSC})/WOOD\ \$_T \geq 50 \quad (5.7.1)$$

Most items are not made entirely of new wood, and therefore the values of the certified new wood and the wood portions are determined as a mass (weight), volume, or cost percent of the total items. Usually the mass percent is used, as in the following equations:

$$MATL\$_{FSCi} = MATL\$_{UNITi}\ (MASS_{FSCi})/MASS_{UNITi} \quad (5.7.2)$$

$$MATL\$_{T} = MATL\$_{UNITi}\ (MASS_{WOODi})/MASS_{UNITi} \quad (5.7.3)$$

And in a similar manner, the total values of the FSC *new* wood and total *new* wood portions needed in Eq. (5.7.1) can be determined from the summation of the individual unit contributions as follows:

$$WOOD\$_{FSC} = \Sigma MATL\$_{FSCi} \quad \text{over all units } i \quad (5.7.4)$$

$$WOOD\$_{T} = \Sigma MATL\$_{WOODi} \quad \text{over all units } i \quad (5.7.5)$$

	Worth	Criterion
MRc7	1 point	$100(\text{WOOD\$}_{FSC})/\text{WOOD \$}_T \geq 50$
Exemplary performance	1 point additional	$100(\text{WOOD\$}_{FSC})/\text{WOOD\$}_T \geq 95$

TABLE 5.7.1 Certified Wood Compliance Equation Summary

Special Circumstances and Exemplary Performance (LEED 2009 MRc7)
An additional EP point can be earned for certified wood if the following is true:

$$100(\text{WOOD\$}_{FSC})/\text{WOOD\$}_T \geq 95 \tag{5.7.6}$$

The point criteria equations for this credit subcategory are summarized in Table 5.7.1.

5.8 Discussion and Overview

The LEED-NC 2.2 Reference Manual lists all the credits in the Materials and Resources category as requiring Construction Phase Submittals. This is so for a variety of reasons. Three of the main reasons are the following:

- Contractors have control over the waste materials and their disposal. It is important to keep track of tipping sheets, etc.
- Many of the material specifications and shop drawings are done or furnished by the contractor. These determine a lot of the material makeup.
- Most of the materials are purchased by the contractor, so the contractor has decision-making capabilities for where the products come from with respect to both local materials and makeup of the materials for recycling, reuse, FSC certification, etc.

It is very important that an effective material tracking program be set up and maintained during the entire construction phase. These requirements should be explained well in the construction documents and periodically reviewed to make sure that the expectations are being met.

References

AISC (1980), *Manual of Steel Construction*, 8th ed., American Institute of Steel Construction Inc., Chicago ©1980, first revised printing 11/84, Second Impression 2/86.

Albermani, F., G. Y. Goh, and S. L. Chan (2006), "Lightweight Bamboo Double Layer Grid System," Engineering Structures, http://www.sciencedirect.com/science/journal/01410296> Volume 29, Issue 7 <http://www.sciencedirect.com/science?_ob=PublicationURL&_tockey=%23TOC%235715%232007%23999709992%23659670%23FLA%23&_cdi=5715&_pubType=J&_auth=y&_acct=C000050221&_version=1&_urlVersion=0&_userid=10&md5=de3b4f08c26c73b1658aee5e404d6a74>, July 2007, pp. 1499–1506.

Cal Recovery (1991), "Conversion Factors for Individual Material Types," Prepared for the California Integrated Waste Management Board by Cal Recover, Inc., December.

CSI (2004), *The Project Resource Manual—CSI Manual of Practice*, 5th ed., Construction Specifications Institute, Alexandria, VA.

FSC (2007), http://www.fsc.org/en/, Forest Stewardship Council website accessed July 18, 2007.

Haselbach, L., and S. Bruner (2006), "Determining Construction Debris Recycling Dumpster Densities," *Journal of Green Building*, Summer, 1(3).

Spiegel, R., and D. Meadows (2006), *Green Building Materials: A Guide to Product Selection and Specification*, 2d ed., CSI, The Construction Specifications Institute, Alexandria, VA.

USGBC (2003), *LEED-NC for New Construction, Reference Guide*, Version 2.1, 2d ed., May, U.S. Green Building Council, Washington, D.C.

USGBC (2005–2007), *LEED-NC for New Construction, Reference Guide*, Version 2.2, 1st ed., U.S. Green Building Council, Washington, D.C., October 2005 with errata posted through Spring 2007.

USGBC (2009), *LEED Reference Guide for Green Building Design and Construction*, 2009 Edition, U.S. Green Building Council, Washington, D.C., April 2009.

Exercises

1. Put together a CPM (critical path method) schedule for items for C&D debris. (See Chap. 8 for information on a CPM.)

2. Make a site plan with staging areas for C&D debris recycling for a project with demolition.
- **A.** Calculate the number of dumpsters needed and sizes.
- **B.** Determine the area needed for the dumpsters and access.
- **C.** Sketch these items on the plan.

3. Make a site plan with staging areas for C&D debris recycling for a project with no demolition.
- **A.** Calculate the number of dumpsters needed and sizes.
- **B.** Determine the area needed for the dumpsters and access.
- **C.** Sketch these items on the plan.

4. Total construction cost of your new building on a site (project A) with no demolition is $10,000,000, and Table 5.E.1 represents the materials tracked as applicable to MR credits 3 through 7. This table does not represent all the materials used, only those tracked for LEED purposes. All new wood and all the furniture items are listed. Include the listed furniture items (wood furniture and workstations) in the calculations.
- **A.** Calculate the eligible points based on Table 5.E.1 for MR credits 3, 4, 5, 6, and 7 (LEED 2009).
- **B.** Is this project eligible for any LEED 2009 EP points as related to MR credits 3, 4, 5, 6, and 7? Why?
- **C.** If this project is in San Francisco, is it also eligible for RP points? Which ones? (See Chap. 7.)

5. Total construction cost of your new building (project A) on a site with no demolition is $10,000,000, and Table 5.E.1 represents the materials tracked as applicable to MR credits 3 through 7. This table does not represent all the materials used, only those tracked for LEED purposes. All new wood and all the furniture items are listed. Do not include the listed furniture items (wood furniture and workstations) in the calculations.
- **A.** Calculate the LEED 2009 eligible points based on Table 5.E.1 for MR credits 3, 4, 5, 6, and 7.
- **B.** Is this project eligible for any LEED 2009 EP points as related to MR credits 3, 4, 5, 6, and 7? Why?
- **C.** If this project is in Honolulu, is it also eligible for RP points? Which ones? (See Chap. 7.)

Product	Company	Postconsumer Recycled %	Preconsumer Recycled %	Content Info Source	Manufacturing Location (mi from Site)	% Raw Materials within 500 mi	% Rapidly Renewable Materials	FSC-Certified Wood (%)	Product Cost ($)
Used brick*	Ali and Allison Salvagers	n.a.	n.a.	n.a.	95	100	n.a.	n.a.	50,000
New brick	Bower's Brick	0	15	Factory letter	600	0	0	n.a.	30,000
Concrete fill*	Bramblett Concrete	n.a.	n.a.	Contractor	69	100	n.a.	n.a.	10,000
Steel	Dorn Metal	60	0	Manufacturer	355	0	0	n.a.	165,000
Ready mix concrete	Amanda Mix	5	0	Brochure	227	100	0	n.a.	32,000
Compost	Tab and Kim Mulch	100	0	Common knowledge	26	100	n.a.	n.a.	5,000
Wheatboard	Martin Walls	0	0	Cut sheet	59	75	75	n.a.	93,000
Metal siding	Neal and Quinn McSiding	25	0	Cut sheet	552	0	0	n.a.	45,000
Metal roofing	Minor Roof	85	0	Brochure	342	25	0	n.a.	47,000
Ceramic tile	Munson Tile	95	0	Brochure	343	45	0	n.a.	12,000
Acoustical tile	Randolph and Roney Ceilings	90	0	Cut sheet	34	0	0	n.a.	15,000
Toilet partitions	Rowland Supply	0	100	Brochure	78	0	n.a.	n.a.	7,000
Carpet	Ashley Carpet	40	0	Cut sheet	115	0	n.a.	n.a.	50,000
Carpet pad	Wilder Padding	0	100	Brochure	56	0	n.a.	n.a.	4,000
Linoleum	Wiseman Floor	0	0	Cut sheet	3,000	0	80	n.a.	18,000

Special finish wood†	Amponsah and Deep Finishes	0	0	Brochure	250	100	0	100	15,000
Misc. finish carpentry†	Sanford Trees	0	0	Brochure	330	100	0	34	32,000
Pine boards†	Gaither Wood	0	0	Cut sheet	45	100	0	50	40,000
Salvaged wood flooring*	Grant Interiors	n.a.	n.a.	Contractor	340	100	n.a.	n.a.	62,000
Wood doors, frames†	Del Porto Accessways	0	0	Brochures	552	0	0	65	120,000
Wood furniture†	Floyd and Frick Furniture	0	0	Manufacturer	123	0	0	30	135,000
Wood roof structure†	Guidt Roof Salvaging	n.a.	n.a.	Brochure	145	100	n.a.	n.a.	115,000
Workstations	Lockard Stations	45	20	Brochure	222	0	0	0	35,000

*Salvaged materials.
†New wood.

Table 5.E.1 Materials and Resources Project A Tracking Summary

6. Total construction cost of your new building on a site (project B) with no demolition is $20,000,000, and Table 5.E.2 represents the materials tracked as applicable to MR credits 3 through 7. This table does not represent all the materials used, only those tracked for LEED purposes. All new wood and all the furniture items are listed. Include the listed furniture items (wood furniture and workstations) in the calculations.

A. Calculate the LEED 2009 eligible points based on Table 5.E.2 for MR credits 3, 4, 5, 6, and 7.

B. Is this project eligible for any LEED 2009 EP points as related to MR credits 3, 4, 5, 6, and 7? Why?

C. If this project is in Austin, TX, is it also eligible for RP points? Which ones? (See Chap. 7.)

7. Total construction cost of your new building on a site (project B) with no demolition is $20,000,000, and Table 5.E.2 represents the materials tracked as applicable to MR credits 3 through 7. This table does not represent all the materials used, only those tracked for LEED purposes. All new wood and all the furniture items are listed. Do not include the listed furniture items (wood furniture and workstations) in the calculations.

A. Calculate the LEED 2009 eligible points based on Table 5.E.2 for MR credits 3, 4, 5, 6, and 7.

B. Is this project eligible for any LEED 2009 EP points as related to MR credits 3, 4, 5, 6, and 7? Why?

C. If this project is in Anchorage, AK, is it also eligible for RP points? Which ones? (See Chap. 7.)

8. You have a small commercial building that is going to be renovated, and you are going for LEED-NC (instead of LEED-EB). The building footprint is 40 ft by 40 ft. It is one story with an exterior wall height of 12 ft and a flat roof. You can assume the following:

- Slab on grade of 12-in depth
- Four exterior doors (7 ft by 3 ft)
- Six exterior windows (4 ft by 3 ft)
- No interior structural walls

You are interested in going for the building reuse credits for the structural items. You plan to keep all the exterior walls and the foundation and slab, but do not plan to keep the roof structure as your client would like a pitched roof, which would provide attic storage. Calculate the percent of building reuse and how many LEED 2009 MR credit 1.1 points are earned.

9. You have a small open-room warehouse that is going to be renovated into a commercial building, and you are going for LEED-NC (instead of LEED-EB). The building footprint is 60 ft by 40 ft. It is one story with an exterior wall height of 12 ft and a flat roof. All interior rooms are 10 ft high. The new interior will consist of the following:

- A 39-ft by 30-ft front retail customer area with a 7-ft by 3-ft exterior door and two 8-ft by 4-ft exterior windows
- A 27-ft by 20-ft office area in the back left with a 3-ft by 6-ft interior window to the customer area
- An 11-ft by 10-ft restroom-utility room on the far back right
- An 11-ft by 16-ft storage area on the middle back right
- A 6-ft by 27-ft hallway down the center of the back part to an existing exit door

These dimensions are interior wall-to-wall measurements. There are four interior doors (3 ft by 7 ft) off the hallway—one to the customer area, one to the office, one to the storage room, and the last one to the restroom-utility room.

You want to see if you can get LEED 2009 MR credit 1.2 by keeping the original hardwood flooring in all areas except the restroom-utility room, keeping the drop ceiling in all areas except the retail customer area, and keeping the gypsum along all the interior sides of the exterior walls.

A. Sketch the design.

B. Determine if you are eligible for MR credit 1.2.

Product	Company	Postconsumer Recycled %	Preconsumer Recycled %	Content Info Source	Manufacturing Location (mi from Site)	% Raw Materials within 500 mi	% Rapidly Renewable Materials	FSC-Certified Wood (%)	Product Cost ($)
Used brick*	Bagley Salvagers	n.a.	n.a.	n.a.	95	100	n.a.	n.a.	100,000
New brick	Koty's Brick	0	15	Factory letter	600	0	0	n.a.	65,000
Concrete fill*	Barwick Rubble	n.a.	n.a.	Contractor	69	100	n.a.	n.a.	20,000
Steel	Burns Metal	60	0	Manufacturer	355	0	0	n.a.	310,000
Ready mix concrete	Kendrick Concrete	5	0	Brochure	227	100	0	n.a.	65,000
Compost	Mogan Mulch	100	0	Common knowledge	26	100	n.a.	n.a.	10,000
Wheatboard	Wall and Wall Walls	0	0	Cut sheet	59	75	75	n.a.	190,000
Metal siding	Byrd Siding	25	0	Cut sheet	552	0	0	n.a.	95,000
Metal roofing	LeCroy Roof	85	0	Brochure	342	25	0	n.a.	94,000
Ceramic tile	Whetstone Tile	95	0	Brochure	343	45	0	n.a.	27,000
Acoustic tile	Powell Ceilings	90	0	Cut sheet	34	0	0	n.a.	32,000
WC partitions	Tristan Supply	0	100	Brochure	78	0	n.a.	n.a.	15,000
Carpet	Bethany Carpet	40	0	Cut sheet	115	0	n.a.	n.a.	110,000
Carpet pad	Sipes Padding	0	100	Brochure	56	0	n.a.	n.a.	24,000
Linoleum	Kimberly Floor	0	0	Cut sheet	3,000	0	80	n.a.	36,000
Special finish wood†	Tune Finishes	0	0	Brochure	250	100	0	100	30,000

TABLE 5.E.2 Materials and Resources Project B Tracking Summary (*Continued*)

Product	Company	Postconsumer Recycled %	Preconsumer Recycled %	Content Info Source	Manufacturing Location (mi from Site)	% Raw Materials within 500 mi	% Rapidly Renewable Materials	FSC-Certified Wood (%)	Product Cost ($)
Misc. finish carpentry†	Wade Trees	0	0	Brochure	330	100	0	34	65,000
Pine boards†	Wagner Wood	0	0	Cut sheet	45	100	0	50	80,000
Reused wood flooring*	Young and Old Interiors	n.a.	n.a.	Contractor	340	100	n.a.	n.a.	130,000
Wood doors, frames†	Gorman Accessways	0	0	Brochures	552	0	0	65	220,000
Wood furniture†	Bridgette Furniture	0	0	Manufacturer	123	0	0	30	260,000
Wood roof structure*	Pearce Roof Salvaging	n.a.	n.a.	Brochure	145	100	n.a.	n.a.	230,000
Workstations	Worthy Stations	45	20	Brochure	222	0	0	0	75,000

*Salvaged materials.
†New wood.

TABLE 5.E.2 Materials and Resources Project B Tracking Summary (*Continued*)

Material	Dumpster Size (yards)	Number of Dumpsters Filled
Cardboard	30	3
Gypsum wallboard	30	1
Steel	6	2
Wood	30	3
Mixed waste to landfill	30	6
Rubble (concrete/brick)	20	3

TABLE 5.E.3 Project C&D Debris Table

10. You set up a C&D debris program at your site and recycle all the cardboard, gypsum wallboard, steel, concrete and brick, and wood for a new construction project. Your dumpster count at the end of the project is as listed in Table 5.E.3. Calculate the percent recycled C&D debris by both volume and weight using the LEED conversion factors. With the weight calculation how many LEED 2009 points might you obtain? With the volume calculations how many LEED 2009 points might you obtain? With the volume calculations how many LEED 2009 points might you obtain?

11. You are putting up a new building at a currently vacant site. You set up a construction debris program at your site and recycle all the cardboard, gypsum wallboard, steel, concrete and brick, and wood for a new construction project. Your dumpster count at the end of the project is as listed in Table 5.E.3. Calculate the percent recycled C&D debris by both volume and weight using the Columbia, S.C., study dumpster densities. With the weight calculation how many LEED 2009 points might you obtain? With the volume calculations how many LEED 2009 points might you obtain?

12. You are planning to start designing a project, and the site engineer asks how and where the C&D debris area should be located. Using the six categories given by LEED for typical conversion factors, estimate dumpster sizes needed and the footprint on the ground needed for this C&D debris recycling area. Make a sketch of the footprint. Make sure that proper access is available to the location for both the recycling contractors and the workers, and sketch in these accessways. Be sure to include fencing as needed.

13. Total construction cost of your new building on a site (project C) with no demolition is $15,000,000, and Table 5.E.4 represents the materials tracked as applicable to MR credits 3 through 7. This table does not represent all the materials used, only those tracked for LEED purposes. All new wood and all the furniture items are listed. Include the listed furniture items (wood furniture and workstations) in the calculations.

A. Calculate the LEED 2009 eligible points based on Table 5.E.4 for MR credits 3, 4, 5, 6, and 7.

B. Is this project eligible for any LEED 2009 EP points as related to MR credits 3, 4, 5, 6, and 7. Why?

14. Total construction cost of your new building on a site (project C) with no demolition is $15,000,000, and Table 5.E.4 represents the materials tracked as applicable to MR credits 3 through 7. This table does not represent all the materials used, only those tracked for LEED purposes. All new wood and all the furniture items are listed. Do not include the listed furniture items (wood furniture and workstations) in the calculations.

A. Calculate the LEED 2009 eligible points based on Table 5.E.4 for MR credits 3, 4, 5, 6, and 7.

B. Is this project eligible for any LEED 2009 EP points as related to MR credits 3, 4, 5, 6, and 7. Why?

Product	Company	Postconsumer Recycled %	Preconsumer Recycled %	Content Info Source	Manufacturing Location (mi from Site)	% Raw Materials within 500 mi	% Rapidly Renewable Materials	FSC-Certified Wood (%)	Product Cost ($)
Used brick*	Bagley Salvagers	n.a.	n.a.	n.a.	95	100	n.a.	n.a.	90,000
New brick	Koty's Brick	0	15	Factory letter	600	0	0	n.a.	65,000
Concrete fill*	Barwick Rubble	n.a.	n.a.	Contractor	69	100	n.a.	n.a.	20,000
Steel	Burns Metal	60	0	Manufacturer	355	50	0	n.a.	310,000
Ready mix concrete	Kendrick Concrete	5	0	Brochure	227	100	0	n.a.	65,000
Compost	Mogan Mulch	100	0	Common knowledge	26	100	n.a.	n.a.	10,000
Wheatboard	Wall and Wall Walls	0	0	Cut sheet	59	75	75	n.a.	190,000
Metal siding	Byrd Siding	55	0	Cut sheet	552	0	0	n.a.	95,000
Metal roofing	LeCroy Roof	85	0	Brochure	342	25	0	n.a.	94,000
Ceramic tile	Whetstone Tile	95	0	Brochure	343	45	0	n.a.	27,000
Acoustic tile	Powell Ceilings	90	0	Cut sheet	34	50	0	n.a.	32,000
WC partitions	Tristan Supply	0	100	Brochure	78	0	n.a.	n.a.	15,000
Carpet	Bethany Carpet	40	0	Cut sheet	115	50	n.a.	n.a.	110,000
Carpet pad	Sipes Padding	0	100	Brochure	56	0	n.a.	n.a.	24,000
Linoleum	Kimberly Floor	0	0	Cut sheet	3,000	0	80	n.a.	36,000
Special finish wood†	Tune Finishes	0	0	Brochure	250	100	0	100	30,000

Misc. finish carpentry†	Wade Trees	0	0	Brochure	330	100	0	34	65,000
Pine boards†	Wagner Wood	0	0	Cut sheet	45	100	0	50	80,000
Reused wood flooring*	Young and Old Interiors	n.a.	n.a.	Contractor	340	100	n.a.	n.a.	130,000
Wood doors, frames†	Gorman Accessways	0	0	Brochures	552	0	0	65	220,000
Wood furniture†	Bridgette Furniture	0	0	Manufacturer	123	0	0	30	160,000
Wood roof Structure*	Pearce Roof Salvaging	n.a.	n.a.	Brochure	145	100	n.a.	n.a.	230,000
Workstations	Worthy Stations	45	20	Brochure	222	0	0	0	75,000

*Salvaged materials.
†New wood.

Table 5.E.4 Materials and Resources Project C Tracking Summary

15. Check the USGBC errata on the USGBC website (www.USGBC.org) for MRc5, Regional Materials, to ensure that the new LEED 2009 EP requirement is 30 percent versus the 40 percent previously listed in LEED 2.2.

16. Which, if any, of the MR credits are available for Regional Priority points in the zip code of your work or school? (See Chap. 7.)

17. Which, if any, of the MR credits are available for Regional Priority points in the following major cities? (See Chap. 7.)

A. New York City
B. Houston, TX
C. Los Angeles, CA
D. Portland, OR
E. Miami, FL
F. Kansas City, MO

18. Your company is purchasing $10,000 worth of windows from a manufacturer who purchases glass panes that are 75 percent postconsumer recycled glass by weight, and aluminum parts that are 55 percent postconsumer recycled by weight. All the wood in the windows are new and are FSC certified. If the glass, metal, and wood portions of the units are on average 40, 25, and 30 percent of the total weight respectively, then calculate both the postconsumer recycled value and the FSC certified new wood value of these windows.

19. Visit hardware or department stores and report the recycle content, FSC, wood, or other wood certification, location of the manufacturer, rapidly renewable content information, and any other environmental information as applicable on at least one wood or natural fiber-based product, one metal product, and one mineral based product such as brick or concrete mix or one plastic based product.

CHAPTER 6

LEED Indoor Environmental Quality

The Indoor Environmental Quality (IEQ) category of the U.S. Green Building Council (USGBC) rating system for new construction version 2.2 (LEED-NC 2.2) and version 2009 (LEED-NC 2009) consists of two prerequisite subcategories and eight credit subcategories that together may earn a possible 15 points (credits) maximum, not including exemplary performance (EP) points in both versions. The notation format for the prerequisites and credits is, for example, EQp1 for Indoor Environmental Quality prerequisite 1 and EQc1 for Indoor Environmental Quality credit 1 in LEED 2.2, whereas in version 2009 the notation has been modified to IEQp1 and IEQc1, respectively. The new LEED 2009 notation will be used throughout this book. A compilation of the two versions, associated points, and a summary of major differences can be found in Table 6.0.0. To facilitate cross-referencing between this text and the USGBC rating system, the second digit in a section heading, equation, table, or figure number represents the credit subcategory number for sections that deal directly with a USGBC LEED-NC 2.2 or 2009 credit subcategory.

The Indoor Environmental Quality (IEQ) portion deals with materials and systems inside the building that affect the health and comfort of the occupants and construction workers.

The two prerequisite subcategories are IEQ Prerequisite 1: Minimum IAQ Performance (IEQp1) and IEQ Prerequisite 2: Environmental Tobacco Smoke (ETS) Control (IEQp2). As is typical for prerequisites, there are no points associated with IEQp1 and IEQp2.

The points for the credits in the Indoor Environmental Quality category are individually related to a specific credit under a subcategory. In both versions (2.2 and 2009), there are eight credit subcategories, five of which have multiple credits and three of which have just one credit and associated point. The eight credit subcategories are IEQ Credit Subcategory 1: Outdoor Air Delivery Monitoring, IEQ Credit Subcategory 2: Increased Ventilation, IEQ Credit Subcategory 3: Construction IAQ Management Plan, IEQ Credit Subcategory 4: Low-Emitting Materials, IEQ Credit Subcategory 5: Indoor Chemical and Pollutant Source Control, IEQ Credit Subcategory 6: Controllability of Systems, IEQ Credit Subcategory 7: Thermal Comfort, and IEQ Credit Subcategory 8: Daylighting and Views.

IEQ credit subcategory 4 has four credits; IEQ credit subcategories 3, 6, 7, and 8 each have two credits; and IEQ credit subcategories 1, 2, and 5 are all single credits. Each credit is worth one point in both versions, not including EP points, which will be discussed later.

Subcategory 2009	Subcategory 2.2	Pts 2009	Pts 2.2	EP Pts 2009	EP Pts 2.2	Major Comments
IEQp1: Minimum Indoor Air Quality Performance	EQp1: Minimum Indoor Air Quality Performance	n.a.	n.a.	n.a.	n.a.	• LEED 2009 is based on ASHRAE 62.1-2007 with errata but without addenda and LEED 2.2 is based on ASHRAE 62.1-2004.* • LEED 2009 allows for natural ventilation in interior spaces without openings by adjoining rooms if unobstructed and double the opening operable area (or 25 sf) is free.
IEQp2: Environmental Tobacco Smoke (ETS) Control	EQp2: Environmental Tobacco Smoke (ETS) Control	n.a.	n.a.	n.a.	n.a.	• LEED 2009 requires signage in smoking and non-smoking areas. • LEED 2009 requires weather-stripping exterior doors and operable windows in residential units.
IEQc1: Outdoor Air Delivery Monitoring	EQc1: Outdoor Air Delivery Monitoring	1	1	–	–	• LEED 2009 is based on ASHRAE 62.1-2007 with errata but without addenda and LEED 2.2 is based on ASHRAE 62.1-2004.* • LEED 2009 corrects the airflow monitoring to within 15% of the set point (instead of 10%). • LEED 2009 gives alternatives for buildings with limited airflow monitoring capabilities.
IEQc2: Increased Ventilation	EQc2: Increased Ventilation	1	1	–	–	• Parts of LEED 2009 are based on ASHRAE 62.1-2007 with errata but without addenda while these LEED 2.2 parts use ASHRAE 62.1-2004.* • LEED 2009 deletes the use of the *Carbon Trust-Good Practice Guide 237*. • LEED 2009 allows either the CIBSE Appl'ns Manuals 10 (2005) or 13 (2000), as applicable to the project space for natural ventilation.
IEQc3.1: Construction Indoor Air Quality Management Plan—during Construction	EQc3.1: Construction Indoor Air Quality Management Plan—during Construction	1	1	–	–	• LEED 2009 uses SMACNA Nov. 2007 instead of the 1995 version and includes the errata (without addenda) of ASHRAE 52.2-1999.

IEQc3.2: Construction Indoor Air Quality Management Plan—Before Occupancy	EQc3.2: Construction Indoor Air Quality Management Plan—Before Occupancy	1	1	–	–	• LEED 2009 requires all finishes are applied prior to the flush-out option. • LEED 2009 Air Testing option has stricter formaldehyde maximum. (27 ppb versus 50 ppb).
IEQc4.1: Low-Emitting Materials—Adhesives and Sealants	EQc4.1: Low-Emitting Materials—Adhesives and Sealants	1	1	–	–	• LEED 2009 has the VOC budget calculation option done separately on total Adhesives and Sealants (Paints/Coatings not included). Also include adhesives from IEQc4.3.
IEQc4.2: Low- Emitting Materials—Paints and Coatings	EQc4.2: Low- Emitting Materials—Paints and Coatings	1	1	–	–	• LEED 2009 has primers adhere to SCAQMD Rule 1113 instead of GS-11 (SCAQMD is slightly less stringent). • LEED 2009 has the VOC budget calculation option done separately on total Paints and Coatings (Adhesives/Sealants not included).
IEQc4.3: Low-Emitting Materials—Flooring Systems	EQc4.3: Low- Emitting Materials—Carpet Systems	1	1	–	–	• LEED 2009 has requirements for flooring systems other than carpet. • LEED 2009 has an alternative testing Option 2 for any of the flooring system types.†
IEQc4.4: Low-Emitting Materials—Composite Wood and Agrifiber Products	EQc4.4: Low-Emitting Materials—Composite Wood and Agrifiber Products	1	1	–	–	–
IEQc5: Indoor Chemical and Pollutant Source Control	EQc5: Indoor Chemical and Pollutant Source Control	1	1	–	–	• The permanent entry systems to capture particulates have increased from 6 ft (2.2) to 10 ft (2009). • Areas needing special exhausting also include science labs, prep rooms, art rooms, and shops of any kind (2009). • Battery banks for temporary back-up must be segregated from other areas (2009). • 2009 requires containment for disposal of hazardous liquid wastes where water and chemical concentrate mixing takes place.

Table 6.0.0 IEQ Summary Table of Major Differences between Versions 2.2 and 2009 and Associated Points (Pts) (*Continued*)

Subcategory 2009	Subcategory 2.2	Pts 2009	Pts 2.2	EP Pts 2009	EP Pts 2.2	Major Comments
IEQc6.1: Controllability of Systems—Lighting	EQc6.1: Controllability of Systems—Lighting	1	1	–	–	–
IEQc6.2: Controllability of Systems—Thermal Comfort	EQc6.2: Controllability of Systems—Thermal Comfort	1	1	–	–	• If provided by operable windows, LEED 2009 is based on ASHRAE 62.1-2007 with errata but without addenda, whereas LEED 2.2 is based on ASHRAE 62.1-2004 for the opening sizes.
IEQc7.1: Thermal Comfort—Design	EQc7.1: Thermal Comfort—Design	1	1	–	–	• LEED 2009 also refers to ASHRAE 55-2004 but with errata and without addenda.
IEQc7.2: Thermal Comfort—Verification	EQc7.2: Thermal Comfort—Verification	1	1	–	–	• LEED 2009 also refers to ASHRAE 55-2004 but with errata and without addenda. • LEED 2009 has three additional requirements: 1. IEQc7.1 is achieved. 2. Not applicable for residential projects. 3. Provide a permanent monitoring system.
IEQc8.1: Daylight and Views—Daylight	EQc8.1: Daylight and Views—Daylight 75% of Spaces	1 (75%)	1	1 (95%)	1 (95%)	• In 2009, the Simulation is now Option 1 instead of 2. The Simulation is at 9 a.m. and 3 p.m. on Sept. 21 instead of noon on either equinox and has a maximum 500 fc if view preserving automated shades not provided. • In 2009, the Calculation Option (formerly Option 1) is replaced by a different Prescriptive Option 2. It analyzes side- and top-lighting zones separately. See Table 6.8.1. • In LEED 2009, the three options—Simulation, Prescriptive, and Measurement—may be used in different areas. (Option 4 Combination) • In all cases, exceptions exist for areas where daylight might hinder tasks, dedicated theater spaces have a min. 10 minimum and appropriate glare control is required.

IEQc8.2: Daylight and Views, Views	EQc8.2: Daylight and Views, Views for 90% of Spaces	1	1	1	1	• LEED 2009 limits sight lines through at most two interior glazing surfaces. • The LEED 2.2 EP point is not specified. • The LEED 2009 EP point is as noted. ‡
	Total	15	15	2	2	

*ASHRAE 62.1-2007 addresses additional ETS items and relative humidity issues with respect to mechanical cooling devices. It also expands the calculations to include variations in the mix of outdoor air and recirculated air in the supplies. Also, in LEED 2009, the ASHRAE 62.1-2007 addenda may be used by project teams but only if consistently used throughout all the LEED credits.

†Based on the California Department of Health Services Standard Practice for the Testing of Volatile Organic Emissions from Various Sources Using Small-Scale Environmental Chambers, including 2004 Addenda.

‡LEED 2009 EP point is based on meeting at a minimum two of the following for 90% or more of regularly occupied spaces:

1. "Have multiple lines of sight to vision glazing in different directions at least 90° apart."
2. "Include views of at least two of the following: vegetation, human activity, objects >70 feet from the exterior of the glazing."
3. "Have access to unobstructed views located within the distance of three times the head height of the vision glazing."
4. "Have access to views with a view factor of three or greater." (Per the Heschong Mahone Group Study, Windows & Offices p. 47)

Table 6.0.0 IEQ Summary Table of Major Differences between Versions 2.2 and 2009 and Associated Points (Pts) (*Continued*)

The prerequisites and credits in the Indoor Environmental Quality category can be grouped into the following two overall sets of goals:

- The two prerequisites and the first five credit subcategories promote good indoor air quality (IAQ). IEQp1, IEQc1, and IEQc2 are all related to indoor air exchange with outside air, a corrective measure for potentially poor indoor quality. IEQp1 sets a minimum air exchange rate. IEQc1 gives alternatives for providing adequate air exchanges by monitoring air flow rates and by utilizing carbon dioxide concentrations as the marker compound. IEQc2 promotes an air exchange rate that is higher than the minimum established in IEQp1. On the other hand, IEQp2, IEQc3.1, IEQc3.2, IEQc4.1, IEQc4.2, IEQc4.3, IEQc4.4, and IEQc5 all represent measures for preventing potential air pollutants from entering into the indoor airspaces.
- The last three credit subcategories promote the well-being and comfort of the occupants with respect to thermal comfort, lighting, and views. IEQc7.1, IEQc7.2, and IEQc6.2 are all related to thermal comfort, with IEQc7.1 related to design, IEQc7.2 related to verification of performance, and IEQc6.2 giving preferred thermal comfort control flexibility to individual or groups of occupants. IEQc6.1 and IEQc8.1 pertain to lighting. IEQc6.1 gives preferred lighting control flexibility to individual or group occupants, and IEQc8.1 addresses increasing natural lighting (daylighting) to supplement the other indoor lighting. IEQc8.2 addresses a minimum amount of views for occupants.

The two prerequisites and the 15 credits in the Indoor Environmental Quality category for both versions 2.2 and 2009 are as follows:

- **IEQ Prerequisite 1** (IEQp1): *Minimum IAQ Performance*
- **IEQ Prerequisite 2** (IEQp2): *Environmental Tobacco Smoke Control*
- **IEQ Credit 1** (IEQc1): *Outdoor Air Delivery Monitoring* (previously referred to as *Carbon Dioxide (CO_2) Monitoring* in LEED-NC 2.1) (EB)
- **IEQ Credit 2** (IEQc2): *Increased Ventilation* (previously referred to as *Ventilation Effectiveness* in LEED-NC 2.1)
- **IEQ Credit 3.1** (IEQc3.1): *Construction IAQ Management Plan—During Construction*
- **IEQ Credit 3.2** (IEQc3.2): *Construction IAQ Management Plan—Before Occupancy* (previously referred to as *Construction IAQ Management Plan—After Construction/Before Occupancy* in LEED-NC 2.1)
- **IEQ Credit 4.1** (IEQc4.1): *Low-Emitting Materials—Adhesives and Sealants*
- **IEQ Credit 4.2** (IEQc4.2): *Low-Emitting Materials—Paints and Coatings*
- **IEQ Credit 4.3** (IEQc4.3): *Low-Emitting Materials—Flooring Systems* (previously referred to as *Low-Emitting Materials—Carpet Systems* in LEED-NC 2.2 and *Low-Emitting Materials—Carpet* in LEED-NC 2.1)
- **IEQ Credit 4.4** (IEQc4.4): *Low-Emitting Materials—Composite Wood and Agrifiber* (previously referred to as *Low-Emitting Materials—Composite Wood* in LEED-NC 2.1)
- **IEQ Credit 5** (IEQc5): *Indoor Chemical and Pollutant Source Control*
- **IEQ Credit 6.1** (IEQc6.1): *Controllability of Systems—Lighting* (previously referred to as *Controllability of Systems—Perimeter Spaces* in LEED-NC 2.1)

- **IEQ Credit 6.2** (IEQc6.2): *Controllability of Systems—Thermal Comfort* (previously referred to as *Controllability of Systems—Non-Perimeter Spaces* in LEED-NC 2.1) (EB)
- **IEQ Credit 7.1** (IEQc7.1): *Thermal Comfort—Design* (previously referred to as *Thermal Comfort—Compliance with ASHARE 55-1992* in LEED-NC 2.1) (EB)
- **IEQ Credit 7.2** (IEQc7.2): *Thermal Comfort—Verification* (previously referred to as *Thermal Comfort—Permanent Monitoring System* in LEED-NC 2.1) (EB)
- **IEQ Credit 8.1** (IEQc8.1): *Daylighting and Views—Daylighting* (EB)
- **IEQ Credit 8.2** (IEQc8.2): *Daylighting and Views—Views* (EB)

LEED-NC 2.2 and 2009 provide tables that summarize some of these prerequisite and credit characteristics, including the main phase of the project for submittal document preparation (Design or Construction). Even though the entire LEED process encompasses all phases of a project, the prerequisites or credits that are noted as Construction Submittal credits may have major actions going on through the construction phase which, unless documented through that time, may not be verifiable from preconstruction documents or the built project. Sometimes these credits have activities associated with them that cannot or need not be performed until or subsequent to the construction phase. All the credits in both the Construction IAQ Management Plan and the Low-Emitting Materials subcategories have been noted in the construction submittal column, because the credits of the former cover construction phase activities and the credits of the latter pertain to materials mentioned that are frequently chosen and used during the construction phase. This designation is noted in Table 6.0.1.

Six of the credit subcategories listed in Table 6.0.1 have the icon EB noted after their title (Outdoor Air Delivery Monitoring, Controllability of System—Thermal Comfort, Thermal Comfort—Design, Thermal Comfort—Verification, Daylighting and Views—Daylight and Daylighting and Views—Views). This icon is a tool to help those who wish to proceed with the continuing LEED-EB certification (Existing Building) after certification of LEED-NC (New Construction or Major Renovation) is obtained. Those credits noted with this icon are usually more cost-effective to implement during the construction of the building than later during its operation. They are also shown in Table 6.0.1. There are many prerequisites or credits which might have variations applicable to the design or implementation of the project related to regional conditions in the United States. All the prerequisites and credits for which regional variations might be applicable are noted in the LEED 2009 Reference Guide and are so listed in Table 6.0.1 for the Indoor Environmental Quality category.

There are EP points under the Innovation and Design category available that relate to several items in other categories. These are available for exceeding each of the respective credit criteria to a minimum level as noted in the respective credit descriptions. EP points are only available for the Daylighting and Views subcategory of the EQ credits. This is noted in Table 6.0.0. Note that only a maximum total of four EP points is available in total for a project in LEED 2.2 and a maximum of three in LEED 2009 and they may be from any of the noted EP options in any of the SS, WE, EA, MR, or IEQ LEED categories (see Chap. 7).

There are many important parameters used to attain and verify the credits. Some are important for multiple credits, such as the determination of the site boundary, which was previously discussed as being an important parameter in many of the credits in the other categories, but is not really directly applicable to any of the IEQ credits. One

Prerequisite or Credit	Construction Submittal	EB Icon	Regional Variations May Be Applicable in LEED 2009
IEQp1: Minimum IAQ Performance			No
IEQp2: Environmental Tobacco Smoke (ETS) Control			Yes
IEQc1: Outdoor Air Delivery Monitoring		Yes	Yes
IEQc2: Increased Ventilation		No	Yes
IEQc3.1: Construction IAQ Management Plan, During Construction	*	No	No
IEQc3.2: Construction IAQ Management Plan, Before Occupancy	*	No	Yes
IEQc4.1: Low-Emitting Materials: Adhesives and Sealants	*	No	No
IEQc4.2: Low-Emitting Materials: Paints and Coatings	*	No	No
IEQc4.3: Low-Emitting Materials: Flooring Systems	*	No	No
IEQc4.4: Low-Emitting Materials: Composite Wood and Agrifiber	*	No	No
IEQc5: Indoor Chemical and Pollutant Source Control		No	Yes
IEQc6.1: Controllability of Systems: Lighting		No	Yes
IEQc6.2: Controllability of Systems: Thermal Comfort		Yes	Yes
IEQc7.1: Thermal Comfort: Design		Yes	Yes
IEQc7.2: Thermal Comfort: Verification		Yes	Yes
IEQc8.1: Daylighting and Views: Daylight		Yes	Yes
IEQc8.2: Daylighting and Views: Views		Yes	Yes

*Construction phase submittal.

Table 6.0.1 IEQ EB Icon, Regional Variation Applicability, and Miscellaneous LEED 2009

important parameter in the Indoor Environmental Quality category is the location of the breathing zone. The measurements are taken within the breathing zone to test for the indoor air quality (IAQ) parameters as specified for IEQc1 if any of the options for carbon dioxide (CO_2) monitoring are used, IEQc2 for the air quality testing option, and IEQc3.2 for the air quality testing option and for the calculations for both IEQp1 and IEQc2. It is important to design a building for improved IAQ in this volume.

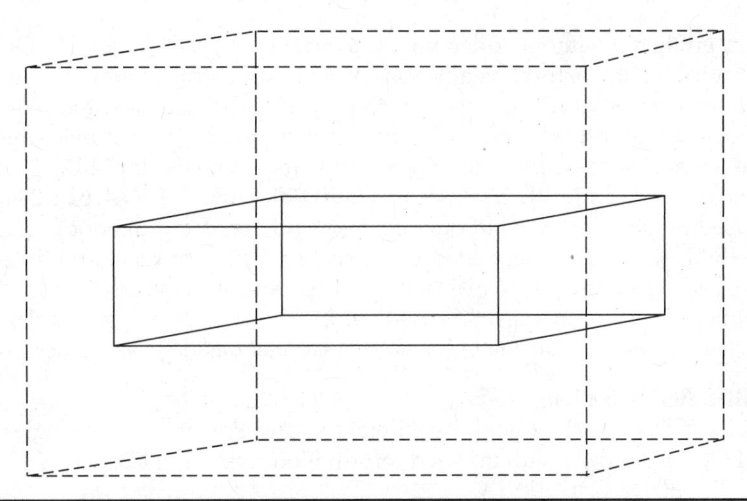

FIGURE 6.0.1 Typical rectangular breathing zone for air quality testing in a room.

The breathing zone is an area defined between two imaginary planes that are located 3 ft and 6 ft above the floor and bounded vertically by planes that are approximately 2 ft off from any wall or fixed air unit. (See Fig. 6.0.1 for a zone in a room approximately 10 ft high.) The measurements are taken within the breathing zone to test for the air quality parameters as specified. However, it may be difficult to locate monitors in this "floating" space; therefore LEED 2009 gives some clarification regarding the location of the sampling devices for credits requiring air quality measurements in the breathing zone in that they are located within the horizontal planes, not necessarily the vertical planes (i.e., they are located between 3 and 6 ft above the floor). The breathing zone represents the volumes that are typically important for good air quality; therefore designing the building such that improved air quality is focused in these volumes is important for a green building. There is no special information in LEED-NC 2.2 for spaces with exceptionally high ceilings, but these should be carefully addressed by the designers.

In summary, most of the Indoor Environmental Quality credits address problems associated with health, comfort, and well-being in the indoor space. Although several of the credits in the Indoor Environmental Quality category might include a capital cost investment, many provide a means to potentially reduce the associated costs over the life of the building, which is done by increasing occupant productivity by maintaining an environment that is more healthful and more comfortable for the occupants. The two prerequisite subcategories *must* be adhered to and verified for each project.

Indoor Environmental Quality Prerequisites

IEQ Prerequisite 1: *Minimum IAQ Performance*

This first prerequisite quantifies a minimum exchange of indoor air with outdoor air to prevent buildup of contaminants in the indoor air. This includes contamination due to moisture such as in residential bathrooms and kitchens and in shower or pool facilities. In LEED 2009, the minimum is set by the recommendations of the American Society of Heating, Refrigerating and Air-Conditioning Engineers in ASHRAE 62.1-2007 (with errata but without addenda) *Ventilation for Acceptable Indoor Air Quality,* Sections 4 through 7, with natural ventilation systems specifically adhering to the minimum requirements as set forth in paragraph 5.1. LEED NC 2.2 used the 2004 version of this same standard. Mixed-mode ventilation, where there are both mechanical and natural systems, adhere to the mechanical ventilation requirements in LEED 2.2, but use Section 6 of the standard for LEED 2009. Also, in LEED 2009, the ASHRAE 62.1-2007 addenda may be used by project teams but only if consistently used throughout all the LEED credits. ASHRAE 62 has been separated into two parts: 62.1 for most buildings and then 62.2 which is for low-rise residential facilities. Therefore, in some cases for low-rise residential facilities, this second part, ASHRAE 62.2-2007 or 2004 as applicable, *Ventilation and Acceptable Indoor Air Quality in Low-Rise Residential Buildings,* should be used instead.

USGBC Rating System

LEED-NC 2.2 lists the Intent, Requirements, and Potential Technologies and Strategies for EQp1 as follows, with updated information from LEED 2009:

Intent

Establish minimum indoor air quality (IAQ) performance to enhance indoor air quality in buildings, thus contributing to the comfort and well-being of the occupants.

Requirements

For mechanically ventilated spaces (Case 1), meet the minimum requirements of Sections 4 through 7 of ASHRAE 62.1-2007 (with errata but without addenda), Ventilation for Acceptable Indoor Air Quality. Mechanical ventilation systems shall be designed using the Ventilation Rate Procedure or the applicable local code, whichever is more stringent. Naturally ventilated buildings (Case 2) shall comply with ASHRAE 62.1-2007 (with errata but without addenda), paragraph 5.1. (Also, in LEED 2009, the ASHRAE 62.1-2007 addenda may be used by project teams but only if consistently used throughout all the LEED credits.)

Potential Technologies and Strategies

Design ventilation systems to meet or exceed the minimum outdoor air ventilation rates as described in the ASHRAE standard. Balance the impacts of ventilation rates on energy use and indoor air quality to optimize for energy efficiency and occupant health. Use the ASHRAE 62 Users Manual for detailed guidance on meeting the referenced requirements.

Calculations and Considerations (LEED 2009)

All ventilation should be designed to prevent the uptake of contaminants into the system. In addition, intakes should be located far from items such as cooling towers, sanitary vents, and vehicular exhaust sources (for example, parking garages, loading docks, and street traffic). Based on LEED 2.2, usually a minimum 25-ft setback is required with a 40-ft setback preferred. ASHRAE 62.1-2007 also requires separation of environmental tobacco smoke (ETS) spaces from non-ETS spaces and clarifies how mechanical cooling systems should be analyzed to limit indoor dampness.

Mechanical Ventilation The ventilation rate procedure is used for the mechanical ventilation areas. This procedure includes both people-related and space-related zone source pollutants. Examples of a person-related air pollutant are dander and cold germs, while a space-related air pollutant may be formaldehyde emitted from a carpet system. The zone is the area of the breathing zone in each interior space *i*. There are two correction factors included in the calculations: one is for the zone distribution effectiveness (EZ_i), and the other is the overall system ventilation effectiveness (EV). The calculations are usually done by experienced HVAC professionals, but the gist can be approximately summarized by the following definitions and sets of equations. In LEED 2009, additional variables, assumptions, and corrections are included for multiple zone recirculating systems which have a mix of outdoor and recirculated return air being supplied to more than one zone.

$OAIZ_i$ Outdoor air intake to zone i (cfm)

BZA_i Breathing zone horizontal area for zone i (ft^2)

BZAR Breathing zone minimum area rate (cfm/ft^2). (It is usually 0.06 cfm/ft^2 as given in the standard for both versions)

$BZOD_i$ Breathing zone occupant density for zone i (people/ft^2)

$BZPR_j$ Breathing zone minimum people rate for use j (cfm/person) as given in the standard. (For example, this is 5 cfm/person for general office spaces, conference rooms, and meeting spaces, and it is 7.5 cfm/person for lecture rooms and classrooms per LEED 2.2.)

EV Overall system ventilation effectiveness

EZ_i Zone distribution effectiveness. (This is typically 1.0 for overhead distribution systems and 1.2 for underfloor distribution systems as given in LEED 2.2.)

For all zones i with mechanical ventilation systems, the standard requires that

$$OAIZ_i \geq BZA_i[(BZOD_i)(BZPR_j) + (BZAR)]/(EZ_i)(EV) \qquad (6.0.1)$$

Natural Ventilation For naturally ventilated areas, there are two main requirements for both versions 2.2 and 2009, with additional scenarios available for LEED 2009:

- There must be an operable window or opening (which is permanently open to this space) within 25 ft of any occupiable area in a room. Note that a roof opening is also acceptable.
- The total operable vertical areas of the openings in the room must be at least 4 percent of the net occupiable horizontal area in that space or room.
- LEED 2009 also addresses interior spaces without direct openings to the outdoors which may be adequately ventilated by the following: ". . . can be ventilated through adjoining rooms if the openings between rooms are unobstructed and at least 8 percent or 25 ft^2 of the area is free."
- LEED 2009 also allows for engineered natural ventilation systems if compliance can be shown through acceptable engineering calculations or simulation.

For the requisite submittals when natural ventilation alone is chosen, floor plans must be submitted showing the minimum 25-ft access to operable windows and both the operable vertical areas of each of the openings and the horizontal occupiable areas of the spaces for the standard option. Summaries of these areas showing compliance should also be included. Figure 6.0.2 shows an example layout of operable windows that does not comply with the prerequisite.

Mixed Mode Ventilation LEED 2009 requires that mixed mode ventilation meet the minimum ventilation rates in ASHRAE 62.1-2007 Section 6, regardless of the ventilation mode based on any acceptable engineering calculation method.

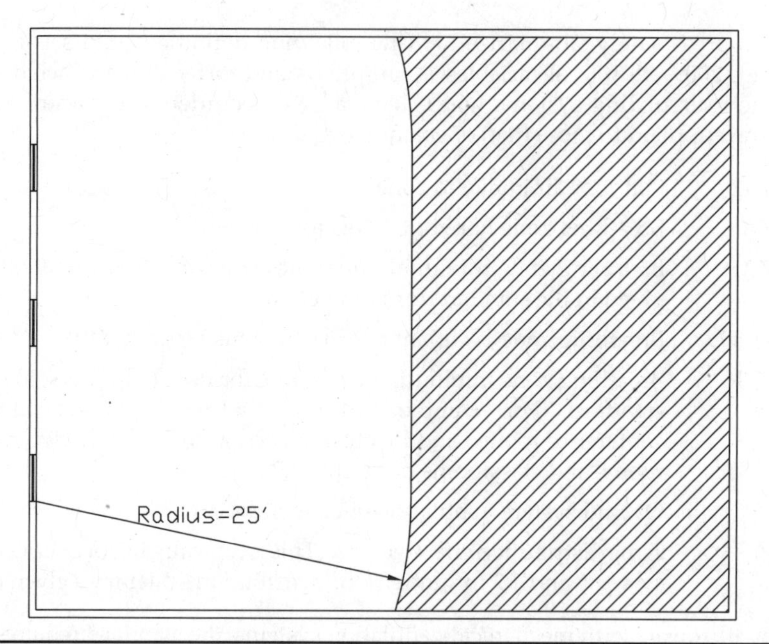

FIGURE 6.0.2 This does not meet IEQ Prerequisite 1 for natural ventilation. Hatched area is not within 25 ft of operable windows.

IEQ Prerequisite 2: *Environmental Tobacco Smoke (ETS) Control*

The purpose of this prerequisite is to prevent secondhand smoke from going into areas other than designated or desired smoking areas of buildings.

USGBC Rating System

LEED-NC 2.2 lists the Intent, Requirements, and Potential Technologies and Strategies for IEQp2 as follows, with modifications as noted to update to LEED 2009:

Intent

Prevent or minimize exposure of building occupants, indoor surfaces, and ventilation air distribution systems to Environmental Tobacco Smoke (ETS).

Requirements

Case 1. All Projects (As per LEED 2009)

OPTION 1

- Prohibit smoking in the building.
- Locate any exterior designated smoking areas at least 25 feet away from entries, outdoor air intakes and operable windows. (LEED 2009 requires signage to allow smoking and/or prohibit smoking in designated areas, or prohibit smoking totally.)

OR

OPTION 2

- Prohibit smoking in the building except in designated smoking areas.
- Locate any exterior designated smoking areas at least 25 feet away from entries, outdoor air intakes and operable windows. (LEED 2009 requires signage to allow smoking and/or prohibit smoking in designated areas, or prohibit smoking totally.)

- Locate designated smoking rooms to effectively contain, capture and remove ETS from the building. At a minimum, the smoking room must be directly exhausted to the outdoors (and as noted in LEED 2009 away from air intakes and building entry paths) with no re-circulation of ETS-containing air to the non-smoking area of the building, and enclosed with impermeable deck-to-deck partitions. With the doors to the smoking room closed, operate exhaust sufficient to create a negative pressure with respect to the adjacent spaces of at least an average of 5 Pa (0.02 inches of water gauge) and with a minimum of 1 Pa (0.004 inches of water gauge).
- Performance of the smoking room differential air pressures shall be verified by conducting 15 minutes of measurement, with a minimum of one measurement every 10 seconds, of the differential pressure in the smoking room with respect to each adjacent area and in each adjacent vertical chase with the doors to the smoking room closed. The testing will be conducted with each space configured for worst case conditions of transport of air from the smoking rooms to adjacent spaces with the smoking rooms' doors closed to the adjacent spaces.

Case 2. Residential and Hospitality Projects Only (As per LEED 2009)

(Listed in LEED 2.2 as OPTION 3-For residential buildings only)

- Prohibit smoking in all common areas of the building.
- Locate any exterior designated smoking areas (including balconies as per LEED 2009) at least 25 feet away from entries, outdoor air intakes and operable windows opening to common areas. (LEED 2009 requires signage to allow smoking and/or prohibit smoking in designated areas, or prohibit smoking totally.)
- Minimize uncontrolled pathways for ETS transfer between individual residential units by sealing penetrations in walls, ceilings and floors in the residential units, and by sealing vertical chases adjacent to the units.
- All doors in the residential units leading to common hallways shall be weather-stripped to minimize air leakage into the hallway. If the common hallways are pressurized with respect to the residential units then doors in the residential units leading to the common hallways need not be weather-stripped provided that the positive differential pressure is demonstrated as in (Case 1) Option 2, considering the residential unit as the smoking room.
- Acceptable sealing of residential units shall be demonstrated by a blower door test conducted in accordance with ANSI/ASTM-E779-03, Standard Test Method for Determining Air Leakage Rate By Fan Pressurization, AND use the progressive sampling methodology defined in Chapter 4 (Compliance Through Quality Construction) of the Residential Manual for Compliance with California's 2001 Energy Efficiency Standards (http://www.energy.ca.gov/title24/2001standards/residential_manual/index.html). Residential units must demonstrate less than 1.25 square inches leakage area per 100 square feet of enclosure area (i.e., sum of all wall, ceiling and floor areas).
- LEED 2009 also requires that all exterior doors and operable windows in the residential units are weather stripped to minimize leakage from outdoors.

Potential Technologies and Strategies

Prohibit smoking in commercial buildings or effectively control the ventilation air in smoking rooms. For residential buildings, prohibit smoking in common areas, design building envelope and systems to minimize ETS transfer among dwelling units.

Calculations and Considerations

Either no smoking is allowed in a building, or if there are smoking areas, such as designated areas in a nonresidential facility, or residential apartments where the occupants may wish to smoke, then these areas are adequately segregated by a combination of physical barriers and negative pressure so that the contaminants from tobacco smoke do not enter the other areas. There is also the potential for secondhand smoke to enter a facility if smoking is allowed near openings to the outdoors and air intakes. Therefore, if outdoor smoking areas are allowed on a site, they must be located away from these potential avenues for indoor contamination. The requirements listed previously in the USGBC LEED-NC 2.2 and 2009 rating system sections give very good explanations of what is required. Project teams should also make sure they abide by local and state smoking bans. A useful summary of the various options and requirements can be found in Fig. 6.0.3. Note the following unit conversions:

$$1 \text{ atm} \sim 101{,}300 \text{ Pa}$$

$$1 \text{ in of water} \sim 249 \text{ Pa}$$

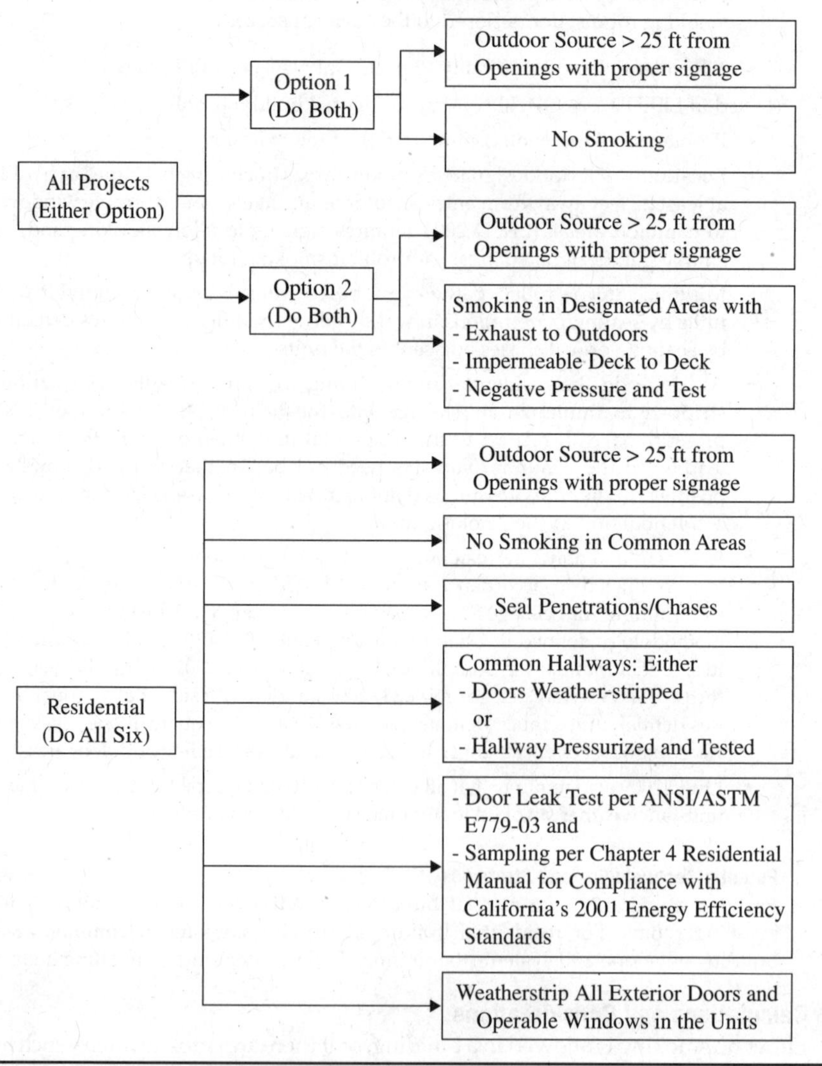

FIGURE 6.0.3 Summary of ETS requirements: all the listed items for each option must be followed for LEED 2009.

6.1 IEQ Credit 1: Outdoor Air Delivery Monitoring

The intent of this credit is to verify the presence of adequate air exchanges with the outside. It is assumed that if there is sufficient air exchange with the outside, then potential contaminants will not build up in the indoor airspace. This credit changed significantly from the requirements in LEED-NC 2.1, with the LEED-NC 2.2 revision, but has not been revised significantly by the LEED 2009 version. LEED 2009 does provide guidance for buildings with limited airflow monitoring capabilities. IEQc1 is worth one point in both LEED 2.2 and LEED 2009.

USGBC Rating System

LEED-NC-2.2 lists the Intent, Requirements, and Potential Technologies and Strategies for IEAQc1 as follows, with additional LEED 2009 revisions included:

Intent

Provide capacity for ventilation system monitoring to help sustain occupant comfort and well-being.

Requirements

Install permanent monitoring systems that provide feedback on ventilation system performance to ensure that ventilation systems maintain design minimum ventilation requirements. Configure all monitoring equipment to generate an alarm when the conditions vary by 10% or more from setpoint (for the carbon dioxide monitors), via either a building automation system alarm to the building operator or via a visual or audible alert to the building occupants.

AND

FOR MECHANICALLY VENTILATED SPACES (Case 1 in LEED 2009)

- Monitor carbon dioxide concentrations within all densely occupied spaces (those with a design occupant density greater than or equal to 25 people per 1000 sq. ft.). CO_2 monitoring locations shall be between 3 feet and 6 feet above the floor.
- For each mechanical ventilation system for which 20% or more of the design airflow is serving non-densely occupied spaces, provide a direct outdoor airflow measurement device capable of measuring the minimum outdoor airflow rate with an accuracy of plus or minus 15% of the design minimum outdoor air rate, as defined by ASHRAE 62.1-2007 (with errata but without addenda). (Also, in LEED 2009, the ASHRAE 62.1-2007 addenda may be used by project teams but only if consistently used throughout all the LEED credits.)

FOR NATURALLY VENTILATED SPACES (Case 2 in LEED 2009)
Monitor CO_2 concentrations within all naturally ventilated spaces. CO_2 monitoring shall be located within the room between 3 feet and 6 feet above the floor. One CO_2 sensor may be used to represent multiple spaces if the natural ventilation design uses passive stack(s) or other means to induce airflow through those spaces equally and simultaneously without intervention by building occupants.

Potential Technologies and Strategies

Install carbon dioxide and airflow measurement equipment and feed the information to the HVAC system and/or Building Automation System (BAS) to trigger corrective action, if applicable. If such automatic controls are not feasible with the building systems, use the measurement equipment to trigger alarms that inform building operators or occupants of a possible deficiency in outdoor air delivery.

Calculations and Considerations (LEED 2009)

The verification of adequate air exchange in spaces where mechanical ventilation systems are used is done in two ways, depending on whether the space will be densely or sparsely occupied. The reason for this segregation is that sparsely occupied areas do not usually need much variation in the airflow, and verification can be with a simple air rate check. Areas that are densely occupied at times will need a much higher air exchange rate during periods of high occupancy than during other times. Many of these occupied spaces are highly occupied for only a few hours during the day, and it would lead to energy inefficiencies to require a higher rate during the times of low occupation. An example might be a conference room or a classroom that may be packed at certain times and either vacant or sparsely used the rest of the time. The definition of *densely occupied* is a maximum occupancy of 25 people per 1000 ft^2 (40 ft^2 per person).

Mechanically Ventilated, Non-Densely Occupied Spaces For areas of the building that are never densely occupied and are mechanically ventilated, the verification is actually a flow measurement of the rate of outdoor airflow in. The rate should not drop more than 15 percent lower than the design rate. For some multiple small spaces, the measuring device can be in the combined return air duct. Note that the rate requirements are based on the air exchange designs of either IEQp1 or IEQc2 if that is sought, whichever has higher air exchange rates. Therefore, for projects not seeking IEQc2, the alarm should be activated when the outdoor air rate into indoor zone i is less than 85 percent of the design; i.e., the alarm should be activated if

$$\text{OAIZ}_i \leq 0.85 \times \text{BZA}_i[(\text{BZOD}_i)(\text{BZPR}_j) + (\text{BZAR})]/(\text{EZ}_i)(\text{EV}) \qquad \text{(not seeking IEQc2)} \quad (6.1.1)$$

or

$$\text{OAIZ}_i \leq 0.85 \times 1.3 \times \text{BZA}_i\,[(\text{BZOD}_i)(\text{BZPR}_j) + (\text{BZAR})]/(\text{EZ}_i)(\text{EV}) \qquad \text{(seeking IEQc2)} \quad (6.1.2)$$

Mechanically Ventilated, Densely Occupied Spaces and Naturally Ventilated Spaces For mechanically ventilated areas that may be crowded with human occupants at times and for all naturally ventilated areas, a special marker compound is used that can be indicative of the number of people in the room and their activity level. This marker compound is carbon dioxide (CO_2). To better understand the significance of the required air exchange control with respect to the levels of CO_2, one should become familiar with the impact that humans have on the concentration of CO_2 in a space.

As we breathe, we intake air laden with oxygen, use a lot of the oxygen, and exhale air that has less oxygen and increased levels of CO_2. If there are more people in a room, there is more oxygen used. Likewise if the people in the room are exercising or performing other high-energy activities, they will need more oxygen for the increased energy levels, and more CO_2 is released.

The uptake of oxygen by a human is related to the metabolic rate of the person. Metabolism is the process of burning energy by living creatures. When an energy source (food) is burned in the human body, oxygen is used to convert the carbon-based compounds in the food to CO_2 and water with a subsequent release of energy. It is important to note that each person burns calories at different rates. The metabolic rate at which a person burns calories while at rest is referred to as 1 *Met*, and it is unique to each person. It may be unique to each person at different stages in his or her life, and as a person becomes more active, the Mets are increased. A metabolic rate of two to four times the resting rate is considered to be low to moderate activity. For very intense activity, the increase may be up to approximately 12 Mets, but usually vigorous exercise results in approximately 6 Mets. So how do we relate this to the CO_2 levels in a room?

When we breathe outside air, we inhale air with oxygen at about 21 percent (volumetric) and CO_2 at around 380 ppm (0.038 percent volumetric). Outdoor levels do vary regionally due to other anthropogenic and natural activities that produce CO_2, and these can vary from 300 to more than 500 ppm, but on average the concentration was approximately 385 ppm in 2007. This average level has been steadily rising over the last century.

When we exhale air, there is a little less oxygen (and more water) and around 4.5 percent (volumetric) CO_2, which is then rapidly diluted by the air around it. (Note that when referring to the concentration of a vapor-phase substance, *ppm* stands for molar parts per million or the number of molecules of the substance with respect to 1 million molecules of the air mixture. Usually ppm is not a weight-based measure in the gaseous phase. At typical temperatures and pressures that humans occupy, the gaseous phase acts approximately like an ideal gas, and molar parts per million can be assumed to be equivalent to volumetric parts per million. Therefore, the concentration ppm as used in this text for vapor-phase concentrations is always molar based.) When we are outdoors, the space is so large that the contribution from our breathing does not noticeably increase the ambient CO_2 concentration; but when we are indoors, the concentration of the CO_2 is elevated in the room since we are adding CO_2 to this set volume. Typically, you can estimate that sedentary activities in a typical space with proper ventilation will *raise* the CO_2 concentrations by about 300 to 600 ppm depending on the human density of the room. Better estimates can be made by doing an air concentration model of the indoor space, but the approximate amounts just stated are a good rule-of-thumb initial estimate for typical indoor CO_2 level increases.

CO_2 is used as a marker compound for several reasons. First, an increase in CO_2 levels usually means a decrease in oxygen levels due to consumption by the occupants. Lower oxygen levels tend to make us more lethargic. Second, if there are higher CO_2 levels, then higher levels of many other air pollutants from the sources of the CO_2 (humans) may be reasonably expected to be in the air. These other pollutants include bacteria and other pathogens that humans cough or breathe out. In addition, more activity tends to increase the potential for settled air pollutants to reentrain into the airspace and for other pollutants to get knocked off of the sources (people) carrying them and into the air. Having monitors that detect higher levels of CO_2 will allow for either automatic or manual increases in the air exchange rate, so that oxygen levels increase and other pollutant levels decrease if the air exchange is made with a less polluted air source. An automatic system is referred to as a *demand-controlled ventilation* (DCV) system. Usually the outdoor air is less polluted and contains more oxygen, but air exchanges through filters can also be designed if there are concerns with the outside air. Usually, at a minimum, mechanical ventilation systems have some sort of a particulate matter filter.

As mentioned previously, one important parameter in the Indoor Environmental Quality category is the location of the breathing zone. This is used for both IEQc1 and IEQc2 for the air quality testing options pertaining to the location of the marker compound monitors. A schematic of a breathing zone as designated for LEED air quality purposes can be seen in Fig. 6.0.1. Here the typical breathing zone is a 3-ft-high rectangle located 3 ft off the floor, and 2 ft from any wall or air handling unit. However, as previously noted, it may be difficult to locate monitors in this "floating" space; therefore LEED 2009 gives some clarification regarding the location of the sampling devices for credits requiring air quality measurements in the breathing zone in that they are located within the horizontal planes, not necessarily the vertical planes (i.e., they are located between 3 and 6 ft above the floor).

Special Circumstances and Exemplary Performance

There is no EP point related to IEQc1 in either the 2.2 or the 2009 version of LEED.

6.2 IEQ Credit 2: *Increased Ventilation*

The intention of this credit is to improve air quality by providing air exchange rates that are higher than the minimums established in the prerequisite. However, increased air exchange rates with outdoor rates may have an impact on energy use, so additional strategies such as countercurrent flows for preheating or cooling may be needed to improve both air quality and energy efficiency. Ventilation strategies should be carefully evaluated in conjunction with energy efficiencies and impacts to maintain an optimum balance between these two goals. IEQc2 is worth one point in both LEED 2.2 and LEED 2009.

USGBC Rating System

LEED-NC 2.2 lists the Intent, Requirements, and Potential Technologies and Strategies for IEQc2 as follows, with modifications to reflect the changes in LEED 2009:

Intent

Provide additional outdoor air ventilation to improve indoor air quality for improved occupant comfort, well-being and productivity.

Requirements

FOR MECHANICALLY VENTILATED SPACES (Case 1 in LEED 2009)

- Increase breathing zone outdoor air ventilation rates to all occupied spaces by at least 30% above the minimum rates required by ASHRAE Standard 62.1-2007, with errata but without addenda (formerly the 2004 version for LEED 2.2) as determined by IEQ Prerequisite 1. (Also, in LEED 2009, the ASHRAE 62.1-2007 addenda may be used by project teams but only if consistently used throughout all the LEED credits.)

FOR NATURALLY VENTILATED SPACES (Case 2 in LEED 2009)
(The following requirement in LEED 2.2 to 'Design natural ventilation systems for occupied spaces to meet the recommendations set forth in the Carbon Trust "Good Practice Guide 237" [1999]' has been deleted in LEED 2009 per addenda dated 12/2/09.) Determine that natural ventilation is an effective strategy for the project by following the flow diagram process shown in Figure 2.8 of the Chartered Institution of Building Services Engineers (CIBSE) Applications Manual 10: 2005, Natural ventilation in non-domestic buildings.

AND

- (Case 2 Option 1 in LEED 2009) Show that the design of the natural ventilation systems meets the recommendations set forth in the CIBSE Applications Manual 10: 2005, Natural ventilation in non-domestic buildings (Path 1) or CIBSE Applications Manual 13:2000, Mixed Mode Ventilation (Path 2), whichever is applicable to the project space.

OR

- (Case 2 Option 2 in LEED 2009) Use a macroscopic, multi-zone, analytic model to predict that room-by-room airflows will effectively naturally ventilate, defined as providing the minimum ventilation rates required by ASHRAE 62.1-2007, with errata but without addenda (formerly the 2004 version for LEED 2.2) Chapter 6, for at least 90% of occupied spaces. (Also, in LEED 2009, the ASHRAE 62.1-2007 addenda may be used by project teams but only if consistently used throughout all the LEED credits.)

Potential Technologies and Strategies

For mechanically ventilated spaces: Use heat recovery, where appropriate, to minimize the additional energy consumption associated with higher ventilation rates. Use public domain software such as NIST's CONTAM, Multizone Modeling Software, along with LoopDA, Natural Ventilation Sizing Tool, to analytically predict room-by-room airflows.

Calculations and Considerations (LEED 2009)

Mechanical Ventilation The required ventilation rate increase of 30 percent for mechanically ventilated spaces (and mixed-mode areas) can be determined by either a modeling method called the *ventilation rate procedure* or a testing method called the *indoor air quality procedure*. These can both be found in the referenced ASHRAE Standard 62.1-2007, with errata but without addenda, Section 6. However, the ventilation rate procedure is the most commonly used and is the recommended method for IEQp1. Alternatively, for low-rise residential buildings ASHRAE 62.2-2007 may be used, as previously mentioned in IEQp1. Therefore, for all the mechanically ventilated zones *i*, the requirements can be summarized by the expanded version of Eq. (6.1.1) as given in Eq. (6.2.1).

$$\mathrm{OAIZ}_i \geq 1.3 \times \mathrm{BZA}_i[(\mathrm{BZOD}_i)(\mathrm{BZPR}_j) + (\mathrm{BZAR})]/(\mathrm{EZ}_i)(\mathrm{EV}) \qquad (6.2.1)$$

An important parameter in the Indoor Environmental Quality category is the location of the breathing zone. This is used for IEQc1 for the air quality testing option and was shown previously in Fig. 6.0.1. The option for mechanically ventilated spaces requires that the 30 percent increase in air exchanges be in the breathing zones.

Natural Ventilation For all naturally ventilated spaces it is necessary to have *both* a special design and a model calculation verification as in the following:

- **Design**
  - Determine if effective per the flow diagram in Fig. 2.8 of the Chartered Institution of Building Services Engineers (CIBSE) Applications Manual 10: *Natural Ventilation in Non-Domestic Buildings*, 2005 (Path 1) or CIBSE Applications Manual 13:2000, Mixed Mode Ventilation (Path 2), whichever is applicable to the project space.
- **Calculations (use one of the following methods)**
  - CIBSE Applications Manual 10 (2005) or CIBSE Applications Manual 13 (2000), whichever is applicable to the project space
  - Macroscopic, multizone analytic model of room-by-room airspaces for >90 percent of occupied spaces per ASHRAE 62.1-2007, with errata but without addenda, Chapter 6. (Also, in LEED 2009, the ASHRAE 62.1-2007 addenda may be used by project teams but only if consistently used throughout all the LEED credits.)

The requirements for naturally ventilated spaces are summarized in Fig. 6.2.1.

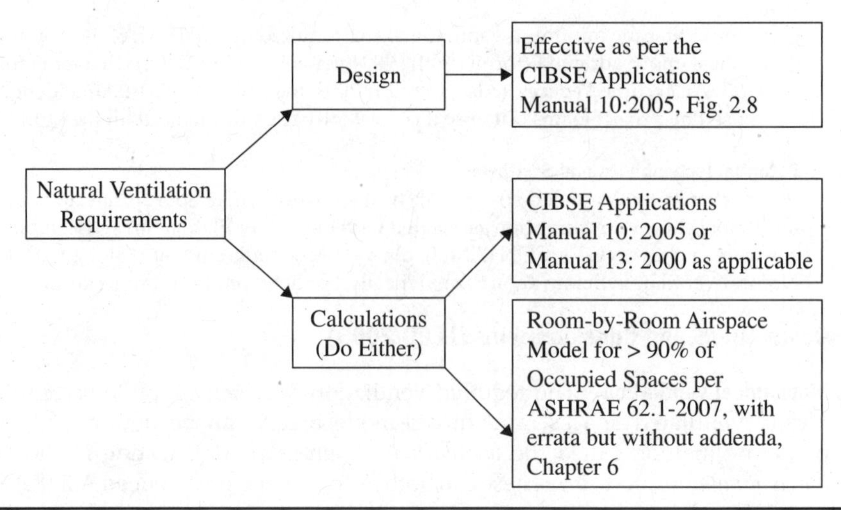

FIGURE 6.2.1 Requirements for increased ventilation in naturally ventilated spaces LEED 2009.

Mixed Mode Ventilation LEED 2009 requires for IEQp1 that mixed mode ventilation meet the minimum ventilation rates in ASHRAE 62.1-2007 Section 6, regardless of the ventilation mode based on any acceptable engineering calculation method, although not specifically restated in IEQc2. LEED 2009 also has the addenda requiring that mixed mode ventilation follow the CIBSE Applications Manual 13:2000, Mixed Mode Ventilation as listed under the natural ventilation requirements.

Special Circumstances and Exemplary Performance (LEED 2009)

At this time, improved indoor air quality is primarily sought by exchanging indoor air with air of better quality from the outside. However, this may not be the best solution in all cases. There are two main concerns that frequently occur. The first is the possibility of poor ambient air quality in some areas during different seasons, and the second is the increased energy use required to condition the additional outside airflow. Therefore, there are also other methods for improved indoor air quality that are being researched and recommended. Many are specific to a certain class of indoor air pollutants such as particulate matter or VOCs and may be important for some applications. These might serve as a special circumstance or an innovative design credit at this time, although future versions of the LEED rating system might add options for special cases. One of the most commonly used indoor air quality mechanisms is filtration, which decreases particulate matter, a pollutant that may carry other contaminants such as microbes or allergens. Improved filtration is partially addressed in the IEQ credit subcategories 3 and 5. Some other methods are mentioned in the Discussion and Overview section of this chapter.

There is no EP point associated with IEQ credit 2 in either the 2.2 or the 2009 version of LEED.

6.3 IEQ Credit Subcategory 3: *Construction IAQ Management Plan*

IEQ Credit 3.1: *Construction IAQ Management Plan—during Construction*

Many activities during construction can have an impact on the indoor air quality of portions of buildings that are occupied during the construction phase and on the indoor air quality of the entire building after construction. To reduce this, there are a series of recommended steps to be taken during construction that will keep these impacts to a minimum. The steps range from keeping construction areas physically segregated from built areas as much as possible, to ensuring that building materials do not get contaminated with moisture or chemicals prior to or after installation. IEQc3.1 is worth one point in both LEED 2.2 and LEED 2009.

USGBC Rating System

LEED-NC 2.2 lists the Intent, Requirements, and Potential Technologies and Strategies for IEQc3.1 as follows, with updated references as per LEED 2009:

Intent

Reduce indoor air quality problems resulting from the construction/renovation process in order to help sustain the comfort and well-being of construction workers and building occupants.

Requirements

Develop and implement an Indoor Air Quality (IAQ) Management Plan for the construction and pre-occupancy phases of the building as follows:

- During construction meet or exceed the recommended Control Measures of the Sheet Metal and Air Conditioning National Contractors Association (SMACNA) IAQ Guidelines for Occupied Buildings under Construction, 2nd edition, Nov. 2007, Chapter 3.
- Protect stored on-site or installed absorptive materials from moisture damage.
- If permanently installed air handlers are used during construction, filtration media with a Minimum Efficiency Reporting Value (MERV) of 8 shall be used at each return air grille, as determined by ASHRAE 52.2-1999 with errata but without addenda. Replace all filtration media immediately prior to occupancy.

Potential Technologies and Strategies

Adopt an IAQ management plan to protect the HVAC system during construction, control pollutant sources and interrupt contamination pathways. Sequence the installation of materials to avoid contamination of absorptive materials such as insulation, carpeting, ceiling tile and gypsum wallboard. Coordinate with Indoor Environmental Quality Credits 3.2 and 5 to determine the appropriate specifications and schedules for filtration media. If possible, avoid using permanently installed air handlers for temporary heating/cooling during construction. Consult the LEED-NC v2.2 and v2009 Reference Guides for more detailed information on how to ensure the well-being of construction workers and building occupants if permanently installed air handlers must be used during construction.

Calculations and Considerations (LEED 2009)

Adherence to this credit is highly dependent on the construction phase activities, and this should be planned for prior to construction. Scheduling and sequencing are critical, as is management of many daily construction activities. An IAQ management plan must be devised and adhered to that incorporates the following three main requirements of this credit: the Sheet Metal and Air Conditioning National Contractors Association

(SMACNA) *IAQ Guidelines for Occupied Buildings under Construction*, 2nd edition Nov. 2007, Chap. 3, must be adhered to; absorptive materials must be protected from moisture; and permanent air handlers must have filtration media installed during construction and replaced prior to occupancy. The SMACNA guidelines are further subdivided into five main categories representing the five SMACNA principles. The SMACNA principles shall be met even if the facility is not occupied during construction as applicable.

SMACNA The five SMACNA principles are

- Protection of permanent HVAC equipment
- Air pollutant source control
- Air pollutant pathway interruption
- Housekeeping to prevent air pollution
- Scheduling to prevent contamination

Protection of permanent HVAC equipment can best be achieved by *not* using permanent HVAC equipment during the construction phase, whenever possible. Temporary heating sources may be a good alternative. In addition, all openings in the system should be sealed, if possible, during construction; and if it is not possible, then filters should be installed and replaced prior to occupancy. Unducted plenums represent other potential sources for contamination into the HVAC system that cannot be easily controlled at point sources. Special consideration should be given to these areas and work scheduled so that the plenums are isolated as quickly as possible with ceilings. Another benefit of protecting HVAC equipment is for continued energy performance. Studies have revealed that fouled HVAC equipment can significantly impact the efficiency of the units.

One additional recommendation for protection of permanent HVAC equipment is not to store construction materials or other potential air pollutant sources in mechanical rooms. These preventive procedures will aid in minimizing particulate matter (PM) contamination of the equipment and ductways. This leads to suggestions for other air pollutant source controls.

The best way to ensure air pollutant source control is to minimize many sources of air pollutants such as toxic finish materials or high-VOC (volatile organic compound or carbon) materials used on the site. However, there are still some materials that probably will be used that might cause a problem if not handled properly. Housekeeping chemicals and used chemical containers should be stored and disposed of responsibly. Idling vehicles and other sources of emissions such as petroleum-fueled construction tools should be reduced if possible and located in areas remote from enclosed areas or accessways to enclosed areas such as HVAC intakes. This leads to suggestions for preventing on-site air pollutants from entering constructed areas.

Air pollutant pathway interruptions can be made by isolating the areas under construction from clean or occupied areas by physical barriers (permanent or temporary) or ventilation and depressurization schemes.

Housekeeping is arguably the most important way to prevent air pollutant contaminants from affecting the facility. Frequent cleaning, vacuuming, and dust prevention measures, such as wetting agents for exposed soils, can prevent much of the PM pollutants from entering the facility and affecting the construction phase workers. Housekeeping also includes properly storing, staging, and protecting absorbent materials so that they do not become moist or absorb unwanted contaminants. Proper housekeeping can be facilitated by scheduling these activities into the construction phase.

Scheduling is a constant theme in the SMACNA principles. Scheduling housekeeping activities into the daily operations of the construction phase is important. Sequencing installation of permanent or temporary barriers between finished and unfinished areas can help in pathway interruption. Scheduling the use of high-VOC or toxic materials or air pollutant-emitting equipment prior to installation of absorbent materials or away from enclosed areas can keep the sources of air pollutants down, as can scheduling protection of the HVAC systems and installation of good filtering media.

Moisture Protection In addition to the SMACNA principles, this credit emphasizes that absorptive material such as sheetrock shall be protected from moisture damage while on-site at all times, from storage through staging, installation, and after installation. Moisture damage can result in other air pollutant problems such as mold after occupancy.

Filtration Media Finally, this credit requires that any air handlers used during construction have filtering media on each return grill, and this medium shall be replaced prior to occupancy. The filtering media used for construction phase IAQ management should meet minimum standards for filtering of particulate matter. In earlier versions of LEED, those with a *minimum efficiency rating value* (MERV) rating of 13 were recommended, but this credit is now based on a MERV rating of 8 and the higher rating of 13 is part of IEQc5. If the project seeks to receive credit for IEQc5, then all the HVAC equipment should be sized to accept filters with the MERV rating of 13.

These filters are used to remove particulate matter from the air. The MERV ratings are based on a combination of removal efficiencies for several different size ranges of particulate matter and a maximum pressure drop. Particulate matter is made up of many different materials and can be either liquid or solid. Most of the larger particles are solids such as dust and soot. Particulate matter comes in all shapes and sizes, and they are classified by listing the size of a particle by the diameter of a sphere of the same volume. The USEPA regulates ambient (outside) concentrations of PM in two different size ranges. The two ranges comprise all particulate matter less than or equal to 10 micrometers (10 μm) in equivalent diameter and all particulate matter less than or equal to 2.5 μm in equivalent diameter. These are commonly referred to as PM10 and PM2.5. Also PM2.5 is sometimes referred to as *fines*. Particulate matter in the air larger than 10 μm in diameter is usually entrained particles that will rapidly settle out and can easily be filtered by our nostrils. Since particulate matter is made up of many substances, both organic and inorganic, it may have a wide range of densities. Usually an average specific gravity is estimated as 2, which is in between the value for many liquids and organics which are closer to 1, and the value for minerals, which is closer to 3. Metals usually have a specific gravity greater than 3.

For the MERV ratings for HVAC equipment filtering relating to indoor air, particulate matter is subdivided into three size ranges: 3 to 10 μm, 1 to 3 μm, and 0.3 to 1 μm (large, medium, and small). For higher MERV values, the removal efficiency will be greater. The MERV ratings also address the allowed pressure drop across a filter. Some typical MERV ratings and removal efficiencies are listed in Table 6.3.1. These were taken from ASHRAE 52.2. Note that there is a minimum size given of 0.3 μm. Usually particles that are smaller than this have so little mass that they are unstable as a solid. [Typical molecules range in size from about 74 picometers (pm) for the smaller hydrogen molecule to 50 angstroms (Å) for large molecules. This converts to about 0.000074 to 0.005 μm. Many of the smaller molecules exist

Group Number	MERV Rating	Average Removal Efficiency (%) 0.3 to 1 µm	Average Removal Efficiency (%) 1 to 3 µm	Average Removal Efficiency (%) 3 to 10 µm	Minimum Final Resistance (in. water gage)
1	1–4	—	—	<20	0.3
2	5	—	—	20–34.9	0.6
	6	—	—	35–49.9	
	7	—	—	50–69.9	
	8	—	—	70–84.9	
3	9	—	<50	≥85	1.0
	10	—	50–64.9	≥85	
	11	—	65–79.7	≥85	
	12	—	80–89.9	≥90	
4	13	<75	≥90	≥90	1.4
	14	75–84.9	≥90	≥90	
	15	85–94.9	≥90	≥90	
	16	≥95	≥95	≥95	

Source: American Society of Heating, Refrigerating and Air-Conditioning Engineers, Inc., *www.ashrae.org*.

Table 6.3.1 MERV Ratings from ASHRAE 52.2-1999: *Method of Testing General Ventilation Air-Cleaning Devices for Removal Efficiency by Particle Size.*

as gases, while most of the larger molecules are liquids or solids at room temperature. There are some larger polymers and supermolecules, but they are not a common ambient air pollution concern. For the sake of a physical comparison, note that a typical textbook may have a page thickness of about 70 µm.]

Special Circumstances and Exemplary Performance

There is no EP point for this credit in either LEED version 2.2 or 2009.

IEQ Credit 3.2: *Construction IAQ Management Plan—before Occupancy*

IEQ credit 3.1 lists a series of measures that can be taken during the construction phase to reduce the potential for air pollution from construction affecting both the workers and the occupants after construction. Then IEQc3.2 lists a series of additional measures which can be taken just before occupancy to further ensure good air quality. IEQc3.2 is worth one point in both LEED 2.2 and LEED 2009.

USGBC Rating System

LEED-NC 2.2 lists the Intent, Requirements, and Potential Technologies and Strategies for IEQc3.2 as follows, with additions noted for LEED 2009 requirements:

Intent

Reduce indoor air quality problems resulting from the construction/renovation process in order to help sustain the comfort and well-being of construction workers and building occupants.

Requirements

Develop and implement an Indoor Air Quality (IAQ) Management Plan for the pre-occupancy phase as follows (LEED 2009 adds that it is implemented after all finishes have been installed and the building has been completely cleaned):

OPTION 1—Flush-Out

- (Option 1: Path 1 for LEED 2009) After construction ends, prior to occupancy and with all interior finishes installed, install new filtration media and perform a building flush-out by supplying a total air volume of 14,000 cu. ft. of outdoor air per sq. ft. of floor area while maintaining an internal temperature of at least 60 degrees F and relative humidity no higher than 60%.

OR

- (Option 1: Path 2 for LEED 2009) If occupancy is desired prior to completion of the flush-out, the space may be occupied following delivery of a minimum of 3,500 cu. ft. of outdoor air per sq. ft. of floor area to the space. Once a space is occupied, it shall be ventilated at a minimum rate of 0.30 cfm/sq. ft. of outside air or the design minimum outside air rate determined in IEQ Prerequisite 1, whichever is greater. During each day of the flush-out period, ventilation shall begin a minimum of three hours prior to occupancy and continue during occupancy. These conditions shall be maintained until a total of 14,000 cu. ft./sq. ft. of outside air has been delivered to the space.

OR

OPTION 2—Air Testing

- Conduct baseline IAQ testing, after construction ends and prior to occupancy, using testing protocols consistent with the United States Environmental Protection Agency Compendium of Methods for the Determination of Air Pollutants in Indoor Air and as additionally detailed in the 2009 Reference Guide.
- Demonstrate that the contaminant maximum concentrations listed 'below' are not exceeded.

CONTAMINANT	MAXIMUM CONCENTRATION
Formaldehyde	27 parts per billion (was 50 ppb in LEED 2.2)
Particulates (PM10)	50 micrograms per cubic meter
Total Volatile Organic Compounds (TVOC)	500 micrograms per cubic meter
4-Phenylcyclohexene (4-PCH)*	6.5 micrograms per cubic meter
Carbon Monoxide (CO)	9 parts per million and no greater than 2 parts per million above outdoor levels

*This test is only required if carpets and fabrics with styrene butadiene rubber (SBR) latex backing material are installed as part of the base building systems.

- For each sampling point where the maximum concentration limits are exceeded conduct additional flush-out with outside air and retest the specific parameter(s) exceeded to indicate the requirements are achieved. Repeat procedure until all requirements have been met. When retesting non-complying building areas, take samples from the same locations as in the first test.
- The air sample testing shall be conducted as follows:
  1) All measurements shall be conducted prior to occupancy, but during normal occupied hours, and with the building ventilation system starting at the normal

daily start time and operated at the minimum outside air flow rate for the occupied mode throughout the duration of the air testing.

2) The building shall have all interior finishes installed, including but not limited to millwork, doors, paint, carpet and acoustic tiles. Non-fixed furnishings such as workstations and partitions are encouraged, but not required, to be in place for the testing.
3) The number of sampling locations will vary depending upon the size of the building and number of ventilation systems. For each portion of the building served by a separate ventilation system, the number of sampling points shall not be less than one per 25,000 sq. ft., or for each contiguous floor area, whichever is larger, and include areas with the least ventilation and greatest presumed source strength.
4) Air samples shall be collected between 3 feet and 6 feet from the floor to represent the breathing zone of occupants, and over a minimum 4-hour period.

Potential Technologies and Strategies

Prior to occupancy, perform a building flush-out or test the air contaminant levels in the building. The flush-out is often used where occupancy is not required immediately upon substantial completion of construction. IAQ testing can minimize schedule impacts but may be more costly. Coordinate with Indoor Environmental Quality Credits 3.1 and 5 to determine the appropriate specifications and schedules for filtration media.

Calculations and Considerations (LEED 2009)

The intention of this credit is to ensure that a building has good air quality after construction. The end of construction is defined as the completion of the stage when all finishes are applied to the base building components and as noted in LEED 2009, after all punch-list items are complete. Furniture and furnishings are not a part of the base building. Also, this credit does not apply to the core and shell portion of LEED until the interior finishes are complete.

There are two main methods of compliance: either there is a substantial flush-out of the building and HVAC systems in either of two prescribed manners, or the team proves that several marker air pollutants are below a certain minimum level after some flush-out. Since it may be costly or unrealistic to delay occupancy after construction is complete for many projects, the second flush-out option allows for the building to be occupied earlier. Also, if there are separate HVAC systems in distinct areas of the building, areas which are completed can be occupied prior to completion of all the areas, if the compliant areas are also isolated and protected by the SMACNA principles as outlined in IEQc3.1. In LEED 2009, it is required that the filter media are replaced prior to flush-out, the filters should be replaced following any flush-out and the filter media should meet the minimum requirements of the design or MERV 13 if IEQc5 is also being sought. There are standards set for the temperature and relative humidity (RH) of the outside air used for flush-out to prevent excessive moisture in the building and also to promote evaporation of the applicable volatile compounds. Volatility increases rapidly with temperature for many organic compounds that are intended to be flushed out of the building. Therefore, to facilitate a rapid flush-out, the temperature of the air used in the flush-out should not be too low.

Prescribed Flush-Out Options The first flush-out option requires a substantial amount of air and may take a few weeks. The second flush-out option requires that a portion (at least 25 percent) of the flush-out be completed prior to occupancy and then can continue at a minimum rate during occupancy until complete.

Air Quality Option The air quality option allows for occupancy as soon as it can be proved that the air quality adheres to a minimum standard. Usually improved air quality is attained by flushing out the spaces. Attaining these minimum standards may require several more days of flush-out if areas are found not in compliance. There are five groups of compounds for this air quality option: formaldehyde, particulate matter of 10-μm diameter or less (PM10), total volatile organic compounds (TVOC), carbon monoxide (CO), and 4-phenylcyclohexene (4-PCH). (The 4-PCH needs only be measured if there are carpets or fabrics installed as part of the base building with styrene butadiene rubber latex backing material.) The maximum allowed concentrations as measured per the prescribed methods are listed in the Requirements section.

Formaldehyde has the chemical formula H_2CO and goes by many other common names, such as *methyl aldehyde*. It is one of the most commonly used chemicals for manufacturing and is commonly used as a preservative for many everyday goods. In buildings, it is emitted from many pressed wood building materials, such as particleboard, as an off-gas from the resins used. Formaldehyde is also one of the simplest organic compounds and is a by-product of many chemical reactions including combustion, such as in cigarette smoking or fuel burning or even from reactions within our own bodies. Therefore, it is found throughout the spaces we occupy at low levels. Formaldehyde is a good compound to test, because its presence at higher levels may mean that either there is still substantial off-gassing of many construction materials and/or there are unacceptable levels of combustion emissions entering the facility. A level of 0.1 ppm is considered to be the level above which there may be eye, throat, or lung irritation. Common levels in homes are usually well below this threshold for human irritation. However, homes with certain insulations or a lot of pressed wood products, and poor ventilation, may reach levels of up to 0.3 ppm. The U.S. Department of Labor Occupational Safety and Health Administration (OSHA) has set a permissible exposure limit (PEL) for formaldehyde of 0.75 ppm [time-weighted average (TWA) in an 8-h workday] and a short-term exposure limit (STEL) of 2 ppm for 15 min as per 29 CFR 1910.1048. The U.S. Department of Housing and Urban Development (HUD) has established a maximum level of 0.4 ppm for mobile homes. The standard in this credit is for formaldehyde to be less than 27 parts per billion (ppb) or, equivalently, 0.027 ppm (it was 50 ppb in LEED 2.2).

TVOC is also commonly referred to as VOC and stands for total volatile organic compounds (or carbons). It consists of many hundreds of organic compounds that are volatilized. Organic compounds are defined as compounds which contain carbon excluding the inorganic carbon compounds such as carbon monoxide, carbon dioxide, carbonic acid, metallic carbides, and carbonates. The term *VOCs* usually includes all the volatilized organic compounds except for a few exclusions due to low reactivity of these specific chemicals. Emissions are regulated for VOCs by the USEPA because many VOCs participate in atmospheric photochemical reactions, such as the smog reaction which causes ozone to develop in the troposphere during the day. However, there are no ambient air quality standards for VOCs as a total group. Some of the VOCs are considered to be carcinogenic and may have individual concentrations set, such as formaldehyde and benzene. Many VOCs are natural, such as limonene emitted by citrus fruits and many emitted by pine trees or cut grass, which give those distinctive smells to the air. VOCs can be by-products of combustion and represent much of the evaporative portion of paints, glues, and other surface-applied products. Since it is not reasonable to evaluate each and every VOC, they are regulated as a group. In addition, their additive impact may be greater than the individual compound concentrations, so minimizing the concentration of VOCs as a group is also helpful to health and property.

Since the VOCs contain a range of many compounds of various molecular weights, the limits given are in the units of mass per volume instead of air molar or volumetric based ppm or ppb (parts per million or billion).

PM10 has been described in detail earlier in IEQc3.1. Particulate matter contains a range of many compounds, and these are in either solid or liquid form. Since particulate matter is not separate gaseous compounds, the limits given are in the units of mass per volume of air instead of ppm. Unacceptable levels of PM10 may mean that there are still many leftover sources from the construction phase in the building or HVAC system and/or that there are pathways from other sources into the occupied areas.

4-Phenylcyclohexene (4-PCH) is applicable only if certain carpet and fabric backing binders are used. Its odor is commonly referred to as "new carpet smell." Each molecule of 4-PCH is made up of 12 carbon atoms and 14 hydrogen atoms.

Carbon monoxide is a commonly known air pollutant that, in high enough quantities, can cause death from asphyxiation. It is a product of incomplete combustion from any carbonaceous fuel source. There are two indoor limits given. The first is a maximum in any case, and the second is a limit for the value above the outdoor levels. Many buildings are located in urban areas where motor vehicle emissions and other emissions from combustion can give an elevated outdoor CO level. In all cases, the maximum value inside should not be more than 2 ppm above the outside. As per the EPA, the national ambient air quality standard (NAAQS) for CO is 9 ppm 8-h nonoverlapping average not to be exceeded more than once per year. (There are some areas in California, Montana, Nevada, Oregon, and Texas which are currently in nonattainment of the standard.)

As outlined in the LEED-NC 2.2 and 2009 Reference Guides, the air quality testing should be performed in the breathing zone (see Fig. 6.0.1) during typical occupancy hours and with the minimum air rates that will occur during occupancy, in addition to several other detailed prescriptive requirements. In this way, the tests will measure what is expected to be the worst background conditions related to the facility but not directly related to human activity in the building. If tests fail, then after corrective action, new tests need to be performed in the same locations.

Special Circumstances and Exemplary Performance

It may be difficult to fully understand the varied units and multiple requirements in this credit, and for other indoor air quality requirements for green facilities. Therefore, some additional information about air pollution calculations and conversions is given at the end of this chapter, and a listing of the units used can be found in App. C.

Several criteria should be considered in the design and construction phase if this credit is sought:

- The flush-out flow rates should be considered, and if the permanent HVAC equipment is not expected to accommodate these flush-out rates in a timely fashion, then some additional temporary units may be needed and/or openings, such as windows, positioned to facilitate the desired level of air exchange.
- If a project is done in phases, then the flush-outs may be phased, but all the SNACNA principles must be adhered to between each phase.
- As mentioned previously, this credit is not directly applicable to core and shell projects until the interiors are complete.
- Chemicals are often used prior to occupancy to prepare and "clean" a facility. If high-VOC chemicals are used, then these may alter the results. They should be avoided.

- Areas designated for the air quality sampling should be representative of areas with the least effective ventilation (worst-case scenarios).
- It may seem difficult to meet the air quality requirements for CO and perhaps PM10 if the facility is built in an area where the outside ambient air has high concentrations, especially since the testing is usually done during occupancy hours that are usually the worst hours for ambient air concentrations. However, there are many ways in which this problem can be addressed. Some of the poor air pollutant concentrations are close to sources such as roadways in urban areas. If the intakes for the HVAC systems or natural ventilation openings are located away from the roadways and potential sources, then the air used for the facility may be adequate to meet good indoor air quality standards.

There is no EP point for IEQ credit 3.2 in either version 2.2 or 2009.

6.4 IEQ Credit Subcategory 4: *Low-Emitting Materials*

The intention of this credit subcategory is to reduce surface coverings or coatings in the interior spaces that might release VOCs into the indoor airspace. In addition to overall VOC concentrations, a few specific organic compounds that are irritable to humans at certain concentrations and some threshold levels are considered. Figure 6.4.1 shows a reception area that has been built with many surfaces that have reduced VOC emissions as compared to other older products.

Figure 6.4.1 Various examples of low VOC-emitting interior surfaces. (*Photograph taken at the Sustainable Interiors ribbon cutting at the Strom Thurmond Building on Fort Jackson, Columbia, S.C., June 2007. The paints were provided by Sherwin-Williams Company and Duron Paints & Wallcoverings. Also shown are carpets and flooring provided by Shaw Industries Group, Lees Carpets and Interface, Inc. These flooring materials may also count toward several MR credits.*)

There are four credits in this subcategory. The four credits (IEQc4.1, IEQc4.2, IEQc4.3, and IEQc4.4) cover the following potential indoor sources in the base building of volatile organic compounds:

- Adhesives and sealants
- Paints and coatings
- Flooring systems (was only carpet systems in LEED 2.2)
- Composite wood and agrifiber products

All four credits in the Low-Emitting Materials subcategory have the same intent as given in the LEED-NC 2.2 and 2009 Reference Guides. IEQc4.1, IEQc4.2, IEQc4.3, and IEQc4.4 are each worth one point in both LEED 2.2 and LEED 2009. They can be applied for independent of each other.

IEQ Credit 4.1: *Low-Emitting Materials—Adhesives and Sealants*

USGBC Rating System

LEED-NC 2.2 lists the Intent, Requirements, and Potential Technologies and Strategies for IEQc4.1 as follows, with the same requirements in LEED 2009:

Intent

Reduce the quantity of indoor air contaminants that are odorous, irritating and/or harmful to the comfort and well-being of installers and occupants.

Requirements

All adhesives and sealants used on the interior of the building (defined as inside of the weatherproofing system and applied on-site) shall comply with the requirements of the following reference standards:

- Adhesives, Sealants and Sealant Primers: South Coast Air Quality Management District (SCAQMD) Rule #1168. VOC limits are listed in Table 6.4.1 and correspond to an effective date of July 1, 2005 and rule amendment date of January 7, 2005.
- Aerosol Adhesives: Green Seal Standard for Commercial Adhesives GS-36 requirements in effect on October 19, 2000 (see Table 6.4.2).

Potential Technologies and Strategies

Specify low-VOC materials in construction documents. Ensure that VOC limits are clearly stated in each section of the specifications where adhesives and sealants are addressed. Common products to evaluate include: general construction adhesives, flooring adhesives, fire-stopping sealants, caulking, duct sealants, plumbing adhesives, and cove base adhesives.

IEQ Credit 4.2: *Low-Emitting Materials—Paints and Coatings*

USGBC Rating System

LEED-NC 2.2 lists the Intent, Requirements, and Potential Technologies and Strategies for IEQc4.2 as follows, with the same requirements in LEED 2009 except as noted:

Intent

Reduce the quantity of indoor air contaminants that are odorous, irritating and/or harmful to the comfort and well-being of installers and occupants.

ADHESIVES					
Architectural Applications	**VOC Limit***	**Specialty Applications**	**VOC Limit***	**Substrate Specific Applications**	**VOC Limit***
Indoor carpet adhesives	50	PVC welding	510	Metal to metal	30
Carpet pad adhesives	50	CPVC welding	490	Plastic foams	50
Wood flooring adhesives	100	ABS welding	325	Porous material (except wood)	50
Rubber floor adhesives	60	Plastic cement welding	250	Wood	30
Subfloor adhesives	50	Adhesive primer for plastic	550	Fiberglass	80
Ceramic tile adhesives	65	Contact adhesive	80		
VCT and asphalt adhesives	50	Special purpose contact adhesive	250		
Drywall and panel adhesives	50	Structural wood member adhesive	140		
Cove base adhesives	50	Sheet applied rubber lining operations	850		
Multipurpose construction adhesives	70	Top and trim adhesive	250		
Structural glazing adhesives	100				

Sealants	VOC Limit*	Sealant Primers	VOC Limit*
Architectural	250	Architectural nonporous	250
Nonmembrane roof	300	Architectural porous	775
Roadway	250	Other	750
Single-ply roof membrane	450		
Other	420		

*VOC limits are given in grams of VOC to liter of the adhesive not including the water component of the adhesive or other exempt components ($g_{voc}/L_{adhesive\ less\ water}$).

TABLE 6.4.1 SCAQMD VOC Limits for Adhesives, Sealants, and Sealant Primers

Adhesive*	% VOC by Weight (Less Water)
General purpose mist spray	65
General purpose web spray	55
Special purpose aerosol adhesives (all types)	70

*Aerosol adhesives: Green Seal Standard for Commercial Adhesives GS-36 requirements in effect on October 19, 2000.

TABLE 6.4.2 Green Seal VOC Limits for Aerosol Adhesives

Requirements

Paints and coatings used on the interior of the building (defined as inside of the weatherproofing system and applied on-site) shall comply with the following criteria:

- Architectural paints and coatings applied to interior walls and ceilings: Do not exceed the VOC content limits established in Green Seal Standard GS-11, Paints, First Edition, May 20, 1993.
    - Flats: 50 g/L (grams of VOCs per liter of product minus water)
    - Non-Flats: 150 g/L (grams of VOCs per liter of product minus water)
    - (Primers: Primers must meet the VOC limit for non-flat paints was part of LEED 2.2, but in LEED 2009 follows the SCAQMD Rule 1113, which varies and is less stringent)
- Anti-corrosive and anti-rust paints applied to interior ferrous metal substrates: Do not exceed the VOC content limit of 250 g/L (grams of VOCs per liter of product minus water) established in Green Seal Standard GC-03, Anti-Corrosive Paints, Second Edition, January 7, 1997.
- Clear wood finishes, floor coatings, stains, and shellacs applied to interior elements (Primers now also included under this standard as per LEED 2009): Do not exceed the VOC content limits established in South Coast Air Quality Management District (SCAQMD) Rule 1113, Architectural Coatings, rules in effect on January 1, 2004. The units are in grams of VOCs per liter of product minus water. The following list of SCAQMD VOC limits are examples. Refer to the standards for complete details.
    - Clear wood finishes: varnish 350 g/L; lacquer 550 g/L
    - Floor coatings: 100 g/L
    - Sealers: waterproofing sealers 250 g/L; sanding sealers 275 g/L; all other sealers 200 g/L
    - Shellacs: Clear 730 g/L; pigmented 550 g/L
    - Stains: 250 g/L

Potential Technologies and Strategies

Specify low-VOC paints and coatings in construction documents. Ensure that VOC limits are clearly stated in each section of the specifications where paints and coatings are addressed. Track the VOC content of all interior paints and coatings during construction.

IEQ Credit 4.3: *Low-Emitting Materials—Flooring Systems (Previously Carpet Systems in LEED 2.2)*

USGBC Rating System

LEED-NC 2.2 lists the Intent, Requirements, and Potential Technologies and Strategies for IEQc4.3 as follows for the carpet systems, which are also applicable in LEED 2009 as an Option 1:

Intent

Reduce the quantity of indoor air contaminants that are odorous, irritating and/or harmful to the comfort and well-being of installers and occupants.

Requirements

Option 1 in LEED 2009 (for carpet systems) All carpet installed in the building interior shall meet the testing and product requirements of the Carpet and Rug Institute's Green Label Plus program.

All carpet cushion installed in the building interior shall meet the requirements of the Carpet and Rug Institute Green Label program.

All carpet adhesive shall meet the requirements of IEQ Credit 4.1: VOC limit of 50 g/L (the units are in grams of VOCs per liter of product minus water).

(Potential Technologies and Strategies for Carpet Systems)

Clearly specify requirements for product testing and/or certification in the construction documents. Select products either that are certified under the Green Label Plus program or for which testing has been done by qualified independent laboratories in accordance with the appropriate requirements. The Green Label Plus program for carpets and its associated VOC emission criteria in micrograms per square meter per hour, along with information on testing method and sample collection developed by the Carpet and Rug Institute (CRI) in coordination with California's Sustainable Building Task Force and the California Department of Health Services (DHS), are described in Section 9, Acceptable Emissions Testing for Carpet, DHS Standard Practice CA/DHS/EHLB/R-174, dated 07/15/04. This document is available at http://www.cal-iaq.org/VOC/Section01350_7_15_2004_FINAL_PLUS_ADDENDUM-2004-01.pdf. [It is also published as Section 01350 Section 9 (dated 2004) by the Collaborative for High Performance Schools (www.chps.net).]

LEED 2009 Option 1 Flooring Systems and Additional Option 2

LEED 2009 Option 1 has the additional requirements for non-carpet flooring systems:

- All hard surface flooring must be certified as compliant with the FloorScore standard (current as of the date of the LEED 2009 rating system or any more stringent version) by an independent third party. Flooring products covered by FLoorScore include vinyl, linoleum, laminate flooring, wood flooring, ceramic flooring, rubber flooring and wall base.
- An alternative compliance path using FloorScore is acceptable for credit achievement: 100% of the non-carpet finished flooring must be FloorScore certified and must constitute at least 25% of the finished floor area. Examples of unfinished flooring include floors in mechanical rooms, electrical rooms and elevator service rooms.
- Concrete, wood, bamboo and cork floor finishes such as sealer, stain and finish must meet the requirements of SCAQMD Rule 1113, Architectural Coatings, rules in effect on January 1, 2004.
- Tile setting adhesives and grout must meet SCAQMD Rule 1168. VOC limits correspond to an effective date of July 1, 2005 and rule amendment date of January 7, 2005.

OPTION 2

LEED 2009 provides a second option for any of the flooring systems such that all flooring elements in the building interior meet the testing and product requirements of the California Department of Health Services Standard Practice for the Testing of Volatile Organic Emissions from Various Sources Using Small-Scale Environmental Chambers, including 2004 Addenda.

IEQ Credit 4.4: *Low-Emitting Materials—Composite Wood and Agrifiber Products*

USGBC Rating System

LEED NC 2.2 lists the Intent, Requirements, and Potential Technologies and Strategies for IEQc4.4 as follows, with additions to reflect the LEED 2009 version:

Intent

Reduce the quantity of indoor air contaminants that are odorous, irritating and/or harmful to the comfort and well-being of installers and occupants.

Requirements

Composite wood and agrifiber products used on the interior of the building (defined as inside of the weatherproofing system) shall contain no added urea-formaldehyde resins. Laminating adhesives used to fabricate on-site and shop-applied composite wood and agrifiber assemblies shall contain no added urea-formaldehyde resins. Composite wood and agrifiber products are defined as: particleboard, medium density fiberboard (MDF), plywood, wheatboard, strawboard, panel substrates and door cores. Fixtures, furniture, and equipment (FF&E) are not considered base building elements and are not included.

Potential Technologies and Strategies

Specify wood and agrifiber products that contain no added urea-formaldehyde resins. Specify laminating adhesives for field and shop applied assemblies that contain no added urea-formaldehyde resins.

Calculations and Considerations IEQ Credit Subcategory 4: *Low-Emitting Materials*

The various referenced rules and organization standards can be found at the following websites:

- The SCAQMD Rule #1168 can be found at http://www.aqmd.gov/rules/reg/reg11/r1168.pdf. The Rule 102 which gives the definition of exempt compounds and other definitions in Rule #1168 can be downloaded from http://www.aqmd.gov/rules/reg/reg01/r102.pdf. The exempt compounds are a group of mainly hydrochlorofluorocarbons (HCFCs), hydrofluorocarbons (HFCs), and chlorofluorocarbons (CFCs). They are considered to be less reactive than most other VOCs with respect to the production of tropospheric ozone, although many are now banned due to their stratospheric ozone depletion potential. (Refer to IEAp3 and IEAc4 for more information on the ozone depletion potential.)
- *Green Seal Standard, GS36, Commercial Adhesives*, October 19, 2000, can be found at http://www.greenseal.org/certification/standards/commercialadhesives.cfm.
- *Green Seal Standard, GS-11, Paints*, 1st ed., May 20, 1993, can be found at http://www.greenseal.org/certification/standards/paints.cfm.
- *Green Seal Standard, GC-03, Anti-Corrosive Paints*, 2d ed., January 7, 1997, can be found at http://www.greenseal.org/certification/standards/anti-corrosivepaints.cfm.
- The SCAQMD *Rule 1113, Architectural Coatings*, as amended beyond the cited date to June 9, 2006, can be found at http://www.aqmd.gov/rules/reg/reg11/r1113.pdf.
- Information about the Carpet and Rug Institute's (CRI's) Green Label and Green Label Plus Programs can be found at http://www.carpet-rug.org/.
- The FloorScore (TM) program is by the Resilient Floor Covering Institute (RFCI) and can be found at http://www.rfci.com/int_FloorScore.htm.

VOC	Maximum Emission Factor [mg/(m²·h)]	
	Carpet Cushion	Carpet
TVOC (total volatile organic compounds)	1.00	0.5
BHT (butylated hydroxytoluene)	0.30	—
Formaldehyde	0.05	0.05*
4-PC (4-phenylcyclohexene)	0.05	0.05
Styrene	—	0.4

*To prove that none is used.

TABLE 6.4.3 Maximum Emission Factors for VOCs from Carpet Cushions and from Carpets per CRI Green Label Website April 11, 2007

- The California Department of Health Services Standard Practice for the Testing of Volatile Organic Emissions from Various Sources Using Small-Scale Environmental Chambers, including 2004 addenda can be found at http://www.cal-iaq.org/VOC/Section01350_7_15_2004_FINAL_ADDENDUM-2004-01.pdf.

Note that the carpet in IEQc4.3 must meet the CRI Green Label Plus requirements, and the carpet cushion must meet the CRI Green Label requirements. The cushion requirement is based on a test of manufacturers' products for emissions of certain VOCs, which are updated on an annual basis. The compounds and the maximum emission factors as given on the CRI Green Label website are listed in Table 6.4.3.

The Green Label Plus requirement in IEQc4.3 is for carpets. This program is listed on the Green Seal website and is a certification that carpets go through a testing for the release of total VOCs (TVOCs) and 13 individual chemicals. These chemicals are

- Acetaldehyde
- Benzene
- Caprolactam
- 2-Ethylhexanoic acid
- Formaldehyde
- 1-Methyl-2-pyrrolidinone
- Naphthalene
- Nonanal
- Octanal
- 4-Phenylcyclohexene
- Styrene
- Toluene
- Vinyl acetate

Material	Requirement
Hard surfaces	Certified as FloorScore* compliant by a third party: • All hard surface flooring or • 100% of the noncarpet finished flooring if it constitutes at least 25% of the finished flooring.
Finishes	Concrete, wood, bamboo, and cork floor finishes (sealer, stain, and other finishes) meet SCAQMD† Rule 1113, Architectural Coatings in effect on January 1, 2004.
Adhesives and grouts for tile	Meet SCAQMD† Rule 1168 VOC limits effective July 1, 2005.

*FloorScore includes Vinyl, Linoleum, laminate, wood, ceramic rubber flooring, and wall base.
†South Coast Air Quality Management District.

TABLE 6.4.4 LEED 2009 IEQc4.3: Low-Emitting Materials Flooring Systems (Option 1 Summary for Materials Other than Carpet)

The specific emissions for each are not given on the website, but the TVOCs and the three individual ones that were previously regulated under the Green Label program are given with their maximum emission levels for the Green Label program in Table 6.4.3. A summary of non-carpet requirements for Option 1 in IEQc3 can be found in Table 6.4.4.

IEQc4.4 refers to a resin used in many products and formerly used in a foam insulation (UFFI) called urea-formaldehyde. Urea-formaldehyde is made in a chemical reaction involving urea [chemical formula CON_2H_4 or $(NH_2)_2CO$] and formaldehyde (H_2CO). Unfortunately, products made with it tend to emit formaldehyde, and their use is discouraged. The requirement states that urea-formaldehyde not be put in any material used for the base building, either made off-site or installed on-site. This requirement is not for furniture or equipment used by the occupants, which perhaps leaves room for another potential innovative credit for these types of materials.

Note again that many of the maximum concentrations given for the products used for surface treatment in this subcategory list these concentrations in milligrams of VOC per liter of the product less the water. In other words, if you purchased a gallon of the paint or adhesive or other listed type product and it contained 50 percent by volume water, then the VOC concentration used to compare to the standard would be double what the actual VOC concentration is in the can. This way, a product cannot be simply watered down to meet the criteria.

Special Circumstances and Exemplary Performance IEQ Credit Subcategory 4: *Low-Emitting Materials (LEED 2009)*

It is really important that all the requirements and material specifications needed for any of the IEQc4 subcategory credits be included in all portions of the construction documents and specifications. It is also essential that information on all the adhesives, paints, and other VOC-containing products used on the site be carefully documented for the project. This may include cut sheets or manufacturers' specifications with the appropriate certification designations, MSDS (Material Safety Data Sheets), and test reports as applicable. Originally, this may have been a large change for construction projects. However, as materials adhering to these requirements and as certified by the organizations become more common, it may be easier for the contractors to maintain the appropriate records without additional effort. If these products are used instead of more polluting products as were previously the standard for a company, then they will become more cost-effective and will also have a positive impact on the environment on all the facilities where they are used, even those not going for green certification.

Two of the credits in this subcategory (IEQc4.1 and IEQc4.2) have alternative compliance paths referred to as the VOC budget. There may be instances when there are no appropriate low-VOC options for certain products, and the team is given the opportunity to do a VOC budget on all the potential sources listed in that particular credit (IEQc4.1 or IEQc4.2) to prove that the overall emissions rate is lower than in a facility that otherwise complied. The VOC budget option allows for a better model of low-VOC product using facilities, but it is not as simple to model or comply with as the prescriptive requirements. This alternative also allows for compliance when there are some mistakes made during the construction phase, such as the inadvertent use of a paint product that is not on the approved list by a subcontractor. Actions that compensate for these mistakes will help attain good indoor air quality and are a way to attain these credits. The calculations use the same amounts of the products in a baseline case as in the design case, but based on the limits in the credits for the baseline and the actual concentrations in the products used (design case). A summation is performed over the amounts of VOCs in the actual products and also over the baseline for that credit. The credit (IEQc4.1 or IEQc4.2) is obtained if the design case total does not exceed the baseline case total.

There are no IEP points for the IEQ Credit Subcategory 4: *Low-Emitting Materials* in LEED 2.2 and LEED 2009.

6.5 IEQ Credit 5: *Indoor Chemical and Pollutant Source Control*

The intention of IEQc5 is to keep potential air pollutants from entering occupied portions of the building after occupancy. Three strategies are employed: preventing dust and grime from entering via pedestrian entrances, keeping indoor chemical sources segregated from occupied areas by physical means, and keeping dust and grime out of the HVAC system by using very high-efficiency filter media. IEQc5 is worth one point in both LEED 2.2 and LEED 2009.

USGBC Rating System

LEED-NC 2.2 lists the Intent, Requirements, and Potential Technologies and Strategies for IEQc5 as follows, with updates for LEED 2009 as noted:

Intent

Minimize exposure of building occupants to potentially hazardous particulates and chemical pollutants.

Requirements

Design to minimize and control pollutant entry into buildings and later cross-contamination of regularly occupied areas:

- Employ permanent entryway systems at least ten feet (was six feet in LEED 2.2) long in the primary direction of travel to capture dirt and particulates from entering the building at entryways that are directly connected to the outdoors and that serve as regular entry points for building users. Acceptable entryway systems include permanently installed grates, grilles, or slotted systems that allow for cleaning underneath. Roll-out mats are only acceptable when maintained on a weekly basis by a contracted service organization.

- Where hazardous gases or chemicals may be present or used (including garages, housekeeping/laundry areas and copying/printing rooms—LEED 2009 also specifically lists science laboratories, prep rooms, art rooms and shops of any kind), exhaust each space sufficiently to create negative pressure with respect to adjacent spaces with the doors to the room closed. For each of these spaces, provide self-closing doors and deck to deck partitions or a hard lid ceiling. The exhaust rate shall be at least 0.50 cfm/sq. ft., with no air recirculation. The pressure differential with the surrounding spaces shall be at least 5 Pa (0.02 inches of water gauge) on average and 1 Pa (0.004 inches of water) at a minimum when the doors to the rooms are closed.
- In mechanically ventilated buildings, provide regularly occupied areas of the building with air filtration media prior to occupancy that provides a Minimum Efficiency Reporting Value (MERV) of 13 or better. Filtration should be applied to process both return and outside air that is to be delivered as supply air.
- LEED 2009 also requires that closed containment be provided for storage of hazardous liquid wastes for appropriate off-site disposal in a regulatory compliant storage area (preferably outside) in places where water and chemical concentrate mixing occurs (e.g. housekeeping, janitorial and science laboratory areas).

Potential Technologies and Strategies

Design facility cleaning and maintenance areas with isolated exhaust systems for contaminants. Maintain physical isolation from the rest of the regularly occupied areas of the building. Install permanent architectural entryway systems such as grills or grates to prevent occupant-borne contaminants from entering the building. Install high-level filtration systems in air handling units processing both return air and outside supply air. Ensure that air handling units can accommodate required filter sizes and pressure drops.

Considerations and Exemplary Performance (LEED 2009)

Most of the prescriptive items in IEQc5 are well detailed in the requirements section already noted and there are many additional suggestions for controlling indoor pollutant sources in the LEED 2009 Reference Guide, such as not locating HVAC intake grills near outdoor sources of pollution. There is also an additional requirement listed in the LEED 2009 Reference Guide for facilities which have temporary backup power battery banks in that these battery banks must be segregated from other areas. Note also that even though there are requirements to treat copier/printing rooms as hazardous chemical areas, this excludes what are referred to as "convenience" printers and copiers, those small units typically used at one's workstation. The requirements for IEQc5 are summarized in Fig. 6.5.1. It should be added that in the design phase, care should be taken to specify HVAC equipment that filter media with an MERV rating of 13 or better can fit, and that these filters are installed and used on the equipment at all times, even prior to occupancy (see IEQc3.1 and 3.2).

There is no EP point associated with this credit for either the 2.2 or the 2009 version of LEED.

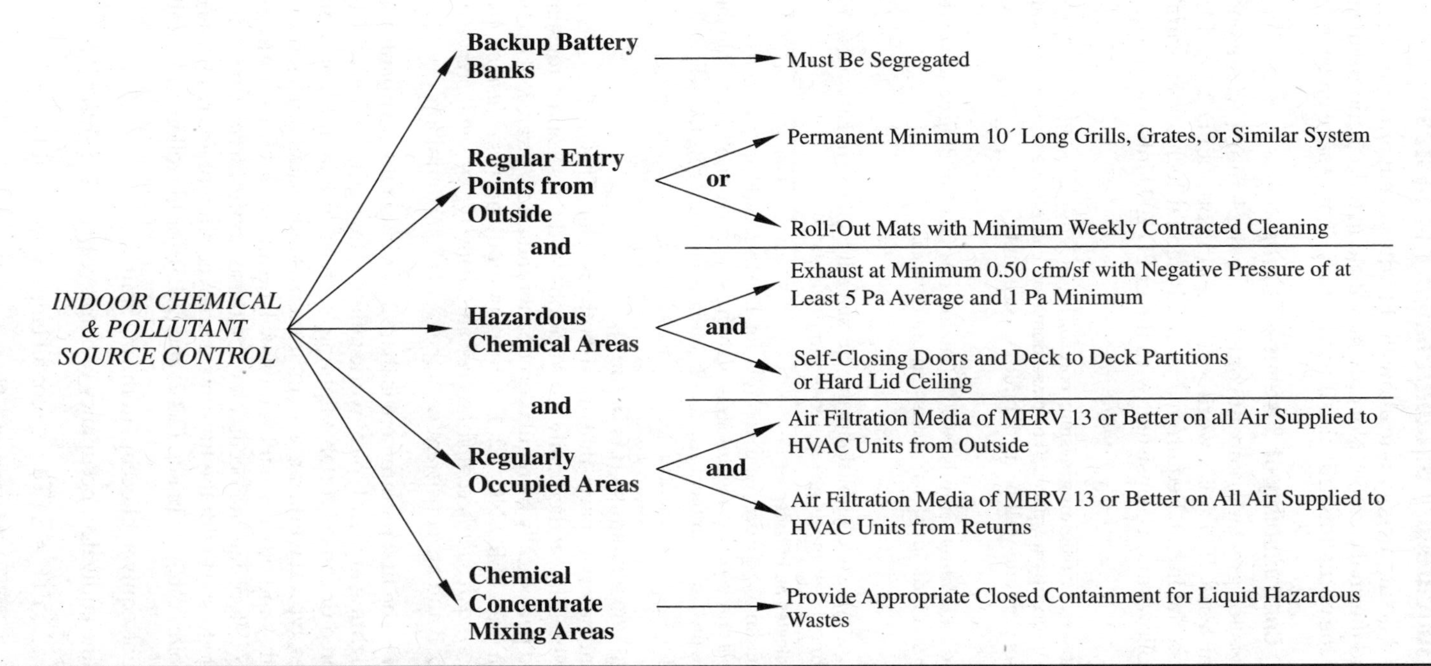

FIGURE 6.5.1 Summary of requirements for IEQ Credit 5: *Indoor Chemical and Source Pollutant Control.*

6.6 IEQ Credit Subcategory 6: *Controllability of Systems*

The intention of this subcategory is to allow the building occupants greater control over both lighting and thermal comfort. Not only may it aid in indoor environmental quality and comfort for the occupants, but also if used wisely, it may save on energy costs.

IEQ Credit 6.1: *Controllability of Systems—Lighting*

This credit allows for greater occupant control over the lighting where one works or performs regular tasks. IEQc6.1 is worth one point in both LEED 2.2 and LEED 2009. LEED-NC 2.2 lists the Intent, Requirements, and Potential Technologies and Strategies for IEQc6.1 as follows and is consistent with the LEED 2009 intent and requirements:

Intent

Provide a high level of lighting system control by individual occupants or by specific groups in multi-occupant spaces (i.e. classrooms or conference areas) to promote the productivity, comfort and well-being of building occupants.

Requirements

Provide individual lighting controls for 90% (minimum) of the building occupants to enable adjustments to suit individual task needs and preferences.

AND

Provide lighting system controllability for all shared multi-occupant spaces to enable lighting adjustment that meets group needs and preferences.

Potential Technologies and Strategies

Design the building with occupant controls for lighting. Strategies to consider include lighting controls and task lighting. Integrate lighting systems controllability into the overall lighting design, providing ambient and task lighting while managing the overall energy use of the building.

Calculations and Considerations (LEED 2009)

This requirement has two parts that must both be met. The first is for the regular occupants of the building at their daily workstations, and the second is for shared spaces that may be used periodically by regular occupants and also by transient occupants. The definitions from the LEED-NC 2.2 Reference Guide for *individual occupant spaces, nonoccupied spaces, nonregularly occupied spaces,* and *shared (group) multioccupant spaces* can be found in App. B and are consistent with the LEED 2009 definitions.

Individual Control The first requirement is that 90 percent of the occupants have control over their task lighting at their typical workstation or living space. Task lighting can be permanently wired or can be portable lighting, such as desk lamps. For commercial or institutional uses, typical workstations can be seen as individual offices, desks or tables in cubicles, booths, and other usual locations where an employee performs most of his or her work. To determine the occupancy for these nonresidential uses, it would be most appropriate to start with the full-time employee equivalents (FTEs) as defined in the Sustainable Sites category in Eqs. (2.4.2) and (2.4.3), for the following definitions:

FTE_j Full-time equivalent occupant during shift j

$FTE_{j,i}$ Full-time equivalent occupancy of employee i during shift j

$$FTE_j = \Sigma\, FTE_{j,i} \quad \text{for all employees in shift } j \tag{2.4.2}$$

$$FTE = \text{Maximum } FTE_j \tag{2.4.3}$$

For occupancies where shift employees share the same workstations as other shift employees depending on the time, and for uses where there is really only one main "shift," providing individual controls at locations equal to 90 percent of FTE should adequately meet the intent of the requirement. If there are alternative regular uses, where many occupants in separate shifts have different workstation locations, then the definition of the required number of individual lighting controls will need to be analyzed and validated differently. However, for most commercial and institutional uses, the minimum number of individual lighting control locations (ILCLs) can be calculated as

$$\text{ILCL} \geq 0.9 \times \text{FTE} \tag{6.6.1}$$

Group Control The second part of this credit is to make sure that there are lighting controls in group spaces, such as conference rooms and classrooms, so that the groups gathered in these areas have greater control over the lighting as appropriate for their group needs. If some of the lighting comes from daylighting, then glare control and room-darkening shades or features should be available too.

Special Circumstances and Exemplary Performance

The individual lighting loads resulting from implementation of the requirements of this credit should be included in the energy calculations as performed in the Energy and Atmosphere category. In many cases there will be a resultant decrease in overall lighting load, as some areas might not need the maximum lighting during all times, but only when the individual prefers the higher lighting intensities. Since individual control may also result in a decrease in energy efficiency if the controls are left on during unoccupied times, education into energy conservation should be a part of this credit. An example might be posted notices to please turn off the individually controlled lights when exiting the area, or automatic sensors that detect when the room is unoccupied. There is no EP point for this credit in both versions 2.2 and 2009.

IEQ Credit 6.2: *Controllability of Systems—Thermal Comfort*

This credit allows for increased occupant control over the thermal environment where one works or performs regular tasks. Thermal comfort is based on many environmental factors and personal factors. Four common environmental factors are air temperature, radiant temperature, humidity, and airspeed, with air temperature being the predominant factor. The personal factors include clothing and activity levels. Clothing is dependent on not only the weather or season, but also the standard attire for the occupants. Thermal comfort is addressed in this credit and in both IEQc71 and IEQc7.2. This credit (6.2) gives flexibility in the thermal control for many individuals or groups so that they can readily change one or more of the environmental factors to suit their preferences. Thermal comfort is considered to be very important, particularly for economic reasons, as comfort is a main factor in productivity and job satisfaction. IEQc6.2 is worth one point in both LEED 2.2 and LEED 2009.

USGBC Rating System

LEED-NC 2.2 lists the Intent, Requirements, and Potential Technologies and Strategies for IEQc6.2 as follows, with updates to LEED 2009 as noted:

Intent

Provide a high level of thermal comfort system control by individual occupants or by specific groups in multi-occupant spaces (i.e., classrooms or conference areas) to promote the productivity, comfort and well-being of building occupants.

Requirements

Provide individual comfort controls for 50% (minimum) of the building occupants to enable adjustments to suit individual task needs and preferences. Operable windows can be used in lieu of comfort controls for occupants of areas that are 20 feet inside of and 10 feet to either side of the operable part of the window. The areas of operable window must meet the requirements of ASHRAE 62.1-2007, with errata but without addenda (was the 2004 version for LEED 2.2) paragraph 5.1 Natural Ventilation.

AND

Provide comfort system controls for all shared multi-occupant spaces to enable adjustments to suit group needs and preferences. Conditions for thermal comfort are described in ASHRAE Standard 55-2004, with errata but without addenda, to include the primary factors of air temperature, radiant temperature, air speed and humidity. Comfort system control for the purposes of this credit is defined as the provision of control over at least one of these primary factors in the occupant's local environment. Project teams can choose to use the addenda to either ASHRAE 62.1-2007 or ASHRAE 55-2004, but only if applied consistently over all the credits applied for the LEED 2009.

Potential Technologies and Strategies

Design the building and systems with comfort controls to allow adjustments to suit individual needs or those of groups in shared spaces. ASHRAE Standard 55-2004 identifies the factors of thermal comfort and a process for developing comfort criteria for building spaces that suit the needs of the occupants involved in their daily activities. Control strategies can be developed to expand on the comfort criteria to allow adjustments to suit individual needs and preferences. These may involve system designs incorporating operable windows, hybrid systems integrating operable windows and mechanical systems, or mechanical systems alone. Individual adjustments may involve individual thermostat controls, local diffusers at floor, desk or overhead levels, or control of individual radiant panels, or other means integrated into the overall building, thermal comfort systems, and energy systems design. In addition, designers should evaluate the closely tied interactions between thermal comfort (as required by ASHRAE Standard 55-2004) and acceptable indoor air quality (as required by ASHRAE Standard 62.1-2007, whether natural or mechanical ventilation).

Calculations and Considerations

Just as in IEQc6.1, this requirement has two parts that must both be met. The first is for the regular occupants of the building at their typical workstations, and the second is for shared spaces that may be used periodically by regular occupants and also by transient occupants. The definitions from the LEED-NC 2.2 Reference Guide for *individual occupant spaces, nonoccupied spaces, nonregularly occupied spaces,* and *shared (group) multioccupant spaces* can be found in App. B and are consistent with the LEED 2009 definitions.

Individual Control The first requirement is that 50 percent of the occupants have control over their thermal comfort at their typical workstation or living space. Thermal comfort controls usually refer to some form of *conditioning* and are usually for both heating and cooling. This conditioning can be active (mechanical HVAC systems) or passive (natural ventilation). A typical type of control may be individual diffusers, or access to an operable window. (The occupant location should be no more than 10 ft sideways from the edge of an operable window and no more than 20 ft away in front of the window.

LEED 2009 also notes that the free-opening of the window must be at least 4 percent of the net floor area, which for the maximum setbacks in IEQc6.2 is a 20 ft by 20 ft net floor area.) As in IEQc6.1, for commercial or institutional uses, typical workstations can be seen as individual offices, desks, or tables in cubicles, booths, and other usual locations where an employee performs most of his or her work. To determine the occupancy for these nonresidential uses, it would be most appropriate to again start with the full-time employee equivalents as defined in the Sustainable Sites category in Eqs. (2.4.1) through (2.4.3), as were used in IEQ credit 6.1.

For occupancies where shift employees share the same workstations as other shift employees at their different work times, and for uses where there is really only one main shift, providing individual controls at locations equal to 50 percent of FTE should adequately meet the intent of the requirement. If there are alternative regular uses, where many occupants in different shifts work in alternative locations, then the definition of the required number of individual thermal controls will need to be analyzed and validated differently. However, for most commercial and institutional uses, the minimum number of individual comfort control locations (ICCLs) can be calculated as

$$\text{ICCL} \geq 0.5 \times \text{FTE} \tag{6.6.2}$$

Group Control The second part of this credit is to make sure that there are thermal comfort controls in group spaces, such as conference rooms and classrooms, so that the groups gathered in these areas have greater control as appropriate for their group needs.

Special Circumstances and Exemplary Performance Thermal comfort control as described in ASHRAE Standard 55-2004 should be balanced with the minimum indoor air quality requirements, as required per ASHRAE Standard 62.1-2007. Also, since many of the features for thermal comfort relate to windows and the HVAC systems, this credit has been labeled with an EB from the USGBC. It is more difficult to change in an existing building. There is no exemplary performance point for this credit.

6.7 IEQ Credit Subcategory 7: *Thermal Comfort*

IEQ Credit 7.1: *Thermal Comfort—Design*

As stated in IEQc6.2, thermal comfort is based on many environmental and personal factors, including air temperature, radiant temperature, humidity, airspeed, clothing, and activity levels. Thermal comfort is also addressed in IEQc6.2 and IEQc7.2. This credit, IEQc7.1, deals with designing the HVAC and other energy-related systems with added emphasis on thermal comfort. Figure 6.7.1 depicts a dessicant wheel used in HVAC systems to help reduce outside humidity levels prior to the cooling coils. Humidity is frequently the factor used for thermal comfort control. IEQc7.1 is worth one point in both LEED 2.2 and LEED 2009.

USGBC Rating System

LEED-NC 2.2 lists the Intent, Requirements, and Potential Technologies and Strategies for IEQc7.1 as follows, with updates to reflect LEED 2009:

Figure 6.7.1 Trane CDQ (Cool, Dry, Quiet) is a dessicant wheel which is placed in series with the cooling coil and can help provide dry air to a space for better humidity control. (*Photograph Courtesy Trane.*)

Intent

Provide a comfortable thermal environment that supports the productivity and well-being of building occupants.

Requirements

Design HVAC systems and the building envelope to meet the requirements of ASHRAE Standard 55-2004, Thermal Environmental Conditions for Human Occupancy with errata but without addenda. Demonstrate design compliance in accordance with the Section 6.1.1 Documentation. Project teams can choose to use the addenda to ASHRAE 55-2004, but only if applied consistently over all the credits applied for the LEED 2009.

Potential Technologies and Strategies

Establish comfort criteria per ASHRAE Standard 55-2004 that support the desired quality and occupant satisfaction with building performance. Design building envelope and systems with the capability to deliver performance to the comfort criteria under expected environmental and use conditions. Evaluate air temperature, radiant temperature, air speed, and relative humidity in an integrated fashion and coordinate these criteria with IEQ Prerequisite 1, IEQ Credit 1, and IEQ Credit 2.

Calculations and Considerations (LEED 2009)

Thermal comfort controls usually refer to some form of conditioning and are for both heating and cooling. This conditioning can be active (mechanical HVAC systems) or passive (natural ventilation) or a combination of both (mixed-mode conditioning). Levels of comfort are different for active and passive systems, and ASHRAE Standard 55-2004 separately addresses these two. For the mechanical systems, there is a *predicted mean vote* (PMV) comfort model which deals with both a thermal balance and a thermal comfort. PMV is based on a thermal sensation scale that ranges from cold (−3) to hot (+3) for low airspeeds (<40 ft/min). Section 5.3 of the standard covers naturally ventilated spaces and criteria vary with mean monthly outdoor temperatures.

Special Circumstances and Exemplary Performance As mentioned in IEQc6.2, thermal comfort control as described in ASHRAE Standard 55-2004 should be balanced with the minimum indoor air quality requirements, as required per ASHRAE Standard 62.1-2004. In addition, lighting can affect thermal comfort, and its impacts should be considered in the designs. Also, since many of the features for thermal comfort relate to windows and the HVAC systems, this credit has been labeled with an EB from the USGBC. It is more difficult to change in an existing building. There is no exemplary performance point for this credit for either LEED version 2.2 or 2009.

IEQ Credit 7.2: *Thermal Comfort—Verification*

Again, thermal comfort is based on many environmental and personal factors, and it is also addressed in IEQc6.2 and IEQc7.1. This credit, IEQc7.2, deals with verifying that the occupants of the building feel that they have an adequate level of thermal comfort. IEQc7.2 is worth one point in both LEED 2.2 and LEED 2009, but in LEED 2009 is only in addition to one point earned in IEQc7.1.

USGBC Rating System

LEED-NC 2.2 lists the Intent, Requirements, and Potential Technologies and Strategies for IEQc7.2 as follows, with modifications to reflect updates in LEED 2009, most notably these three; the requirement to also achieve IEQc7.1, ineligibility for residential projects, and the requirement to provide a permanent monitoring system to ensure compliance with IEQc7.1:

Intent

Provide for the assessment of building thermal comfort over time.

Requirements

Achieve IEQc7.1: Thermal Comfort-Design (LEED 2009)

AND

Provide a permanent monitoring system to ensure performance meets the desired comfort criteria in the IEQc7.1 design (LEED 2009).

AND

Residential projects are not eligible (LEED 2009)

AND

Agree to implement a thermal comfort survey of building occupants within a period of six to 18 months after occupancy. This survey should collect anonymous responses about thermal comfort in the building including an assessment of overall satisfaction with thermal performance and identification of thermal comfort-related problems. Agree to develop a plan for corrective action if the survey results indicate that more than 20% of occupants are dissatisfied with thermal comfort in the building. This plan should include measurement of relevant environmental variables in problem areas in accordance with ASHRAE Standard 55-2004, with errata but without addenda. Project teams can choose to use the addenda to ASHRAE 55-2004, but only if applied consistently over all the credits applied for the LEED 2009.

Potential Technologies and Strategies

ASHRAE Standard 55-2004 provides guidance for establishing thermal comfort criteria and the documentation and validation of building performance to the criteria. While the standard is not intended for purposes of continuous monitoring and maintenance of the thermal environment, the principles expressed in the standard provide a basis for design of monitoring and corrective action systems.

	Very Dissatisfied −3	Dissatisfied −2	Somewhat Dissatisfied −1	Neutral 0	Somewhat Satisfied 1	Satisfied 2	Very Satisfied 3
Are you satisfied with the thermal comfort of your workspace?							

Table 6.7.1 Thermal Comfort Questionnaire

Calculations and Considerations

The survey requirement of this credit is well described in the LEED write-up, but little guidance is given for the permanent monitoring system requirements. The survey should be based on a variation of a seven-point Likert-type scale where the lowest is −3, representing very dissatisfied, and the highest is +3, representing very satisfied. A typical question may look like the one in Table 6.7.1.

Anyone responding with a number less than zero (neutral) should be included in the dissatisfied category. Additional questions should allow the respondents to explain how they are dissatisfied (or satisfied) so that ideas for corrective action can be developed. Since any answer less than neutral is included in the rating, why even have seven distinct rating levels? This is to help in the determination of corrective action. For instance, a person who is only mildly dissatisfied, perhaps on hot summer days, may become satisfied with a simple policy change, such as a more casual clothing requirement during heat waves or the use of a personal fan. Similarly, corrective action for a very dissatisfied individual may require installing additional controls or items in the main HVAC system. Additional questions may be asked about factors such as humidity ratings and lighting levels, etc., to aid in designing means for improved satisfaction with the workspace.

Special Circumstances and Exemplary Performance

As mentioned in IEQc6.2 and IEQc7.1, many of the features for thermal comfort relate to windows and the HVAC systems, so this credit has been labeled with an EB from the USGBC. It may be more difficult to take corrective action in an existing building. There is no exemplary performance point for this credit for either LEED version 2.2 or 2009.

6.8 IEQ Credit Subcategory 8: *Daylighting and Views*

The intentions of these two credits are to allow for more natural lighting during the day, known as daylighting, and for more views of the outside for the occupants in the regularly used areas in a building. Wall surfaces which are part of the inside of an exterior wall that are less than 2.5 ft from the floor surface are not usually included in view or daylighting calculations. These same wall surfaces from 2.5 ft above the floor to 7.5 ft above the floor are considered to be "view" surfaces and can also count toward daylighting. Any glazed wall or ceiling surface (sky or toplights, light tubes, etc.)

above 7.5 ft from the floor are usually designed for daylighting purposes (but not for views). Many birds are injured each year by flying into glazed surfaces. Consideration should be given to using etched glazing or other means of modifying glazing so that the modified exterior reflectances aid in reducing these mishaps.

IEQ Credit 8.1: *Daylighting and Views—Daylight (75% of Spaces)*

IEQc8.1 is worth one point in both LEED 2.2 and LEED 2009. IEQc8.1 has major differences between the two LEED versions (2.2 and 2009) and has some of the options renumbered even though the intent remains similar between the two versions. Therefore, a separate listing of the requirements for each is given. The intent is first given in the version 2.2 section.

USGBC Rating System (LEED 2.2)

LEED-NC 2.2 lists the Intent, Requirements, and Potential Technologies and Strategies for IEQc8.1 as follows:

Intent (LEED 2.2 and LEED 2009)

Provide for the building occupants a connection between indoor spaces and the outdoors through the introduction of daylight and views into the regularly occupied areas of the building.

Requirements (LEED 2.2)

LEED 2.2 OPTION 1—CALCULATION

Achieve a minimum glazing factor of 2% in a minimum of 75% of all regularly occupied areas.

The glazing factor is calculated as follows:

Glazing Factor = Window Geometry Factor × (Window Area [SF]/Floor Area [SF]) × (Actual T_{vis}/ Minimum T_{vis}) × Window Height Factor

OR

LEED 2.2 OPTION 2—SIMULATION

Demonstrate, through computer simulation, that a minimum daylight illumination level of 25 footcandles has been achieved in a minimum of 75% of all regularly occupied areas. Modeling must demonstrate 25 horizontal footcandles under clear sky conditions, at noon, on the equinox, at 30 inches above the floor.

OR

LEED 2.2 OR LEED 2009 OPTION 3—MEASUREMENT

Demonstrate, through records of indoor light measurements, that a minimum daylight illumination level of 25 footcandles has been achieved in at least 75% of all regularly occupied areas. Measurements must be taken on a 10-foot grid for all occupied spaces and must be recorded on building floor plans. In all cases, only the square footage associated with the portions of rooms or spaces meeting the minimum illumination requirements can be applied towards the 75% of total area calculation required to qualify for this credit. In all cases, provide daylight redirection and/or glare control devices to avoid high-contrast situations that could impede visual tasks. Exceptions for areas where tasks would be hindered by the use of daylight will be considered on their merits.

Potential Technologies and Strategies

Design the building to maximize interior daylighting. Strategies to consider include building orientation, shallow floor plates, increased building perimeter, exterior and interior permanent shading devices, high performance glazing and automatic photocell-based controls. Predict daylight factors via manual calculations or model daylighting strategies with a physical or computer model to assess footcandle levels and daylight factors achieved.

USGBC Rating System (LEED 2009)

The requirements for the updated LEED 2009 version of IEQc8.1 can be summarized by the following four options.

Requirements LEED 2009

LEED 2009 OPTION 1—SIMULATION (This was previously Option 2 under LEED 2.2 with some modifications.)

LEED 2009 IEQc8.1 Option 1 requires a computer simulation demonstrating that at least 75 percent of regularly occupied areas achieve a minimum daylight illuminance of 25 footcandles (fc) and a maximum daylight illuminance of 500 fc at 30 inches above the floor level in a clear sky condition at both 9 a.m. and 3 p.m. on September 21st (March 31st is given as an alternative in the Reference Guide). The grid lines used in the simulation must not be more than 5 feet apart.

LEED 2009 OPTION 2—PRESCRIPTIVE (This prescriptive option replaces the previous Calculation Option #1 in LEED 2.2.)

LEED 2009 IEQc8.1 Option 2 separates all daylit areas into two different types of zones, one for side-lighting and one for top-lighting. These should be separately analyzed and then summed to determine compliance. A summary of the requirements can be found in Table 6.8.1.

LEED 2009 OPTION 3—MEASUREMENT

LEED 2009 IEQc8.1 Option 3 is the same as LEED 2.2 IEQc8.1 Option 3.

Side-Lighting Daylight Zones	Top-Lighting Daylight Zones
The product of the visible light transmittance (T_{vis} or VLT) and the window to floor area ratio (WFR) must be between 0.150 and 0.80 for all window areas >30 in above the floor.	Skylights with a minimum T_{vis} of 0.5 should cover between 3 and 6% of the roof area.
The floor (bay) area used for each window is defined by the bay width and twice the window height into the room.	Skylights should not be separated by >140% of the ceiling height.
The ceiling cannot obstruct into the area under a line drawn from the height of the window-head to a distance two window-head heights away from the plane of the window for any portion of the window or the floor area included in the WFR for that particular bay.	The daylit zones under a skylight are its plan view plus in every direction the lesser of 70% of the ceiling height, half the distance to the nearest skylight or any opaque partition farther away from the plan edge than 70% of the distance between the top of the partition and the ceiling.
Provide sunlight redirection and/or glare control as needed.	Any skylight diffuser must have an ASTM D1003 tested haze value >90% and not be in direct line of sight.
The LEED 2009 Reference Guide provides some typical visible transmittance values as 0.86 for N/S and E/W bays and 0.45 for corner bays.	

Table 6.8.1 LEED 2009 IEQc8.1: Daylight and Views—Daylight Option 2. Prescriptive Requirements

LEED 2009 OPTION 4—COMBINATION

LEED 2009 IEQc8.1 allows for any combination of LEED 2009 Options 1, 2, and 3 that meets the minimum 75 percent daylight illumination requirement.

For all the options in LEED 2009 glare control is required to avoid high-contrast situations that could impede visual tasks. Some allowance (minimum 10 fc) is given for dedicated theater spaces and other spaces where daylight might hinder tasks may be given exceptions on a case-by-case basis. Regularly occupied areas also include those areas occupied by animals for veterinary, and animal boarding and shelter facilities.

Calculations and Considerations

The definitions for *glazing, glazing factor, daylighting, incident light, nonoccupied spaces, nonregularly occupied spaces, regularly occupied spaces,* and *visible light transmittance* T_{vis} are listed in App. B. The most important definition is that of *regularly occupied areas.* Regularly occupied areas are spaces where most occupants actually work on a regular basis. Maintenance work is not included. In residential facilities, they are the living or family areas. In these areas, having daylighting from more "indirect" sources is promoted.

Figure 6.8.1 depicts some of the typical types of daylighting glazed surfaces and their locations. To obtain this credit, there are restrictions on these various locations for daylighting and views so that glare is controlled. Typical glare control practices are listed in the LEED-NC 2.2 and 2009 Reference Guides.

In both versions of LEED, there are three main options for credit compliance, with a forth option which combines any of the three compliance paths in LEED 2009. In both versions, the first two are by some form of lighting model, and the third is based on actual measurements. The first in LEED 2.2 and the second in LEED 2009 are based on glazing factors, which are dependent on the window types and orientations. The second (version 2.2) or first (LEED 2009) and third options in either are based on an estimate (model) or measurement, respectively, of horizontal footcandles (illumination) in the regularly occupied areas.

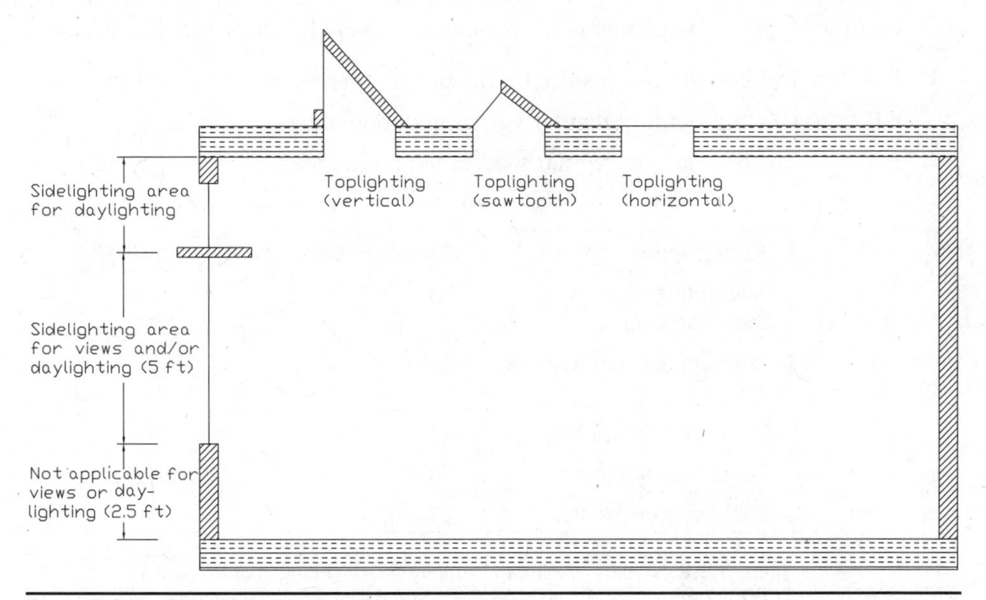

FIGURE 6.8.1 Sidelighting and toplighting.

In either version, all the options require having 75 percent of the regularly occupied floor areas in the building meet a minimum daylight illumination level. However, there are some differences in how the 75 percent is determined. The applicable regularly occupied floor areas can be segregated into different modeling units. They can be in modeling units of "rooms" or "areas." For the modeling options in LEED 2.2, most small rooms are taken as one unit, and large rooms can be segregated into applicable areas. In LEED 2009, the prescriptive modeling option uses special areas defined by the window locations and sizes. The square footage of all the modeling units that meet the minimum daylight illumination level can be used for the 75 percent determination. For the third option, the one based on measuring illumination levels, the applicable regularly occupied floor areas are segregated into 10-ft by 10-ft grids. All the 100-ft^2 squares that meet the minimum daylighting levels count toward the 75 percent determination. Only the calculation model in LEED 2.2 will be explained in greater detail here, as the second model requires more complex simulations and the third option requires actual measurements after construction. In LEED 2009, the prescriptive option (LEED 2009 Option 2) methodology is summarized in Table 6.8.1.

LEED 2.2 Option 1 (Calculation) The first option gives a simple method that can be used in the design phase to quickly evaluate options that may or may not meet the credit criteria. As can be seen in the LEED-NC 2.2 Requirements, it is based on determining if the glazing factor for at least 75 percent of the regularly occupied areas meets a minimum value of 0.02 (2 percent). The equation given in the Requirements section is based on the window's location (height), angle (geometry factor), and a minimum visible transmittance T_{vis} based on the window location. Together, these parameters can be combined to give a *combined daylight factor* for each window type *i*, denoted by CDF_i. Typical minimum values for each combined daylight factor as specified by the parameters set in LEED-NC 2.2 are listed in Table 6.8.2. Note that each window type may vary, and the manufacturers' information should be used for the final calculations.

By using these factors, the area of each window type, and the areas of the regularly occupied spaces, compliance can be calculated using the following definitions:

DC Daylighting compliance for the project (in percent)

GF_j Glazing factor of regularly occupied room area *j*

RA_j The floor area of regularly occupied room area *j*

Glazing Type	Combined Daylighting Factor CDF_i
Sidelighting (2.5–7.5 ft above the floor)	0.20
Sidelighting (>7.5 ft above the floor)	0.20
Toplighting, vertical	0.50
Toplighting, sawtooth	0.83
Toplighting, horizontal	1.25

Table 6.8.2 Typical Minimum Combined Daylighting Factors (CDF_i) LEED 2.2

RAT The total floor area of all regularly occupied room areas

WA_{ij} The sum of the areas of window type i in room area j

The glazing factor of regularly occupied room area j can be determined as

$$GF_j = \frac{\sum_i (CDF_i) \times (WA_{ij})}{RA_j} \quad \text{for all window types } i \text{ in regularly occupied area } j \quad (6.8.1)$$

The total floor area of all regularly occupied room areas can be determined as

$$RAT = \sum_j RA_j \quad \text{for all regularly occupied room areas } j \quad (6.8.2)$$

In LEED 2.2 Option 1, daylighting compliance for the project can be calculated as

$$DC = 100 \times \left(\frac{\sum_j RA_j}{RAT} \right) \quad \text{for all regularly occupied areas } j \text{ where } GFj \geq 0.02 \quad (6.8.3)$$

For either version and all options, the project complies with IEQc8.1 if

$$DC \geq 75 \quad (6.8.4)$$

Simulation and Measurement Options These options are based on a horizontal illumination level. The requisite amount is at least 25 horizontal footcandles by natural lighting in a minimum of 75 percent of the regularly occupied areas. Illumination engineers are experienced in providing *isoluxes* of areas. Isoluxes are plan views giving horizontal illumination contours. For many exterior applications, these illumination contours are usually given at ground level, but for the simulation options (Option 1 in LEED 2009 and Option 2 in LEED 2.2) and either version of the measurement option (Option 3), the isoluxes would be on the horizontal plane at 30 in above the floor. This elevation corresponds more closely to the level of a worktable or desktop. Isolux stands for "same light," and the term *lux* (abbreviated lx) is also the SI unit for illuminance. One footcandle equals approximately 10.764 lx. In the 2.2 version, these simulations or measurements are based on natural light that would be expected or is measured on a clear day at noon on the equinox (either vernal or autumnal). In LEED 2009, the time has been changed to both 9 a.m. and 3 p.m. on the equinox for the simulation option, but is not specified for the measurement option, so it is assumed that the times for the measurements have also been changed to 9 a.m. and 3 p.m. to be consistent with the simulation option.

Special Circumstances and Exemplary Performance

Since the location of windows and other daylighting features is an intricate part of the architecture of a building, this credit has been labeled with an EB from the USGBC. It is more difficult to change in an existing building. Figure 6.8.2 shows an area with natural daylighting at a LEED-NC-certified building in Columbia, S.C. Figure 6.8.3a and b shows the outdoor and indoor views of skylights at a mall in South Carolina.

FIGURE 6.8.2 Natural lighting at Cox and Dinkins, Engineers and Surveyors, Columbia, S.C. (*Photograph taken June 2007.*)

IEQc8.1 has also been listed as available for an EP credit in LEED-NC 2.2 and LEED 2009. This additional point can be achieved if at least 95 percent of the regularly occupied spaces are daylighted by the minimum glazing factor as in LEED 2.2 Option 1, the prescriptive option in LEED 2009 or the minimum 25 fc as simulated or measured for either version. This corresponds to the following requirement:

$$DC \geq 95 \quad \text{additional EP point criterion} \tag{6.8.5}$$

Since Option 3 in either version is based on measurements that can only be taken on two days of the year, and under restrictive weather conditions, it may cause a scheduling conflict for certification if the weather is not favorable. In that case, a description of special circumstances regarding the measurements taken on alternate dates may need to be included in the submittals.

IEQ Credit 8.2: Daylight and Views—Views (for 90% of Spaces)

IEQc8.2 is worth one point in both LEED 2.2 and LEED 2009

USGBC Rating System

LEED-NC 2.2 lists the Intent, Requirements, and Potential Technologies and Strategies for IEQc8.2 as follows and is essentially identical for LEED 2009:

Intent

Provide for the building occupants a connection between indoor spaces and the outdoors through the introduction of daylight and views into the regularly occupied areas of the building.

Requirements

Achieve direct line of sight to the outdoor environment via vision glazing between 2 feet 6 inches and 7 feet 6 inches above finish floor for building occupants in 90% of all regularly

(a)

(b)

FIGURE 6.8.3 Skylights and celestories at Columbiana Mall in Irmo, S.C. (a) Exterior view; (b) interior view. (*Photograph taken July 2007.*)

occupied areas. Determine the area with direct line of sight by totaling the regularly occupied square footage that meets the following criteria:

- In plan view, the area is within sight lines drawn from perimeter vision glazing.
- In section view, a direct sight line can be drawn from the area to perimeter vision glazing. Line of sight may be drawn through interior glazing. For private offices, the entire square footage of the office can be counted if 75% or more of the area has direct line of sight to perimeter vision glazing. For multi-occupant spaces, the actual square footage with direct line of sight to perimeter vision glazing is counted.

Potential Technologies and Strategies
Design the space to maximize daylighting and view opportunities. Strategies to consider include lower partition heights, interior shading devices, interior glazing, and automatic photocell-based controls.

Calculations and Considerations (LEED 2009)

Compliance with IEQc8.2 requires a minimum number of regularly occupied areas in the building to have direct lines of sight to exterior views from a specified height above the floor level. This height is usually given as approximately 3.5 ft, which is the average seated eye height of most occupants, but it can vary if there is reasonable justification for a lower or higher height such as for younger students. Since this is a three-dimensional requirement, lines must be drawn on a plan view showing that there are direct sight lines to a view from the horizontal planar area, and these areas must also be drawn in elevation to show that there are sight lines from the eye height that reach the vision glazing. (The definition of vision glazing as given by LEED-NC 2.2 can be found in the definitions section in App. B and is the glazed exterior wall space between 2.5 and 7.5 ft from the floor as depicted in Fig. 6.8.1.) The sight lines can go through interior glazing, but in LEED 2009, this is limited to at most two interior glazing surfaces for any sight line to be compliant.

The 90 percent compliance rate is for all regularly occupied areas, with one main exception. Private offices need only have 75 percent of their floor area within the sight line criterion for the entire office space to count toward the credit. Why? Private offices are usually more spacious (in square feet per occupant) and are occupied by a sole occupant who usually has the opportunity to move his or her desk or work area to a place of choice. Therefore, there need not be views from all corners of the room. (In contrast, employees in multioccupied office areas, conference rooms, or cafeterias usually do not have a choice where they sit or work.) If private offices have a minimum of 75 percent of the floor area compliant, then the entire floor area counts. If not, then only the fraction of the floor area that is compliant counts. This encourages designers and owners to put private offices in the interior spaces with glazing through interior walls to exterior walls. In this way, privacy in private offices can be maintained with shades and blinds, while views are still available and not obstructed from other occupants.

A direct consequence of this requirement is the challenge to provide cubicle areas with views. In response, many companies are designing alternative cubicle designs that have some attributes that allow for views from the work areas. Figure 6.8.4a through d shows some newer cubicle designs.

View compliancy (VC) with IEQc8.2 can be determined in the following way using the following variables:

VA_j Area compliant with both view criteria in regularly occupied room area j (including private offices)

(a) (b) (c) (d)

FIGURE 6.8.4 Examples of innovative cubicle designs that may aid in addressing views. The manufacturer is (a) Steelcase, Inc., (b) Herman Miller, Inc., (c) Kimball International, Inc., and (d) Knoll, Inc. The chairs in front were provided by Haworth, Inc., and may optionally count toward other MR credits if furniture is included in the calculations. (*Photographs taken at the Strom Thurmond Building on Fort Jackson, Columbia, S.C., in June 2007.*)

$VAPO_k$	Area compliant with both view criteria in private office k (the $VAPO_k$ values represent a subset of the VA_j values)
VC	View compliance for the project (in percent)
RA_j	Floor area of regularly occupied room area j (including private offices)
$RAPO_k$	Floor area of private office k ($RAPO_k$ values represent a subset of RA_j values)
RAT	Total floor area of all regularly occupied room areas

The total floor area of all regularly occupied room areas, as previously listed in IEQc8.1, is

$$RAT = \sum_j RA_j \quad \text{for all regularly occupied room areas } j \tag{6.8.2}$$

And therefore VC for the project can be calculated as

$$VC = 100 \times \left[\left(\frac{\sum_j VA_j}{RAT} \right) + \left(\frac{\sum_k (RAPO_k - VAPO_k)}{RAT} \right) \right] \quad \text{for all } k \text{ where } (VAPO_K)/(RAPO_K) \geq 0.75 \tag{6.8.6}$$

The term to the left of the addition sign in Eq. (6.8.6) represents the total floor areas in regularly occupied spaces that comply with both view criteria, and the terms to the right of the addition sign represent the additional amount that may be included for private offices where at least 75 percent of the private office area meets both criteria. The project then complies with IEQc8.2 if

$$VC \geq 90 \tag{6.8.7}$$

Special Circumstances and Exemplary Performance

Since the location of windows and large work areas is an intricate part of the architecture of a building, this credit has been labeled with an EB from the USGBC. It is more difficult to change in an existing building.

There is an EP point available for IEQc8.2 for both versions, although no specific criterion is set for it in LEED 2.2. LEED 2009 has the criteria for the IEQc8.2 EP point specified as also meeting at least two of the following four options for 90 percent or more of the regularly occupied spaces:

1. "Have multiple lines of sight to vision glazing in different directions at least 90 degrees apart."
2. "Include views of at least two of the following: vegetation, human activity, or objects >70 ft from the exterior of the glazing."
3. "Have access to unobstructed views located within the distance of three times the head height of the vision glazing."
4. "Have access to views with a view factor of three or greater as per the Heschong Mahone Group Study, Windows and Offices, p. 47."

6.9 Discussion and Overview

This category covers indoor air quality, indoor thermal comfort, indoor lighting, and views. In the future, these and other important indoor environmental concerns will be expanded upon. For example, in the USGBC LEED for Schools rating system released in 2007, both a prerequisite and a credit have been added that relate to acoustics or sound pollution. Currently, however, the major focus is on indoor air quality as the impacts on the human occupants of the buildings have been shown to be important.

There are other alternatives for promoting good indoor air quality. There may be the potential for many other types of monitoring other than CO_2, and other ways to "clean" indoor air than through outdoor air exchanges, which is very important in dense urban areas or other areas where outside air quality is not always up to the National Ambient Air Quality Standards (NAAQS). These technological alternatives may eventually become other supplemental options in the indoor air quality credit subcategories or earn innovation credits. Two new technologies that are being developed to improve air quality with respect to microbes and pathogens in indoor environments are UV lighting and copper components such as heat exchanger fins, cooling coils, and condensate drip pans in the HVAC systems. More information about the possible advantages of copper can be found on the website of the Copper Development Association (CDA) (*http://www.copper.org/*). Figure 6.9.1 depicts a factory installation of UV lights in a Trane air handler.

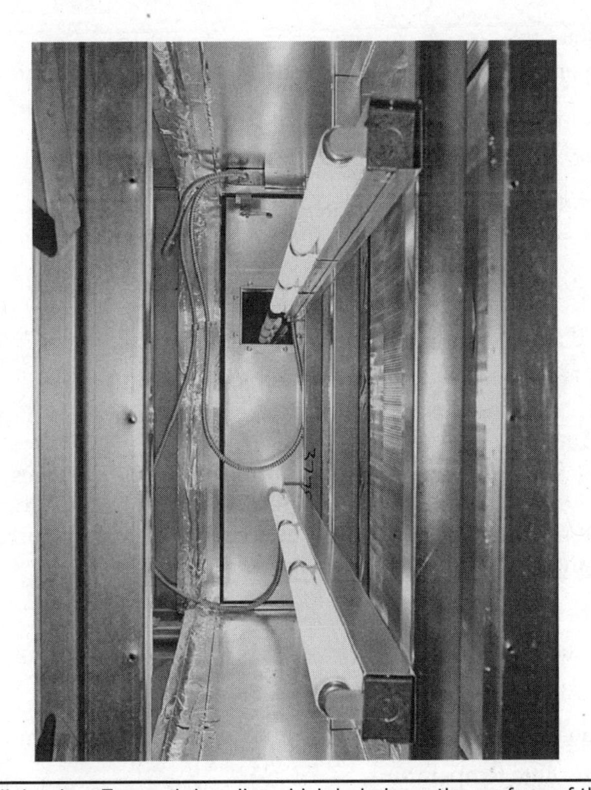

Figure 6.9.1 UV lights in a Trane air handler which help keep the surface of the coil and drain pan clean. (*Photograph Courtesy Trane.*)

ASHRAE was awarded a grant in 2006 from the EPA to develop a special publication entitled *Advanced Indoor Air Quality (IAQ) Design Guide for Non-Residential Buildings.* The document was expected to be complete in late 2008, and the work was being done in conjunction with many other organizations including the USGCB, AIA, and SMACNA. This document should be useful in comparing air quality alternatives and designing innovative control methods.

Also, ASTM International is developing a standard that addresses methods for the evaluation of residential indoor air quality. It is Standard D7297 and is under the jurisdiction of Subcommittee D22.05 on Indoor Air, which is part of ASTM Committee D22, Air Quality. The Center for Disease Control and Prevention, headquartered in Atlanta, provides many other references for healthy indoor and outdoor environments. A website with resources entitled "Designing and Building Healthy Places" can be found at http://www.cdc.gov/healthyplaces/.

Air Pollution Calculation Conversions and Concentrations

Air pollutant concentrations and emission rates are typically reported in many units. Conversion of these units is not always straightforward and is also dependent on the temperature and pressure of the readings. Typically the values given from standard testing are converted to standard temperature and pressure, which is at 1 atmosphere (atm) (14.7 psia) and between 20 and 25°C. The following items can aid in understanding these concepts and conversions to estimate air pollution levels.

Temperature

The SI temperature scale is in degrees Celsius (°C), and the absolute SI scale is in kelvins (K) such that

$$\mathrm{K} = {}^\circ\mathrm{C} + 273 \tag{6.9.1}$$

The temperature scale in the United States, degrees Fahrenheit (°F), can be converted to its absolute scale, degrees Rankine (°R), by the following:

$${}^\circ\mathrm{R} = {}^\circ\mathrm{F} + 460 \tag{6.9.2}$$

To convert between Celsius and Fahrenheit, the following equation is used:

$${}^\circ\mathrm{C} = 1.8({}^\circ\mathrm{F} - 32) \tag{6.9.3}$$

Ideal Gas Law

The ideal gas law is a model that quantifies how the volumes and concentrations of gases change with temperature and pressure. It can be used to model most gases and mixtures of gases at typical ambient temperatures and pressures. The ideal gas law is Eq. (6.9.4) and is based on the following variables:

P	absolute pressure
V	volume of gas
n	moles of gas
R_{Ideal}	ideal gas law constant; one value is 0.0821 L · atm · (mol · K)
T	absolute temperature

$$PV = nR_{\mathrm{Ideal}}T \tag{6.9.4}$$

If the concentration of a gas or a mixture of gases is given in moles per volume (n/V), then the changed concentration due to temperature variations can readily be estimated by multiplying the initial concentration by the ratio of the two temperatures. Note that the volume increases with an increase in temperature, so the concentration decreases with an increase in temperature, and the temperatures used must be based on one of the absolute scales (Kelvin or Rankine).

If the concentration of a gas or a mixture of gases is given in moles per volume (n/V), then the changed concentration due to pressure variations can readily be estimated by multiplying the initial concentration by the ratio of the two pressures. Note that the volume increases with a decrease in pressure, so the concentration decreases with a decrease in pressure, and the pressures used must be absolute pressures, not gauge pressures.

Some Typical Properties of Ambient Air

The following properties of ambient air are useful in performing indoor air pollutant concentration calculations:

- At room temperature and 1 atm, there is approximately 24 liters per mole (L/mol) of air.
- At the earth's surface, dry air is predominantly composed of 78 percent nitrogen, 21 percent oxygen, and 1 percent argon. This is a mole or volume percent.
- Dry air composed of approximately 78 percent nitrogen, 21 percent oxygen, and 1 percent argon has an average molecular weight of approximately 29 g/mol.

Concentrations

Air pollutant concentrations are typically given either by mass of the pollutant per volume of air or by a molar ratio. The first unit type (mass/volume) can be used for gaseous, gaseous mixtures, liquid, or solid air pollutants. One common unit is micrograms per meter cubed ($\mu g/m^3$).

The molar ratios can only be used for the gaseous pollutants. They are usually given in parts per million (ppm) or parts per billion (ppb). When these units are used for gaseous phase concentrations, they are in parts per million or billion mole. They are not based on mass. Since most of the air pollution concentrations that are of interest inside buildings and in ambient air can be modeled by the ideal gas law, the ppm or ppb on a molar basis is equal to the ppm or ppb on a volumetric basis.

If masses of air pollutants are known, the molecular weights must be used to convert the units to ppm or ppb. This cannot be done for the air pollutant categories that combine several types of pollutants, such as total VOCs, even though the VOCs are gases. Therefore, these concentrations are always given in mass per volume.

Box Models

Simple air pollutant box models based on the conservation of mass can be used to estimate indoor air pollutant concentrations (see Chap. 10 for a description of simple box models). Although the examples given in Chap. 10 are based on water and water pollutant concentrations, similar box models can also be used for both air mass balances and air pollution mass balances. However, care must be taken when balancing the air masses and air pollutant masses so that variations in flows, volumes, and concentrations due to temperature and pressure changes are addressed.

References

ASHRAE (1999), ASHRAE 52.2-1999: *Method of Testing General Ventilation Air-Cleaning Devices for Removal Efficiency by Particle Size*, American Society of Heating, Refrigerating and Air-Conditioning Engineers, Atlanta, GA.

ASHRAE (2004), ASHRAE 62.1-2004: *Ventilation for Acceptable Indoor Air Quality*, American Society of Heating, Refrigerating and Air-Conditioning Engineers, Atlanta, GA.

ASHRAE (2004), ASHRAE 62.2-2004: *Ventilation and Acceptable Indoor Air Quality in Low-Rise Residential Buildings*, American Society of Heating, Refrigerating and Air-Conditioning Engineers, Atlanta, GA.

ASHRAE (2007), ASHRAE 62.1-2007: *Ventilation for Acceptable Indoor Air Quality*, American Society of Heating, Refrigerating and Air-Conditioning Engineers, Atlanta, GA.

ASHRAE (2007), ASHRAE 62.2-2007: *Ventilation and Acceptable Indoor Air Quality in Low-Rise Residential Buildings*, American Society of Heating, Refrigerating and Air-Conditioning Engineers, Atlanta, GA.

ASTM (2003), ANSI/ASTM-E779-03: *Standard Test Method for Determining Air Leakage Rate By Fan Pressurization*, ASTM, West Conshohocken, PA.

ASTM (2007), ASTM D1003: Standard Test Method for Haze and Luminous Transmittance of Transparent Plastics, ASTM, West Conshohocken, PA.

ASTM (2007), "Indoor Air Quality," Global Notebook Section of *ASTM Standardization News*, July, 35(7): 4.

ASTM (2007), "New Practice Will Consistently Evaluate Indoor Air Quality Problems," Technical News Section of *ASTM Standardization News*, July, 35(7): 10.

California DHS (2004), The California Department of Health Services Standard Practice for the Testing of Volatile Organic Emissions from Various Sources Using Small-Scale Environmental Chambers, including 2004 addenda, http://www.cal-iaq.org/VOC/Section01350_7_15_2004_FINAL_ADDENDUM-2004-01.pdf.

Carbon Trust (1999), "Natural Ventilation in Non-Domestic Buildings—A Guide for Designers; Developers and Owners," *Carbon Trust Good Practice Guide 237*, The Carbon Trust, London, United Kingdom.

CDA (2007), http://www.copper.org/health/, Copper Development Association website accessed July 19, 2007.

CDC (2010), "Designing and Building Healthy Places" Center for Disease Control and Prevention website, http://www.cdc.gov/healthyplaces/, accessed February 4, 2010.

CEC (2001), *Residential Manual For Compliance with California's 2001 Energy Efficiency Standards*, California Energy Commission Pub. No. P400-01-022, June 1, 2001, Sacramento, CA

CIBSE (2000), CIBSE Applications Manual 13: *Mixed Mode Ventilation*, Chartered Institution of Building Services Engineers, London, United Kingdom.

CIBSE (2005), *CIBSE Applications Manual 10: Natural Ventilation in Non-Domestic Buildings*, Chartered Institution of Building Services Engineers, London, United Kingdom.

Cooper, C. D., and F. C. Alley (2002), *Air Pollution Control, A Design Approach*, Waveland Press, Inc., Prospect Heights, IL

CRI (2006), http://www.carpet-rug.com/, Carpet and Rug Institute website accessed July 24, 2007,

De Nevers, N. (2000), *Air Pollution Control Engineering,* 2d ed., McGraw-Hill, New York.

EPA OAR (1991), *Building Air Quality: A Guide for Building Owners and Facility Managers,* EPA Document 402-F-91-102, December, U.S. Environmental Protection Agency, Office of Air and Radiation, Washington, D.C.

EPA OAR (1998), *Building Air Quality Action Plan,* EPA Document 402-K-98-001, June, U.S. Environmental Protection Agency, Office of Air and Radiation, Washington, D.C.

EPA OAR (2006), http://www.epa.gov/iaq/formalde.html, U.S. Environmental Protection Agency, Office of Air and Radiation, Washington, D.C., website accessed October 19, 2006.

EPA OAR (2006), http://www.epa.gov/iaq/voc.html, U.S. Environmental Protection Agency, Office of Air and Radiation, Washington, D.C., website accessed October 19, 2006.

EPA OAR (2006), http://www.epa.gov/oar/oaqps/greenbk/o3co.html, U.S. Environmental Protection Agency, Office of Air and Radiation, Washington, D.C., website accessed October 19, 2006.

EPA OAR (2006), http://www.epa.gov/pmdesignations/basicinfo.htm, U.S. Environmental Protection Agency, Office of Air and Radiation, Washington, D.C., website accessed October 19, 2006.

FloorScore (2009), Resilient Floor Covering Institute (RFCI), http://www.rfci.com/int_FloorScore.htm.

Green Seal (1993), *The Green Seal Standard GS-11, Paints,* 1st ed., May 20, 1993, http://www.greenseal.org/certification/standards/paints.cfm, Green Seal, Washington, D.C.

Green Seal (1997), *The Green Seal Standard GC-03, Anti-Corrosive Paints,* 2d ed., January 7, 1997, http://www.greenseal.org/certification/standards/anti-corrosivepaints.cfm, Green Seal, Washington, D.C.

Green Seal (2000), *Green Seal Standard, GS36, Commercial Adhesives,* October 19, 2000, http://www.greenseal.org/certification/standards/commercialadhesives.cfm, Green Seal, Washington, D.C.

Heinsohn, R. J., and R. L. Kabel (1999), *Sources and Control of Air Pollution,* Prentice-Hall, Upper Saddle River, NJ.

Heschong Mahone Group (2003), Windows and Offices; A Study of Office Worker Performance and the Indoor Environment, Heschong Mahone Group Inc., Gold River, CA. http://www.h-m-g.com/, accessed January 3, 2010.

Likert, R. (1932), "A Technique for the Measurement of Attitudes," *Archives of Psychology,* 140: 55.

Michels, H. T. (2006), "Anti-Microbial Characteristics of Copper," *ASTM Standardization News,* October, 34(10): 3–6.

SCAQMD (2004), *Rule 1113, Architectural Coatings,* January 1, 2004, South Coast Air Quality Management District, Diamond Bar, CA

SCAQMD (2005), *Rule 1168, Adhesive and Sealant Applications,* as Amended January 7, 2005, South Coast Air Quality Management District, Diamond Bar, CA

Siegel, J., I. Walker, and M. Sherman (2002), "Dirty Air Conditioners: Energy Implications of Coil Fouling," Lawrence Berkeley National Laboratory Paper LBNL-49757, March 1.

SMACNA (1995), *IAQ Guidelines for Occupied Buildings under Construction,* Chap. 3, Sheet Metal and Air Conditioning National Contractors Association, Chantilly, VA.

SMACNA (2008), *IAQ Guidelines for Occupied Buildings under Construction*, Chap. 3, Sheet Metal and Air Conditioning National Contractors Association, Chantilly, VA.

USGBC (2003), *LEED-NC for New Construction, Reference Guide*, Version 2.1, 2d ed., May, U.S. Green Building Council, Washington, D.C.

USGBC (2005–2007), *LEED-NC for New Construction, Reference Guide*, Version 2.2, 1st ed., U.S. Green Building Council, Washington, D.C., October 2005 with errata posted through Spring 2007.

USGBC (2007), *LEED for Schools for New Construction and Major Renovations*, Approved 2007 Version, April, U.S. Green Building Council, Washington, D.C.

USGBC (2009), *LEED Reference Guide for Green Building Design and Construction*, 2009 Edition, U.S. Green Building Council, Washington, D.C., April 2009.

Varodompun, J., and M. Navvab (2007), "HVAC Ventilation Strategies: The Contribution for Thermal Comfort, Energy Efficiency, and Indoor Air Quality," *Journal of Green Building*, Spring, 2(2): 131–150.

Wark, K., C. F. Warner, and W. T. Davis (1998), *Air Pollution, Its Origin and Control*, Addison Wesley Longman, Menlo Park, CA.

Wikipedia (2006), http://en.wikipedia.org/wiki/Metabolic_equivalent, as accessed October 16, 2006.

Exercises

1. Put together a CPM schedule for items for SMACNA and the other requirements for IEQc3.1 and IEQc3.2. (See Chap. 8 for information on a CPM.)

- **A.** With procurement staggered
- **B.** With on-site protected storage
- **C.** For sequencing installation of absorbent materials

2. Make a site plan with staging areas for material protection.

- **A.** Calculate the size of the covered areas needed.
- **B.** Determine the area needed for the staging and access.
- **C.** Put these items on the plan.

3. When we breathe outside air, we inhale air with oxygen at about 21 percent (volumetric) and carbon dioxide (CO_2) at around 380 ppm (0.038 percent volumetric) and exhale air that has a little less oxygen (a bunch more water) and around 4.5 percent CO_2, which is then rapidly diluted by the air around it. Typically, you can estimate that sedentary activities in a typical space with proper ventilation will raise the CO_2 concentrations by about 350 ppm. Sedentary activities are said to be 1 Met (metabolic rate), which is a measure of the rate at which we expend energy, which can also represent the rate at which we exhale CO_2. Vigorous exercise might be about 6 Met.

- **A.** In a room with computer programmers and typical ventilation, estimate what you might expect the concentration of CO_2 to be (in ppm and also percent volumetric).
- **B.** Estimate what this concentration might be in a jazzercise room if you do not increase the ventilation from the level in the computer room (in ppm and percent volumetric). (Many times the CO_2 alarm is set at 1000 ppm. CO_2 is considered toxic at about 5 percent volumetric.)

4. For IEQc3.1 and IEQc5 there are requirements to have particulate filtering capabilities on the air handlers based on the ASHRAE MERV (minimum efficiency reporting value) ratings. For these ratings, typical airborne particulates are categorized into the following size ranges: 0.3 to 1 μm, 1 to 3 μm, and 3 to 10 μm. The micrometer (abbreviated μm) is given as an average diameter measurement and is 10^{-6} m. If you assume that these particles are round and have a specific gravity of 2, what are their volume ranges and mass ranges?

5. IEQ credit 4.3 limits total VOC emissions and also several specific compounds emissions, including formaldehyde from carpets and carpet cushioning. It also limits the VOC level (concentration) in the carpet adhesive as per IEQc4.1. You are carpeting a room that is 4.5 m by 6 m wide and 3.2 m tall, and it is totally sealed so that there are no air exchanges. If you install carpet cushioning and carpet on the floor that all are at the emissions limits as established by the Carpet and Rug Institute (CRI) for formaldehyde and as listed in Table 6.4.3., how long (in hours) will it take until the concentration in the room of formaldehyde reaches the threshold for human irritation (around 0.1 ppm)? Remember, we are assuming a perfectly sealed room. You can assume that both the formaldehyde and the air are ideal gases.

6. You are carpeting a room that is 4.5 m by 6 m wide and 3.2 m tall, and the room changes out its full volume of air every 4 h. If you install carpet cushioning and carpet on the floor that all are at the emissions limits as established by the Carpet and Rug Institute (CRI) for formaldehyde as listed in Table 6.4.3, what is the steady-state concentration of formaldehyde in the room? Please note the following:

- *Steady state* means that the mass rate of formaldehyde leaving the room in the air which is leaving equals the mass rate of the formaldehyde which is being emitted into the room from the carpet system as noted. (See Chap. 10 for mass balance box models.)
- You can assume that both the formaldehyde and the air are ideal gases.

7. You are carpeting a room that is 4.5 m by 4.5 m wide and 3.0 m tall, and the room changes out its full volume of air every 2 h. If you install carpet cushioning and carpet on the floor that all are at the emissions limits as established by the Carpet and Rug Institute for formaldehyde as listed in Table 6.4.3, what is the steady-state concentration of formaldehyde in the room? Please note the following:

- Steady state means that the mass rate of formaldehyde leaving the room in the air which is leaving equals the mass rate of the formaldehyde which is being emitted into the room from the carpet system as noted. (See Chap. 10 for mass balance box models.)
- You can assume that both the formaldehyde and the air are ideal gases.

8. Your room has equal mass concentrations of particulate matter in the 1- to 3-μm range and in the 3- to 10-μm range and an insignificant amount of the smaller particulate matter on a mass basis. What is the overall efficiency of a MERV 8 filter and a MERV 13 filter for this room based on Table 6.3.1?

9. Your room has approximately 25 percent on a mass basis of the particulate matter in the 1- to 3-μm range, approximately 20 percent in the 3- to 10-μm range, and the remainder in the 0.3- to 1.0-μm range. What is the overall efficiency of a MERV 8 filter for this room based on Table 6.3.1? What is the overall efficiency of a MERV 13 filter for this room based on Table 6.3.1?

10. Ambient air PM2.5 1997 maximum standards are 15 $\mu g/m^3$ on an average annual basis and 65 $\mu g/m^3$ on an average 24-h basis in 1997. The 24-h standard was tightened to 35 $\mu g/m^3$ in 2006. You have a room that is 4 m by 4 m by 3 m. If you flush out the volume of this room and it is at the maximum annual ambient air standard for PM2.5, what mass of PM2.5 would you expect to collect on a MERV 13 filter?

11. Estimate the time it would take to flush out a building after occupancy for the minimum rates set in LEED 2009 IEQ credit 3.2 Option 1, Path 1. The standard gives a total of 14,000 ft^3 of outside air per square foot. The Path 2 alternate flush-out option allows for occupancy after 3500 ft^3 is complete. At the minimum rate of 0.30 $ft^3/(min \cdot ft^2)$ for a typical 9-h day occupancy (8 h plus lunch), how many extra days will this take as compared to Path 1? (Note that the system must start at least 3 h prior to occupancy.)

12. The maximum allowed concentration for formaldehyde for LEED 2009 IEQ credit 3.2 based on air quality testing is 27 ppb (parts per billion). What is the concentration in micrograms per cubic meter? Formaldehyde is CH_2O. You can assume that both the formaldehyde and the air are ideal gases.

13. The maximum allowed concentration for 4-phenylcyclohexene for IEQ credit 3.2 based on air quality testing is 6.5 μg/m³. What is the concentration in ppm? You can assume that both the 4-PCH and the air are ideal gases.

14. The maximum allowed concentration for carbon monoxide for IEQ credit 3.2 based on air quality testing is 9 ppm. What is the concentration in micrograms per cubic meter? You can assume that both the CO and air are ideal gases.

15. A typical diameter of a mechanical pencil lead is 0.5 mm. How many PM particles of 10-μm size would need to be lined up side by side to form this diameter?

16. Look at your classroom or lecture hall. Make a sketch of the room and determine if it meets the natural ventilation requirement for operable windows as per IEQ prerequisite 1. If it does not, then make a plan sketch and elevation of the room which would meet the requirements if this is feasible. If it is not, then explain why.

17. Look at your classroom or lecture hall. Make a sketch of the room and determine if it meets the view requirements of IEQ credit 8.2. If it does not, then make a plan and an elevation sketch of the room which would meet the requirements. If this is not feasible, explain why.

18. Look at your classroom or lecture hall. Make a sketch of the room and determine if it meets the daylighting requirements of LEED 2.2 IEQ credit 8.1, using the typical minimum values of the combined daylighting factors in Table 6.8.2. If it does not, then make a plan and an elevation sketch of the room with the additional features which would meet the requirements. If this is not feasible, explain why.

19. You are designing the window arrangement in a new office space in a single-story flat-roof building. The room is 30 ft wide by 80 ft long with an exterior wall along one length. Design windows and openings so that both the natural ventilation requirements in IEQ prerequisite 1 and the view requirements of IEQ credit 8.2 are met. Show this with both a plan and an elevation sketch of the room.

20. You are designing the window arrangement in a new office space in a single-story flat-roof building. The space is 40 ft wide by 90 ft long with an exterior wall along one length. There need to be six private 10-ft by 10-ft offices in this office space, with the remainder open for cubicles and meeting areas. Design windows and openings so that both the daylighting requirements of LEED 2.2 IEQ credit 8.1 and the view requirements of LEED 2.2 IEQ credit 8.2 are met. Show this with plan and elevation sketches of the spaces. (Use the typical minimum values of the combined daylighting factors in Table 6.8.2.)

21. You have performed a survey of all your employees and have collected the number of responses as listed in Table 6.E.1. Do you qualify for a point for IEQc7.2 if IEQc7.1 is achieved and you also provide a permanent monitoring system?

22. Perform a comfort survey on the students in your class or another group as specified by your lecturer. Summarize your findings.

	Very Dissatisfied −3	Dissatisfied −2	Somewhat Dissatisfied −1	Neutral 0	Somewhat Satisfied 1	Satisfied 2	Very Satisfied 3
Are you satisfied with the thermal comfort of your workspace?	0	18	32	94	85	8	2

TABLE 6.E.1 Thermal Comfort Survey Results

23. Based on the workforce in Chap. 3 Exercise 9, determine the LEED 2009 IEQc6.1 minimum number of individual lighting control locations (ILCL) required for the project.

24. Based on the workforce in Chap. 3 Exercise 9, determine the LEED 2009 IEQc6.2 minimum number of individual comfort control locations (ICCL) required for the project.

25. Based on the workforce in Chap. 2 Exercise 6, determine the LEED 2009 IEQc6.1 minimum number of individual lighting control locations (ILCL) required for the project.

26. Based on the workforce in Chap. 2 Exercise 6, determine the LEED 2009 IEQc6.2 minimum number of individual comfort control locations (ICCL) required for the project.

27. Based on the workforce in Chap. 2 Exercise 7, determine the LEED 2009 IEQc6.1 minimum number of individual lighting control locations (ILCL) required for the project.

28. Based on the workforce in Chap. 2 Exercise 7, determine the LEED 2009 IEQc6.2 minimum number of individual comfort control locations (ICCL) required for the project.

29. Look at your classroom or lecture hall. Make a sketch of the room and determine if it meets the daylighting requirements of LEED 2009 IEQ credit 8.1, using Table 6.8.1 and an assumed visible transmittance of 0.86 for N/S or E/W bays and 0.45 for corner bays. If it does not, then make a plan and an elevation sketch of the room with the additional features which would meet the requirements. If this is not feasible, explain why.

30. You are designing the window arrangement in a new office space in a single-story flat-roof building. The space is 40 ft wide by 90 ft long with an exterior wall along one length. There need to be six private 10-ft by 10-ft offices in this office space, with the remainder open for cubicles and meeting areas. Design windows so that both the daylighting requirements of LEED 2009 IEQ credit 8.1 and the view requirements of LEED 2009 IEQ credit 8.2 are met. Show this with plan and elevation sketches of the spaces. (Use the information in Table 6.8.1 and a visible transmittance of 0.86.)

31. Which, if any, of the IEQ credits are available for Regional Priority points in the zip code of your work or school? (See Chap. 7.)

32. Which, if any, of the IEQ credits are available for Regional Priority points in the following major cities? (See Chap. 7.)

A. New York City
B. Houston, TX
C. Los Angeles, CA
D. Portland, OR
E. Miami, FL
F. Kansas City, MO

CHAPTER 7

LEED Innovation in Design Process and Regional Priorities

The Innovation in Design category deals with issues otherwise not included in the other categories, that have alternative strategies to obtain goals of other categories, or which exceed to a specified degree some of the intents from the other credit categories. This category of the U.S. Green Building Council (USGBC) rating system for new construction version 2.2 (LEED-NC 2.2) consisted of two credit subcategories which together may earn a possible five points (credits) maximum. Up to four of this maximum of five could come from exemplary performance (EP) points earned in the other categories. In LEED 2009, this has been slightly modified to allow for an additional Innovation in Design point, but restricting these points from EP to a maximum of three. The notation format for credits is, for example, IDc1.1 for Innovation in Design credit 1.1 in LEED 2.2 and IDc1 for Innovation in Design credit subcategory 1 in LEED 2009. Also, in this chapter, to facilitate easier cross-referencing between this text and the USGBC rating system, the second digit in a section heading, equation, table, or figure number represents the credit subcategory number for sections that deal directly with a USGBC LEED-NC 2.2 and 2009 credit subcategories.

There are two credit subcategories in both the 2.2 and the 2009 versions. The two credit subcategories are ID Credit Subcategory 1: Innovation in Design and ID Credit Subcategory 2: LEED Accredited Professional. The ID credit subcategories, credits, and available points are summarized for both LEED 2.2 and 2009 in Table 7.0.0.

Basically, the credits in the Innovation in Design category can be grouped into the following three overall sets of goals:

- In LEED 2009, up to three points in the Innovation in Design subcategory can be obtained for projects which substantially exceed the intent of certain credits in the other categories. These are the EP options as listed in each section and described in Chaps. 2, 3, 4, 5, and 6. The LEED 2009 Building Design and Construction Reference Guide also mentions that some credits may earn an EP point if there are two compliance pathways which are both attained and are additive.
- In LEED 2009, up to five points (less the EP points awarded) in the Innovation in Design subcategory can be used for other innovative strategies that are not included in the previous categories, but which can be shown to address environmental issues by strategies not otherwise addressed in LEED.

Subcategory 2009	Subcategory 2.2	Points 2009	Points 2.2	Major Comments
IDc1: Innovation in Design	IDc1.1: Innovation in Design	1–5	1	LEED 2009 allows from 1 to 5 Innovation in Design points of which 3 may be EP points (Path 2) and any or all may be Path 1, *Innovation in Design*. LEED 2.2 allowed from 1 to 4 for each with a maximum total of 4 for both.
	IDc1.2: Innovation in Design		1	
	IDc1.3: Innovation in Design		1	
	IDc1.4: Innovation in Design		1	
IDc2: LEED Accredited Professional	IDc2: LEED Accredited Professional	1	1	
	Total	6	5	

TABLE 7.0.0 ID Summary Table of Major Differences between Versions 2.2 and 2009 and Associated Points (Pts)

- The one point in the LEED Accredited Professional subcategory can be used to ensure that one of the team members is familiar with the overall LEED process. The intent is to have someone who can understand the systematic approach needed in sustainable design.

The LEED 2009 notations in the Innovation in Design category are as follows:

- **ID Credit 1** (IDc1): *Innovation in Design*
- **ID Credit 2** (IDc2): *LEED Accredited Professional*

LEED-NC 2.2 provides a table which summarizes whether the credit is considered a design submittal or a construction submittal. IDc2 is a construction submittal as the professional is expected to participate throughout the project phases. The five possible Innovation in Design points in LEED 2009 can be either design or construction submittals depending on their intent and requirements. Likewise, any of or all the project team members may be the main decision makers for these credits depending again on their content. In the same manner it can only be determined by the particular intent and requirements of an innovation credit as to whether it should have the EB icon associated with it. IDc2 does not have an EB icon, as a LEED Accredited Professional can be used in any subsequent phases of a project life.

The Regional Priority category of the LEED New Construction rating system is new to LEED 2009. It allows for additional points based on environmental or energy priorities specific to a geographic region. Each small region of the country, divided by zip code, can allocate up to six credits from the five main categories of LEED to be of special priority and each project may obtain up to four points from this list of six possible credits to count toward certification. The notation in the Regional Priority category of LEED 2009 is:

- RP Credit 1 (RPc1): First Regional Priority Credit

7.1 ID Credit 1: *Innovation in Design*

USGBC Rating System

LEED-NC 2.2 lists the Intent, Requirements, and Potential Technologies and Strategies for IDc1 as follows, with modifications as noted for LEED 2009 (LEED 2.2 numbered these from IDc1.1 through IDc1.4):

Intent

To provide design teams and projects the opportunity to be awarded points for exceptional performance above the requirements set by the LEED-NC Green Building Rating System and/or innovative performance in Green Building categories not specifically addressed by the LEED-NC Green Building Rating System.

Requirements

LEED 2009 IDc1 Path 1 Innovation in Design (1–5 points): In writing, identify the intent of the proposed innovation credit, the proposed requirement for compliance, the proposed submittals to demonstrate compliance, and the design approach (strategies) that might be used to meet the requirements.

LEED 2009 IDc1 Path 2 Exemplary Performance (1–3 points): One point is awarded for each exemplary performance point earned from other categories.

(Note: The sum of the points earned from Paths 1 and 2 cannot exceed five in LEED 2009.)

Potential Technologies and Strategies

Substantially exceed a LEED-NC performance credit such as energy performance or water efficiency. Apply strategies or measures that demonstrate a comprehensive approach and quantifiable environment and/or health benefits.

Calculations and Considerations

As mentioned previously, there are two main avenues for obtaining the five available innovation points in LEED 2009. Either provide an innovative strategy not specifically addressed by LEED-NC 2009 or exceed one of the LEED credit criteria from the previous categories.

In the first case (Path 1), many different options can be pursued. The LEED-NC 2.2 and 2009 Reference Guides give three specific criteria. The environmental performance associated with the innovation credit must be quantifiable, this performance must be comprehensive throughout the project, and it must be such that it could be applicable to other projects. If an innovation credit is awarded to one project, it does not mean that it will automatically be awarded to other projects; and conversely, if an innovation credit is not awarded for a specific project, it does not mean that it cannot be awarded for another project with other circumstances. A good source for information about some possible innovation credits for a particular project can be found in the other LEED rating systems and the LEED application guidances which were listed in Tables 1.1.1 and 1.2.1. The five innovation LEED 2009 credits on a specific project must be substantially different from one another, and none can be technically similar to other credits (except as noted for the EP pathway).

In the second case (Path 2), the exceedance is usually double the standard or the next incremental step. Any previous credit subcategory that may be available for one of these ID credits is listed with the availability of EP noted in previous chapters. Table 7.1.1 gives a summary of the subcategories for which exemplary performance IDc1 points may be given in LEED 2009. Although there are 17 points listed in Table 7.1.1, only three can be applied to any one project.

Credit	Exemplary Performance and Other ID Point Availability	Exemplary Performance Point Criterion
SSc2: Development Density and Community Connectivity	Yes, 1 point for Option 1 only	Site density is double the area density or Area density is doubled for double the radius
SSc4.1: Alternative Transportation: Public Transportation Access	Yes, 1 point for either option (Not available if obtain SSc4 category plan EP point)	Double and also a minimum 200 transit rides daily Option 1: 2 stations and minimum 200 transit rides/day Option 2: 4 stops and minimum 200 transit rides/day
SSc4.1: Alternative Transportation: Public Transportation Access	Yes, 1 point for entire SSc4 subcategory (Not available if obtain SSc4.1EP point)	Comprehensive plan
SSc4.2: Alternative Transportation: Bicycle Storage and Changing Rooms		
SSc4.3: Alternative Transportation: Low-Emitting and Fuel-Efficient Vehicles		
SSc4.4: Alternative Transportation: Parking Capacity		
SSc5.1: Site Development: Protect or Restore Habitat (For previously developed sites, criterion is maximum of 50% of site less building footprint or 20% of the total site)	Yes, 1 point for previously developed sites	Greater of 75% of site less building footprint or 30% of the total site
SSc5.2: Site Development: Maximize Open Space: Option 1: Exceed Open Space Requirement by 25% Option 2: Match the Building Footprint with Open Space if No Zoning Option 3: 20% Open Space When None Required	Yes, 1 point	Double Option 1: Exceed by 50% Option 2: Double Building Footprint with Open Space Option 3: 40% Open Space When None Required
SSc6: Stormwater Design	Yes, 1 point for entire SSc6 subcategory	Capture and treat beyond requirements
SSc7.1: Heat Island Effect: Non-Roof (50%)	Yes, 1 point	100%

Table 7.1.1 Summary of Exemplary Performance Point Availability in LEED 2009

Credit	Exemplary Performance and Other ID Point Availability	Exemplary Performance Point Criterion
SSc7.2: Heat Island Effect: Roof Option 2: 50% Green Roof	Yes, 1 point for Option 2 only	100% Vegetated Roof
WEc2: Innovative Wastewater Technologies (50%)	Yes, 1 point for either option	1 point for 100%
WEc3: Water Use Reduction	Yes, 1 point	1 point for 45% EPAct
EAc1: Optimize Energy Performance	Yes, 1 point	50% for new construction or 46% for existing buildings
EAc2: On-Site Renewable Energy	Yes, 1 point	15%
EAc3: Enhanced Commissioning	Yes, 1 point	Comprehensive Envelope Commissioning
EAc6: Green Power	Yes, 1 point	Doubling the minimum amount of electricity to 70% or Doubling the minimum contract length to 4 years*
MRc2: Construction Waste Management	Yes, 1 point	95%
MRc3: Materials Reuse	Yes, 1 point	15%
MRc4: Recycled Content—(postconsumer + ½ pre-consumer)	Yes, 1 point	30%
MRc5: Regional Materials—Extracted, Processed, and Manufactured Regionally	Yes, 1 point	30% (was 40% in LEED 2.2)
MRc6: Rapidly Renewable Materials (>2.5%)	Yes, 1 point	5%
MRc7: Certified Wood (>50%)	Yes, 1 point	95%
EQc8.1: Daylighting and Views: Daylight 75% of Spaces	Yes, 1 point	95%
EQc8.2: Daylighting and Views: Views for 90% of Spaces	Yes, 1 point	Meet at least two of four special view criteria

*Not specifically addressed in the LEED 2009 Reference Guide, but noted in the LEED 2.2 Reference Guide.

Table 7.1.1 *(Continued)*

Since education is considered to be an extremely important avenue for the development and implementation of sustainable construction, there are many projects which choose an educational outlet for one of the innovation in design credits. This type of activity must be unique and sustainable, not simply a one-time lecture or event. The importance of education has been further emphasized in the LEED for Schools rating system for K–12 schools as adopted in 2007. In this rating system a credit point had been added to the Innovation in Design category which specifically addresses continuing education opportunities.

Special Circumstances and Exemplary Performance

One of the variables which must be determined for each project and which must be kept consistent throughout the credits in a LEED submittal is the size and location of the project site. It has been noted that there may be exceptions to this for special circumstances such as campus settings. Since campus and multiple-building projects may have varying boundaries depending on the phases of construction, additional LEED guidances have been developed to aid in the project interpretations for several credits. Likewise, many projects have other unique criteria that may not fit well into the credit intents and requirements as listed in LEED-NC 2009.

There is no EP point available for this credit subcategory as this credit subcategory is the repository of the EP points from other credit subcategories.

7.2 ID Credit 2: *LEED Accredited Professional*

USGBC Rating System

LEED-NC 2.2 lists the Intent, Requirements, and Potential Technologies and Strategies for IDc2 as follows, and is essentially not changed for LEED 2009:

Intent

To support and encourage the design integration required by a LEED-NC green building project and to streamline the application and certification process.

Requirements

At least one principal participant of the project team shall be a LEED Accredited Professional (AP).

Potential Technologies and Strategies

Educate the project team members about green building design and construction and application of the LEED Rating System early in the life of the project. Consider assigning the LEED AP as a facilitator of an integrated design and construction process.

Considerations and Exemplary Performance

IDc2 has been designated as a construction phase submittal. Mainly, this is done because the LEED Accredited Professional should participate in all aspects of the project including construction. There is no EP point available for this credit.

7.3 Regional Priorities

The regional priorities category allows for additional points for obtaining certain credits that are considered to be priorities in the region where the project is located. These regional priorities have been established with input from the USGBC chapters around

City	Zip Code	Credit Available for Regional Priority Point (Minimum Threshold Level and/or Option)					
Seattle, WA	98101	SSc1	SSc5.1	SSc6.1	EAc1 (48%/44%)	EAc2 (13%)	MRc7
Boston, MA	02108	SSc3	SSc6.1	SSc7.1	SSc7.2	EAc2 (1%)	MRc1.1 (75%)
Phoenix, AZ	85003	SSc2	SSc4.4	SSc7.1	WEc1 (Opt.2)	WEc3 (30%)	EAc2 (3%)
Columbia, SC	29201	SSc4.1	SSc6.1	WEc3 (40%)	EAc1 (28%/24%)	EAc2 (1%)	IEQc7.1
Columbus, OH	43201	SSc6.1	EAc2 (1%)	MRc1.1 (95%)	MRc2 (75%)	MRc3 (5%)	IEQc8.1

Table 7.3.1 Example of Credits and Threshold Levels and/or Required Options Available for Regional Priority (RP) Points in Five Cities for LEED-NC 2009

the country. They are listed on the USGBC website by state and are further subdivided by zip code (www.usgbc.org/LEED2009/). The listings are available for the LEED-NC rating system as well as other LEED rating systems such as Existing Buildings (EB), Schools, Core and Shell (CS), and Commercial Interiors (CI). In each case, six credits from the main LEED categories (SS, WE, EA, MR, or IEQ) were designated for possible RP credit, from which four may count toward the point total. Many credit subcategories award different numbers of points depending on the option, level of achievement, or a threshold value. Therefore, the designation for Regional Priority credit also gives the option or level to which the specific credit must be attained in order to be awarded the additional point. Currently, the regional priority listings only pertain to zip codes within the United States. Other regions of the world may be included in the future.

Eligible regional priority credits for five zip codes within some major US cities are listed in Table 7.3.1. Note the varying priorities. For example, Boston priorities focus on improving existing infrastructure and stormwater issues, while Phoenix has a water efficiency focus. Both also give priority to urban heat island mitigation and on-site renewable energy production.

7.4 Discussion and Overview

As previously mentioned in Chap. 1, the USGBC is developing guidances specific to many special subsets of project application types, to facilitate LEED certification and sustainable practices for these special types of projects. There are many special options for Innovation in Design credits as outlined in the various guidances being developed by the USGBC and its partners for specific project type categories. In addition, sometimes EP points are handled differently in these guidances and in the other LEED rating sytems. There are options in some where a credit point that is listed as an EP in LEED-NC 2.2 or 2009 may be an additional listed credit in the guidance or system for the particular project type. If what was once an EP point is counted as a credit, then this sometimes leaves open the opportunity for more innovation in design credits to be incorporated into the project.

References

Katz, G. (2006), *Greening America's Schools Costs and Benefits*, October, A Capital E Report, Capital E, Washington, D.C.

USGBC (2003), *LEED-NC for New Construction, Reference Guide*, Version 2.1, 2d ed., May, U.S. Green Building Council, Washington, D.C.

USGBC (2004–2005), *LEED-CI Green Building Rating System for Commercial Interiors*, Version 2.0, November 2004, updated December 2005, U.S. Green Building Council, Washington, D.C.

USGBC (2004–2005), *LEED-EB Green Building Rating System for Existing Buildings, Upgrades, Operations and Maintenance*, Version 2, October 2004, updated July 2005, U.S. Green Building Council, Washington, D.C.

USGBC (2005), *LEED-NC Application Guide for Multiple Buildings and On-Campus Building Projects, for Use with the LEED-NC Green Building Rating System*, Versions 2.1 and 2.2, October, U.S. Green Building Council, Washington, D.C.

USGBC (2005–2007), *LEED-NC for New Construction, Reference Guide*, Version 2.2, 1st ed., U.S. Green Building Council, Washington, D.C., October 2005 with errata posted through Spring 2007.

USGBC (2006), *LEED Green Building Rating System for Core and Shell Development*, Version 2.0, July, U.S. Green Building Council, Washington, D.C.

USGBC (2007), *LEED for Homes Program Pilot Rating System*, Version 2.11a, January, U.S. Green Building Council, Washington, D.C.

USGBC (2007), *LEED for Neighborhood Development Rating System*, Pilot, Congress for the New Urbanism, Natural Resources Defense Council, U.S. Green Building Council, Washington, D.C.

USGBC (2007), *LEED for Retail—New Construction and Major Renovations*, Pilot Version 2, April, U.S. Green Building Council, Washington, D.C.

USGBC (2007), *LEED for Schools for New Construction and Major Renovations*, Approved 2007 Version, April, U.S. Green Building Council, Washington, D.C.

USGBC (2009), *LEED Reference Guide for Green Building Design and Construction*, 2009 Edition, U.S. Green Building Council, Washington, D.C., April 2009.

USGBC (2010), LEED 2009 Website, www.usgbc.org/LEED2009/, accessed February 4, 2010.

Exercises

1. You are redeveloping an acre of land that was originally an asphalt parking lot. Along with the building and parking, you plan to turn 10,000 ft^2 of the lot into landscaped areas with natural and adapted plantings. What credits might this aid you in earning? What EP points might you consider?

2. True or false? The following action may be eligible for EP credit: A stormwater management plan is made for the site that reduces the rate and quantity of the existing runoff (1- and 2-year, 24-h storm) by 20 percent. The site currently has an imperviousness of approximately 35 percent. Why or why not?

3. True or false? The following action may be eligible for EP credit: The fire suppression systems in the new building will not use any CFCs, any halons, or any HCFCs. Why or why not?

4. True or false? The following action may be eligible for EP credit: Of the materials for the construction of the new building, except for plumbing, mechanical, furniture, and electrical materials, 50 percent (based on their cost value) are extracted or harvested and manufactured within 500 mi of the site. Why or why not?

5. True or false? The following action may be eligible for EP credit: Individual lighting controls will be available in every office and cubicle and at every desk in the new office building. Why or why not?

6. True or false? The following action may be eligible for EP credit: A covered staging area will be provided at the construction site to store materials away from the rain. Why or why not?

7. Research a LEED certified project in your area and find out which ID credits were obtained.

8. There are several projects that have used education as an ID credit. Find a specific example and summarize it. Do you think this will be applicable to all projects, or is it only specific to certain projects? Explain your opinion.

9. Look at a nearby construction project, and think of a possible innovative credit, other than EP, that you might like to consider for that project. Describe it and explain why you think it is innovative.

10. Go to the USGBC website and find out if regional priority credits are now available for regions outside the United States.

11. Go to the USGBC website and see if any of the regional priorities for LEED-NC for the cities listed in Table 7.3.1 have changed.

12. Go to the USGBC website and find the regional priority credits and associated thresholds for your zip code for LEED-NC, LEED-EB, and LEED for schools.

13. Go to the USGBC website and find the regional priority credits and associated thresholds for your work or school zip code for LEED-NC, LEED-CS, and LEED-CI.

CHAPTER 8

A Systematic View of Green and Minimum Program Requirements

Minimum Program Requirements

In version 3.0 of LEED (LEED-NC 2009), a series of Minimum Program Requirements (MPRs) was established by the USGBC in order to apply for registration of a project. When released in April 2009, there were eight MPRs, but one was removed in the listing and there are now seven. They are all shown in Table 8.0.0 with a summary of many of their important features. Although the requirements for *reasonable timetables and rating system sunset dates* are no longer listed as an MPR, some of these requirements might still be part of the registration documents.

8.1 Green Building from the Project Viewpoint

How to Get Started

The trick to getting started is to meet, then meet again, and then meet again. Integrated building design is the cornerstone for creating sustainable projects. If the systems and components that make up the facility are only looked at independently, then the result usually will not be optimal for performance, sustainability, and cost. It is important that all aspects of the project be viewed as interrelated.

Initial meetings with all the project team members who have already been identified are crucial to the success of the project. At these initial meetings, basic project scopes and requirements are reviewed. With respect to the incorporation of green concepts in the project, it is beneficial to start reviewing the prerequisites and credit opportunities of the LEED rating system. Project members should be asked to start identifying the necessary steps and design features needed to meet all the prerequisites. They should also be asked to review the list of possible credits and start identifying credits that may or may not be feasible to incorporate into the design while maintaining the project scope and objectives.

Many large projects hold early formal meetings commonly referred to as *charrettes.* Invited participants are the owner, stakeholders, identified project team member representatives, and other potential participants in the design and construction community, in addition to public

MPR	Overview
1: Must Comply with Environmental Laws	• Design and construction must comply with all environmental laws for the time and place of construction. • This includes all real property within the LEED project boundary.
2: Must Be a Building	• At least one building in its entirety must be included for new construction or major renovation except as allowed for mixed-use and Core and Shell. • The project must be on already existing land. • A structure that is designed to be moved in its lifetime is not included. (Note that the guidance does not define what a building is.) • If two or more buildings under the same contract are both pursuing certification and are in a contiguous area, then they must certify as one project.
3: Must Use a Reasonable Site Boundary	• All land that was or will be disturbed during the project must be included. • The maximum area is contiguous land of the building owner with exceptions for noncontiguous parcels if they are within a 0.4 km walkable distance, owned by the same owner, in the same jurisdiction, comply with all the prerequisites and same credits, etc. • Segmentations of large campuses into project areas must not overlap and overall, must include all the campus land.
4: Must Comply with Minimum FTE and Floor Area Requirements	• If the annual average FTE is less than 1, the project may use LEED, but not obtain any of the IEQ credits. All prerequisites must still be met. • The indoor enclosed gross floor area must be at least 1000 ft^2.
5: Must Comply with Minimum Occupancy Rates	This is only for the Existing Building rating system, but if a new building seeks recertification after its first certification, then it must adhere to this. Basically, there are some formulas to verify that a certain proportion of the design occupancy is fulfilled for a certain time period prior to recertification.
(was 6)*: Registration and Certification Activity Must Comply with Reasonable Timetables and Rating System Sunset Dates*	• After registration under LEED 2009, a significant amount of certification application activity must occur within 4 years or the registration may be canceled by the USGBC (subject to proper notifications). • There is a 6 year maximum time in which to certify under a prior version of a rating system after a newer version is released. • The LEED certification must be applied for no longer than 2 years after project completion (typically as per the Certificate of Occupancy).
6 (was 7)*: Must Allow USGBC Access to Whole-Building Energy and Water Usage Data	All LEED 2009 certified projects must allow the USGBC access to actual whole-project energy and water usage data if such meters are in place for research purposes starting within 1 year of certification. This requirement must be carried forward to future owners.
7 (was 8)*: Must Comply with the Minimum Building Area to Site Area Ratio	The gross floor area (GFA) to project site area ratio must be 0.02 or greater. (This is typically referred to as the Floor Area Ratio—FAR.)

*Former requirement number 6 is no longer listed and former requirements 7 and 8 are renumbered 6 and 7, respectively.

TABLE 8.0.0 Minimum Program Requirements (MPRs) in LEED 2009 (April 2009)

and regulatory participation as appropriate. Charrettes are intensely focused, collaborative efforts. They can help create momentum to meet project goals in addition to fostering trust and buy-in from the regulatory agencies and community. Collaborative agreement early in a project can help avoid costly design changes and also identify potential partners and opportunities for increased sustainability and performance from the project. Education is a vital component.

Another set of obstacles can involve the local and state regulatory agencies if they are not familiar with some of the practices used. Likewise local statutes may prohibit, discourage, or not address some of the techniques suggested for more sustainable construction. Therefore, it is important to work closely with local and state officials and see which practices may be incorporated and what steps may be taken to overcome some of these obstacles. As the sustainable practices become more common, some of these obstacles will disappear.

Several states have started to adopt green building policies. This serves to facilitate the process in two ways. First, adoption of green policies by states helps direct state and local agencies in the manner that green projects will be permitted and evaluated. Second, requiring green practices for many state projects helps improve the learning curve and familiarity in the region with respect to sustainable construction. Understanding and knowledge are always some of the best tools for implementation of new practices. By 2005 two states, Washington and Nevada, had passed legislation relating to sustainable construction; and nine other states—California, Arizona, Colorado, Michigan, Pennsylvania, New York, New Jersey, Maryland, and Maine—had Executive Orders in place relating to either sustainable construction or energy efficiency. Most also referred to the USGBC LEED rating systems in their policy directives.

How to Keep Track

At the initial meetings and charrettes, project objectives are outlined. It is important that these objectives be referred to repeatedly throughout the project life and reevaluated and revised as needed. One portion of the objectives is to establish the green design criteria. LEED issues should be incorporated into a project as early as possible, to facilitate an effective and efficient project completion and to minimize costs and changes. Just as in other project and construction management philosophies which over the years have been successfully incorporated into design and construction, such as Quality Assurance and Quality Control (QA/QC), the incorporation of green building practices may at first seem awkward. However, they should soon become routine and part of an established procedure with their benefits accepted and appreciated. In similar manner to QA/QC, green building issues are continually reevaluated and improved upon during the life of a project.

So where should it all start? It starts first with the owner and project conception. In some cases the owner may voluntarily opt to seek LEED certification; in other cases it may be required by local or state mandate. Either way, the intention to seek LEED certification is usually first identified in an early draft of the specifications set by the owner for a project. The LEED-NC process refers to the specifications set by the owner as the "Owner's Project Requirements" or OPR. As the OPR evolve, the specific LEED credits preferred or recommended by the owner eventually get listed.

These green criteria can be easily kept track of in checklists that are based on the LEED credit lists. The checklists are dynamic documents that change and improve with the project. They can be readily downloaded from the USGBC website. A blank checklist for LEED 2.2 is shown as Fig. 1.2.3. For each credit the question is asked whether the project should attempt to achieve it. The answers can be yes, maybe (?), or no. Each credit is checked off in the Yes, the Maybe (?), or the No column. "Yes" means that at this point in the project, it is the intention that this credit will be sought. The question mark is for those credits that are still being evaluated for cost and project effectiveness, and "no" means that the associated credit will probably not be considered for certification of the subject project. Throughout the life of a project, any of the notations may be moved into different columns for each credit based on more detailed design, information, and project team decisions.

These estimated credit checklists are very good tools for directing all the project team members during the project. However, it may be difficult to fill out the early versions. This may be particularly difficult if the project is the first for an owner, or the first of its kind in a region, since it may be difficult to determine which credits are most applicable and effective for the particular project type and for the regional resources. Various members of the USGBC and researchers have started to develop ranked lists with columns expressing the "ease" or possibility of obtaining each credit based on local experience. Table 8.1.1 gives examples of three types of rankings that have been used to start the initial checklist. One ranking is a subjective ranking of ease of obtaining, the second ranking is the percent of regional projects that have received the credit, and the third ranking is a ranking that gives an initial estimate of viability ranked by the certification levels in the LEED-NC rating system. The items in Table 8.1.1 are for illustrative purposes only and do not represent the actual ease based on any scientific or region-specific analyses. Prerequisites have been excluded from the list as they are required.

The ranked list used is subject to change readily as the demand for green products changes and the knowledge or experience with green construction evolves in an area. This ranked list is also partially based on projects adhering to earlier versions of LEED, and subsequent versions may make certain credits more or less applicable for similar projects in the future. Regardless, past experiences can still give a reasonable starting point for initial estimates of LEED-NC certification feasibility.

The University of South Carolina (USC) has worked on many LEED registered projects. One of the projects was unique in that it represented a proposed baseball stadium, which was significantly different from the types of projects that the LEED-NC 2.2 rating system was usually based on. With this project, it was not easy to readily identify the sustainable opportunities as outlined in LEED-NC 2.2, but USC wanted to explore the possibility of LEED certification. To deal with this, the project team first developed a checklist from its current ranked list based on prior experience with other projects. One month later, after some preliminary design was complete, the list was revised based on further information and reviews of CIRs. Table 8.1.2 depicts the changes in the credit checklist for these two periods in the early design stage for the baseball stadium project. As the project proceeded beyond this early design phase, there were fewer and fewer changes. Table 8.1.2 shows how a checklist can help initiate the process even with a project for which there is much uncertainty as to the best sustainable route, and for which there are few previous similar projects to use for initial estimates.

Project Methods and Scheduling

Project Methods and Phases

Project schedules and interactions differ depending on the type of project and the type of construction process used. For example, some projects follow the more traditional design-bid-build process, whereas others might have one design-build contract. Since continuity and a systems approach are important for effective incorporation of sustainable construction into projects, the different project delivery methods do impact the manner in which these are handled. The method of interaction between the project participants

Credit	Ease of Earning			Local Project Experience (% Earned)	Probable Certification Level			
	Easy	Neutral	Hard		Certified	Silver	Gold	Platinum
SSc1	✓			80	✓			
SSc2*		✓		60		✓		
SSc3*		✓		40			✓	
SSc4.1*		✓		50			✓	
SSc4.2	✓			95	✓			
SSc4.3	✓			80	✓			
SSc4.4	✓			85	✓			
SSc5.1			✓	30				✓
SSc5.2		✓		65		✓		
SSc6.1	✓			95	✓			
SSc6.2		✓		65		✓		
SSc7.1	✓			80	✓			
SSc7.2	✓			80	✓			
SSc8			✓	45				✓
WEc1†	✓ (2)			85(2)	✓ (2)			
WEc1†		✓ (4)		45(4)			✓ (4)	
WEc2	✓			95	✓			
WEc3†	✓ (2)	✓ (3)	✓ (4)	50(3)	✓ (2)		✓ (3)	✓ (4)
EAc1†	✓ (5)	✓ (10)	✓ (18)	50 (10)	✓ (5)	✓ (8)	✓ (12)	✓ (18)
EAc2†		✓ (3)	✓ (7)	50 (3)		✓ (3)	✓ (5)	✓ (7)
EAc3		✓		60			✓	
EAc4		✓		65		✓		
EAc5			✓	15				✓
EAc6			✓	25				✓
MRc1.1†		✓ (2)		60(2)		✓ (2)		
MRc1.2*		✓		40			✓	
MRc2 (50%)	✓			100	✓			
MRc2 (75%)		✓		60		✓		
MRc3 (5%)		✓		55			✓	

Table 8.1.1 Example of Three Types of Ranked List of Credits for Initial Estimates (LEED 2009). This list is an example only and should be appropriately altered based on local experience (*Continued*)

Credit	Ease of Earning			Local Project Experience (% Earned)	Probable Certification Level			
	Easy	Neutral	Hard		Certified	Silver	Gold	Platinum
MRc3 (10%)			✓	25				✓
MRc4 (10%)	✓			100	✓			
MRc4 (20%)		✓		55			✓	
MRc5 (10%)	✓			90	✓			
MRc5 (20%)		✓		60		✓		
MRc6			✓	5				✓
MRc7*		✓		50			✓	
IEQc1			✓	30				✓
IEQc2	✓			85	✓			
IEQc3.1		✓		55			✓	
IEQc3.2			✓	20				✓
IEQc4.1	✓			90	✓			
IEQc4.2	✓			95	✓			
IEQc4.3	✓			95	✓			
IEQc4.4	✓			85	✓			
IEQc5		✓		60		✓		
IEQc6.1			✓	20				✓
IEQc6.2			✓	10				✓
IEQc7.1		✓		40			✓	
IEQc7.2		✓		60				✓
IEQc8.1	✓			90	✓			
IEQc8.2		✓		60			✓	
IDc1†	✓ (2)	✓ (3)	✓ (5)	50 (3)	✓ (2)	✓ (3)		✓ (5)
IDc2	✓			100	✓			
RP†	✓ (2)	✓ (3)	✓ (4)	✓ (2)	✓ (1)	✓ (2)	✓ (3)	✓ (4)

*Very site specific. Each has been given a median rating for illustrative purposes only.
†May earn varying points based on the level of performance attained. The total cumulative number of points earned as ranked is in the parentheses.

Table 8.1.1 Example of Three Types of Ranked List of Credits for Initial Estimates (LEED 2009). This list is an example only and should be appropriately altered based on local experience (*Continued*)

Credit	September 2006			October 2006		
	Yes	?	No	Yes	?	No
SSc1	✓				✓[*]	
SSc2		✓			✓	
SSc3		✓		✓[*]		
SSc4.1	✓			✓		
SSc4.2	✓				✓[*]	
SSc4.3		✓				✓[*]
SSc4.4		✓			✓	
SSc5.1		✓			✓	
SSc5.2	✓			✓		
SSc6.1		✓			✓	
SSc6.2	✓			✓		
SSc7.1	✓				✓[*]	
SSc7.2		✓			✓	
SSc8		✓				✓[*]
WEc1.1	✓				✓[*]	
WEc1.2		✓			✓	
WEc2			✓			✓
WEc3.1	✓			✓		
WEc3.2	✓			✓		
EAc1[†]	✓ (2–4)			✓[*] (2–4)	✓[*] (2–4)	
EAc2[†]			✓			✓
EAc3	✓			✓		
EAc4	✓			✓		
EAc5	✓			✓		
EAc6			✓			✓
MRc1.1			✓			✓
MRc1.2			✓			✓
MRc1.3			✓			✓
MRc2.1	✓			✓		
MRc2.2		✓		✓[*]		
MRc3.1			✓			✓
MRc3.2			✓			✓

Table 8.1.2 LEED-NC Version 2.2 Registered Project Checklist with Credit Estimates for a USC Baseball Stadium in Columbia, S.C., September 2006 and October 2006 (*Continued*)

	September 2006			October 2006		
Credit	**Yes**	**?**	**No**	**Yes**	**?**	**No**
MRc4.1	✓			✓		
MRc4.2		✓		✓*		
MRc5.1	✓			✓		
MRc5.2		✓		✓*		
MRc6			✓			✓
MRc7			✓			✓
EQc1	✓			✓		
EQc2		✓			✓	
EQc3.1	✓			✓		
EQc3.2		✓			✓	
EQc4.1	✓			✓		
EQc4.2	✓			✓		
EQc4.3	✓			✓		
EQc4.4		✓		✓*		
EQc5	✓				✓*	
EQc6.1		✓			✓	
EQc6.2		✓			✓	
EQc7.1	✓			✓		
EQc7.2	✓			✓		
EQc8.1	✓			✓		
EQc8.2		✓			✓	
IDc1.1–1.4†	✓ (1–2)	✓ (3–4)		✓ (1–2)	✓ (3–4)	
IDc2	✓			✓		
Total	29–31	20	18–20	27–31	18–22	20

*Represents a change.
†May earn more than one point. The range of points earned as ranked is in the parentheses.

TABLE 8.1.2 LEED-NC Version 2.2 Registered Project Checklist with Credit Estimates for a USC Baseball Stadium in Columbia, S.C., September 2006 and October 2006 (*Continued*)

and their responsibilities with respect to sustainable construction will be different for the different methods. It may be easier to have a single LEED lead for the design-build method, whereas, with the design-bid-build process, it may be necessary to transfer the lead from perhaps an architectural firm designee to a construction firm designee as the project proceeds. Depending on the prerequisite and credit, the importance of the impact of the delivery method chosen will differ.

Regardless of the project delivery method, the stages of a project can usually be summarized into a few main categories:

- Project conceptualization
- Schematic design
- Planning and zoning (P&Z) and other municipal planning organization (MPO) reviews
- Design development (detailed design)
- Permits
- Construction document development
- Bid and procurement
- Construction
- Closeout
- Operations

As previously mentioned, many of the credits must be considered in the early stages of project conception to be effective for the project, whereas others may slowly evolve as the design phase progresses into greater detail. It is important that there be interaction regarding the LEED process early on and frequently throughout each step.

CPM Schedules

The term *schedule* actually refers to two different project design and management tools. One type of schedule is a tally of specific items. An example is a door schedule on an architectural plan. A door schedule is the list of the number of types of doors needed for the project. This type of schedule usually includes information on sizes, materials, other characteristics, and manufacturers. The other tool called a schedule refers to a listing or diagram of activities with relationship to time. It is this type of schedule, a time-activity schedule, that is being further evaluated in this section. Note that both types of schedules, item-related and time-activity-related, are important to projects. For instance, in Chap. 10 where erosion and sedimentation control measures are further explored, the design package might require both an itemized schedule of materials or plants needed for erosion control and a time-activity schedule to sequence the erosion control activities so that the soils are adequately protected from erosion during all the construction stages.

Activity-time scheduling is a critical tool in any design-construction process. There are many types of scheduling techniques used, but most can be summarized in the following two formats:

- Bar charts (Gantt chart)
- Critical path method (CPM)

Bar or Gantt charts are very common and represent activities in a vertical list with time on the horizontal and bars along the horizontal which give some idea when these activities will take place. Table 8.1.3 depicts a simple bar chart of some typical project phases.

CPMs are similar listings of activities except that a CPM schedule also shows a dependency of activities on other activities. For instance, a building should be designed

Project Phase	Year 1												Year 2										
Project concept	x																						
Schematic design		x	x	x																			
MPO reviews				x	x	x	x	x															
Design development					x	x	x	x	x	x													
Permits									x	x	x	x	x										
Construction document development							x	x	x	x	x												
Bid and procure										x	x	x	x	x									
Construction												x	x	x	x	x	x	x	x	x	x		
Closeout																					x	x	x
Operations																							x

TABLE 8.1.3 Simple Bar Chart of Typical Project Phases

before it is constructed, not constructed before it is designed. CPMs are usually depicted as diagrams with interconnected arrows. If the activity is listed on the arrow, then it is referred to as an "activity on the arrow" CPM, and if the activity is depicted in a box between arrows, then it is referred to as an "activity on the node" CPM. Time can be included in the CPM schedule by either listing time requirements with the activity or placing the activity on a figure similar to the bar chart where time is depicted on the horizontal axis. The term *critical path method* comes from the act of adding time along any path of arrows and determining the path which would take the longest time. The activities along this path are considered critical since any delay in them will potentially delay the completion of the entire project scheduled.

CPMs can be evaluated manually or with many established project scheduling programs. Some programs are simple, and others are quite complex and include additional opportunities to relate many features to each activity such as the workforce needs or material needs. Adding these is usually referred to as *resource loading*. Other versions add opportunities for delayed actions or only partial completion of activities before the next starts. CPMs are frequently used and are considered to be valuable tools in project management. Figure 8.1.1 depicts a very simplified CPM of the activities listed in Table 8.1.3 in the activity on the node format with some delayed or early starts and finishes on some of the activities. This is done because some of the activities within a node may start before a predecessor is totally complete. Determination of the critical path would need to include these delayed or early start activity estimates.

Scheduling is even more important for several reasons in sustainable construction. First, the additional activities needed to design, collect data, and verify systems, etc., which are a part of the LEED process must be integrated into the design and construction schedules. There may be cost or schedule conflicts if many of the LEED related activities are not performed in a timely fashion. The second reason pertains to a more detailed scheduling level. As noted in Chap. 6 on indoor environmental quality, if construction activities are not scheduled in special ways, there are opportunities for impacts to indoor air quality such as with mold or particulate matter. Scheduling is one of the five principles of SMACNA for good indoor air quality. This is the same for many other activities such as erosion and sedimentation control. If the sediment fence is not installed on a graded site prior to a storm event, then major

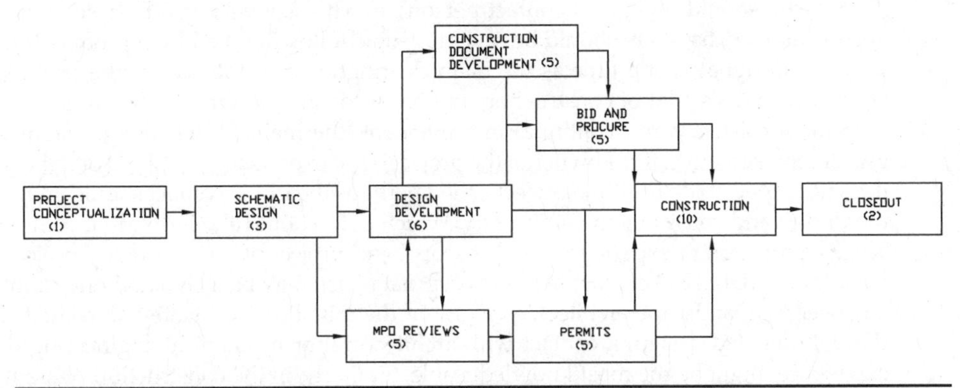

Figure 8.1.1 Simplified CPM schedule for some major construction project activities with some overlapping starts and finishes. (*Overall activity time in months is given.*)

erosion may result. Usually the best scheduling technique for these types of activities is the CPM.

Because of the need for detailed schedules with sequenced activities related to several of the LEED credits, there is also a need for staging areas for these activities to be noted on the plans. Some examples are as follows:

- Staging areas need to be noted for areas where materials may need to be covered from the elements.
- Areas need to be designated for C&D debris separation and storage prior to removal.
- Erosion and sediment control measures need to be designated on the site plans.

Lead People and LEED

Throughout the project it is important to make sure that there is some consistency in the incorporation of LEED into the project. All summary information relating to the LEED aspects should be filtered through one entity established for this, and frequent contact should be made between the central LEED clearinghouse person and all the project team members and interested parties. This entity may change through the course of a project and the LEED information be passed to another, but there should always be a central clearinghouse established at all times.

The inclusion of parties in LEED decisions outside of the core design team may be crucial even at the onset of the project. For instance, modifications may impact earlier zoning approvals, and many zoning-related changes may take months for approval, severely impacting a project schedule. During the zoning stage, the lead for the project may be, for instance, an attorney or planner, and later during detailed design, the lead may pass to an architectural or other design firm. Nonetheless, there should be a central contact for the green design aspects in contact with the project lead person who can see if any changes have major impacts on the green process. In addition, many projects have specialty designers and occupants or owner subsets, who should be involved. The baseball stadium project at the University of South Carolina represents a case where there are specialty designers for the baseball field as well as special needs during operations of the team and managers which should be considered in decisions, including many LEED related decisions.

So who should this green contact person be? This is a very good question, and the answer may be that there should be several people following the LEED process, but with only one designated at a time as the main clearinghouse. If the owner of a project is an entity which does a lot of construction and has a construction management department, then for consistency and continual improvement, the main LEED contact might be an owner representative. If the owner of the project is not experienced in LEED or if the owner does not expect to do multiple projects, then perhaps the LEED contact should come from the design and construction community. Which field should the individual come from? This is a function of experience, project scope, and project phase. For many projects, the most influential LEED contact in the conceptual phase may be a civil and environmental engineer or a landscape architect, as many of the initial decisions are site related. Then, through detailed design, a contact with architectural or mechanical engineering design experience might be the most knowledgeable. Similarly, in the construction phase it may be easiest for a commissioning or contractor representative to oversee the main part of the

LEED process. Regardless, as long as there is consistency and a smooth transition from the main contacts throughout the project life, the process should work well.

So what needs to be done to make sure that the LEED clearinghouse contact is adequately involved in all stages of the project? Communication, communication, communication! It is not necessary to call special meetings with the green contact all the time that might add to the project schedule and costs. It is necessary, however, that this person be included on all meeting and important decision correspondence. Then on a periodic basis it is helpful to have a portion of a meeting or a special charrette dedicated to the LEED process. This can be done simply by adding the green contact person to the project team meetings and including a section on LEED in construction and other meetings, such as is done with QA/QC. The project checklist with credit estimates as portrayed in Tables 1.2.3 or 8.1.2 is a good example of a document that should be updated by the green contact person and distributed to all project team members frequently throughout a project.

The really important part is that the entire project be looked upon as a system and the green credits be interpreted as a package. It has been found that if individual entities each evaluate specific credits, there is a tendency to add or delete credits based on limited economic and/or environmental perspectives. This may lead to going green becoming an added cost to a project. However, if credits are evaluated in a systemwide manner with many project features clustered to obtain multiple credits and environmental goals, then this reduces the costs of construction and the project. The only way that this can be done effectively is to always have one green lead, or "director of sustainability" as others may wish to call the person, for a project at any time. Any transfers of personnel for this position must be done well and comprehensively.

Miscellaneous

There will be many additional rules of thumb developed to aid in the green process as the green rating systems become more prevalent. However, already several suggestions can be made based on past experiences. The best way to learn about many of these is to stay informed about green building and, with respect to the LEED rating system, become familiar with the CIRs and other documents available. Some additional suggestions are listed in the following sections.

Preliminary Site Assessment

Many existing site features can impact the green decisions that are made. Having a standard questionnaire or template for preliminary site assessment that includes questions about these features can be a useful tool. Including a summary of these features on the documents in all phases of the project, including project conceptualization, may facilitate a more efficient and economical incorporation of sustainability into the project. Some of the items include

- Prevailing wind direction
- Building rotation
- Neighboring shading features such as foliage and buildings
- Floodplain locations
- Wetlands, streams, rock outcroppings, and other natural features
- Soil types
- Existing on-site vegetation

- Grade considerations
- Existing area and on-site utilities
- Prior site uses and any existing facilities such as pavement, pads, and buildings

Predesign Site Selection

Predesign site selection can have a large impact on the ease of obtaining many of the credits. Some of the credits may not seem attainable at first glance, or may be obtained if there is some forethought about how they may be accomplished. For instance,

- The development density option in SSc2 may be more applicable than expected for campuses and other projects where large areas of land may be used for recreational or community purposes.
- The community connectivity option in SSc2 may be more applicable if there are community programs that the proposed project can support, which will enhance this credit, such as contributing or supporting "rails to trails" projects or donating land or funds for the municipality to improve pedestrian access, etc.
- The brownfields credit may be obtained for projects that have some contamination in only small areas, such as an old UST.
- The project team should check for possible planned SSc4.1 public transportation access or perhaps develop a unique plan for project-provided access to these systems.
- All three of the remaining Alternative Transportation credits (SSc4.2, 4.3, and 4.4) may be fairly easy to accomplish if parking is looked at from a holistic view, especially in campus situations, where a comprehensive parking plan may be applicable to multiple projects.

Project Area and Project Boundary

As mentioned in earlier chapters, both the project area and the project boundary are important variables in many of the credit determinations for the USGBC rating system. For many projects these are easily set as the lot in question. However, for campus projects and projects on land areas which also contain other buildings or facilities, there is some flexibility in determining the project area and the project boundary, or extent of disturbance. Careful consideration of the project area and extent of disturbance can aid in obtaining LEED credits and also have a positive influence on the impact of the project. There is no one good solution. Sometimes it is useful to have greater area included if the additional space contains beneficial features such as open space or natural habitat. Sometimes it is useful to have a smaller project area if a goal is to maintain dense urban communities to better use existing infrastructure. It is important that the project area be viewed as a variable that can have a positive impact on sustainability and that the decisions include the various project participants.

There are some cases in which the project area may be made up of land portions that are not contiguous. This may be common on campuses or other land areas where there are existing facilities that are to remain or there are utility sources that are shared by many buildings. It is more difficult to differentiate whether renewable energy sources or other resources are part of the project when they may be shared by many facilities, both existing and proposed. The University of South Carolina was recently working on a library

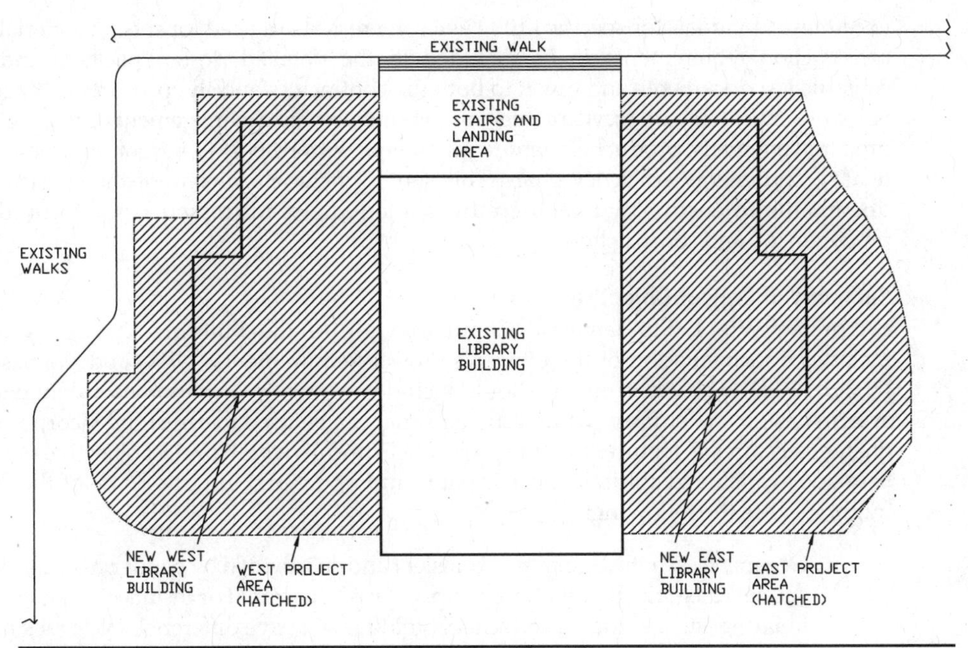

Figure 8.1.2 Example project where portions of the site area are not contiguous.

addition project where the project area was bisected by some of the existing facilities. Due to the existing layout, it was determined that the addition should consist of two new wings, one on the west side of the existing library and one on the east side. Figure 8.1.2 is a rough sketch of what a design for the additions might look like. As can be seen, the project area consists of two noncontiguous parts. This makes the determination of some of the LEED criterion variables such as the density radius and the density boundary more difficult. However, the goal in sustainable construction is to meet the intent of a credit, and extenuating circumstances can always be explained in the submittals.

Designating the project area does not relieve the project from responsibility for certain actions which may be outside of the boundary. One example includes the ESC measures that are part of the SS requirements. These must be implemented anywhere that the project might impact, which may include protecting existing catch basins off-site. For most projects, there will be off-site work relating to the project such as drives and utilities in the state DOT or local right-of-way. Appropriate ESC measures must be taken there too.

Designation of Responsible Parties

As a project develops, it is also useful to keep a table listing the responsible party for many of the specific data collection, reporting, and oversight needs for sustainable construction. The simplest way to start is to designate responsible parties for each of the credits listed on the project checklist. This can then be further subdivided into actual activities as needed for accurate project control. The preliminary designation of responsible parties should also be made a part of the bid package and construction documents so that the contractors are well aware of their responsibilities from the onset. This helps prevent costly change orders and also lets the green criteria become an initial part of the project quality control and quality assurance program. To facilitate this transfer of responsibility to certain individuals for specific LEED items, there is also a need for special material and construction requirements to be included in the detailed design package and the construction documents and given to both the contractors and the procurement agents. Some example items are recycle content requirements, waste management needs, regional procurement goals, product content requirements with respect to indoor environmental quality, and energy efficiency goals. This should include a review of the construction documents to ensure that each credit is adequately addressed prior to bidding, procurement, and construction.

Summary of Project Objectives

EA prerequisite 1, Fundamental Commissioning of the Building Energy Systems, includes documentation of the Owner's Project Requirements (OPR) and the Basis of Design (BOD). These documents should include not only information useful for energy system design and efficiency, but also items that can be used to direct the incorporation of other sustainable features into the project. A listing of overall project objectives should be made available to all project participants and updated regularly. Some of the items might include the following:

- Occupancies, both regular and transient and subdivided by gender, age, and other special categorization that may impact project design. For instance, a high school building and a kindergarten school building will have different bicycle rack needs for the same number of students. A nursing home and an apartment building will have different access and water needs. Occupancies determine many of the needs for sustainable site, water, energy, and room or furniture requirements.
- Transportation and areal access needs including parking and expected trip generation.
- Interior space needs.
- Exterior space objectives.
- Hours and frequencies of operation.
- Maintenance needs.
- Special needs including
  - Security
  - Special chemical use or storage needs
  - Special facilities such as laboratories or examination rooms

This list is in no way complete, but can aid in asking initial questions of the owner or decision makers so that design and construction can proceed in a more efficient and sustainable manner.

8.2 Summary of Relevant Codes and Credits

Summary of Relevant Codes

The LEED-NC 2009 rating system and Reference Guide list many codes, standards, and other guidances that are used in the various prerequisite and credit determinations. Table 8.2.1 lists many of these by prerequisite or credit as a summary tool for obtaining any supplemental information needed to aid in the project execution.

Prerequisite or Credit	Referenced Standards/Codes, etc.
SSp1: Construction Activity Pollution Prevention	EPA Construction General Permit (CGP for ESC)
SSc1: Site Selection	• 7 CFR 657.5 (Prime Farmland) • FEMA 100-year flood definition • U.S. Fish and Wildlife Service Threatened or Endangered Species List • 40 CFR Parts 230–233 and Part 22 (Wetlands)
SSc2: Development Density and Community Connectivity	None
SSc3: Brownfield Redevelopment	• ASTM E1903-97 Phase II Environmental Site Assessment • EPA brownfields definition
SSc4.1: Alternative Transportation: Public Transportation Access	None
SSc4.2: Alternative Transportation: Bicycle Storage and Changing Rooms	None
SSc4.3: Alternative Transportation: Low-Emitting and Fuel-Efficient Vehicles	• CARB definition of zero emissions vehicles (ZEV) or • ACEEE Annual Vehicle Rating Guide
SSc4.4: Alternative Transportation: Parking Capacity	Maximum parking allowed by Local zoning code minimum
SSc5.1: Site Development: Protect or Restore Habitat	None
SSc5.2: Site Development: Maximize Open Space	Local zoning code
SSc6.1: Stormwater Management: Quantity Control	None
SSc6.2: Stormwater Management: Quality Control	TARP (Washington State) Data Protocol for BMP monitoring
SSc7.1: Heat Island Effect: Non-Roof	• SRI: ASTM E1980-01 • Emittance: ASTM E408-71(1996)e1 or C1371-04 • Reflectance: ASTM E903-96, E1918-97, or C1549-04
SSc7.2: Heat Island Effect: Roof	• SRI: ASTM E1980-01 • Emittance: ASTM E408-71(1996)e1 or C1371-04 • Reflectance: ASTM E903-96, E1918-97, or C1549-04
SSc8: Light Pollution Reduction	ASHRAE/IESNA Standard 90.1-2007, Section 9 (Exterior)

Table 8.2.1 Listing of Referenced Standards—LEED 2009 (*Continued*)

Prerequisite or Credit	Referenced Standards/Codes, etc.
WEp1: Water Use Reduction Required	• EPAct: Energy Policy Act of 1992, 2005 • UPC1-2006 • IPC 2006 Section 604
WEc1: Water Efficient Landscaping	None
WEc2: Innovative Wastewater Technologies	• EPAct: Energy Policy Act of 1992, 2005 • UPC1-2006 • IPC 2006 Section 604
WEc3: Water Use Reduction	• EPAct: Energy Policy Act of 1992, 2005 • UPC1-2006 • IPC 2006 Section 604
EAp1: Fundamental Commissioning of the Building Energy Systems	None
EAp2: Minimum Energy Performance	• ASHRAE/IESNA Standard 90.1-2007 Appendix G And Mandatory of Sections 5.4, 6.4, 7.4, 9.4, 10.4) Or • ASHRAE Advanced Energy Design Guide for Small Office Buildings, 2004 Or • ASHRAE Advanced Energy Design Guide for Small Retail Buildings, 2006 Or • ASHRAE Advanced Energy Design Guide for Warehouses and Self Storage Buildings, 2008 Or • NBI Advanced Buildings™ Core Performance™ Guide
EAp3: Fundamental Refrigerant Management	EPA Clean Air Act, Title VI, Section 608, Refrigerant Recycling
EAc1: Optimize Energy Performance	• ASHRAE/IESNA Standard 90.1-2007 Appendix G Or • ASHRAE Advanced Energy Design Guide for Small Office Buildings, 2004 Or • ASHRAE Advanced Energy Design Guide for Small Retail Buildings, 2006 Or • ASHRAE Advanced Energy Design Guide for Warehouses and Self Storage Buildings, 2008 Or • NBI Advanced Buildings™ Core Performance™ Guide

Table 8.2.1 Listing of Referenced Standards—LEED 2009 (*Continued*)

Prerequisite or Credit	Referenced Standards/Codes, etc.
EAc2: On-Site Renewable Energy	ASHRAE/IESNA Standard 90.1-2007 Appendix G (if EAc1 Option 1)
EAc3: Enhanced Commissioning	None
EAc4: Enhanced Refrigerant Management	None
EAc5: Measurement and Verification	International Performance Measurement and Verification Protocol (IPMVP), Vol. III: January 2006; Option D (whole-building calibration simulation) or Option B (energy conservation measure isolation)
EAc6: Green Power	• ASHRAE/IESNA Standard 90.1-2007, Appendix G (if EAc1 Option 1) • Center for Resource Solutions Green-e program technical requirements
Materials and Resources Category Overall	CSI MasterFormat™ Division List 2004 Edition
MRp1: Storage and Collection of Recyclables	None
MRc1.1–1.2: Building Reuse	None
MRc2: Construction Waste Management	None
MRc3: Materials Reuse	None
MRc4: Recycled Content	ISO 14021—Environmental labels and declarations—Self-declared environmental claims (Type II environmental labeling)
MRc5: Regional Materials	None
MRc6: Rapidly Renewable Materials	None
MRc7: Certified Wood	Forest Stewardship Council's Certification
IEQp1: Minimum IAQ Performance	For mechanically ventilated areas: • ASHRAE 62.1-2007, Ventilation for Acceptable Indoor Air Quality, Sections 4–7 For naturally ventilated areas: • ASHRAE 62.1-2007, paragraph 5.1
IEQp2: Environmental Tobacco Smoke (ETS) Control	Residential facilities if not weather-stripped: • ANSI/ASTM-E779-03, Standard Test Method for Determining Air Leakage Rate by Fan Pressurization and • Residential Manual for Compliance with California's 2001 Energy Efficiency Standards, Chapter 4
IEQc1: Outdoor Air Delivery Monitoring	For mechanically ventilated areas: ASHRAE Standard 62.1-2007 with errata

Table 8.2.1 *(Continued)*

Prerequisite or Credit	Referenced Standards/Codes, etc.
IEQc2: Increased Ventilation	For mechanically ventilated areas: • ASHRAE Standard 62.1-2007 with errata For naturally ventilated areas: • *Carbon Trust Good Practice Guide 237* (1998) and • Chartered Institution of Building Services Engineers (CIBSE) Applications Manual 10: 2005, natural ventilation in nondomestic buildings, effective as per Fig. 1.18 And model by either • Chartered Institution of Building Services Engineers (CIBSE) Applications Manual 10: 2005, natural ventilation in nondomestic buildings or CIBSE Applications Manual 13:2000, Mixed Mode Ventilation (as applicable) Or • ASHRAE 62.1-2007, Chapter 6, for at least 90% of occupied spaces
IEQc3.1: Construction IAQ Management Plan, During Construction	• Sheet Metal and Air Conditioning National Contractors Association (SMACNA) IAQ Guidelines for Occupied Buildings under -Construction, 2007, Chapter 3 ASHRAE 52.2-1999 for MERV ratings
IEQc3.2: Construction IAQ Management Plan, Before Occupancy	Option 2: EPA Compendium of Methods for the Determination of Air Pollutants in Indoor Air
IEQc4.1: Low-Emitting Materials: Adhesives and Sealants	Adhesives, sealants, and Sealant Primers: • South Coast Air Quality Management District (SCAQMD) Rule #1168, through 01/07/2005 Aerosol adhesives: • *Green Seal Standard for Commercial Adhesives GS-36* requirements 10/19/2000
IEQc4.2: Low-Emitting Materials: Paints and Coatings	Architectural paints, coatings, and primers applied to interior walls and ceilings: -VOC content limits in *Green Seal Standard GS-11, Paints*, 1st ed., May 20, 1993. • Anti-corrosive and antirust paints applied to interior ferrous metal substrates: • VOC content limit of 250 g/L established in *Green Seal Standard GC-03, Anti-Corrosive Paints*, 2d ed., January 7, 1997. • Clear wood finishes, floor coatings, stains, and shellacs applied to interior elements: • VOC content limits established in South Coast Air Quality Management District (SCAQMD) Rule 1113, Architectural Coatings 10/1/2004

Table 8.2.1 Listing of Referenced Standards—LEED 2009 (*Continued*)

Prerequisite or Credit	Referenced Standards/Codes, etc.
IEQc4.3: Low-Emitting Materials: Flooring Systems	• Carpet Systems: CRI's Green Label Plus program for carpet CRI's Green Label program for cushion See IEQc4.1 for adhesives • Hard Flooring Systems: FloorScore certified for flooring See IEQc4.2 for finishes See IEQc4.1 for adhesives/grout Or • All flooring CA Dept. of Health Services Std. Practice for the Testing of Volatile Organic Emissions form Various Sources using Small-Scale Environmental Chambers, including 2004 addenda
IEQc4.4: Low-Emitting Materials: Composite Wood and Agrifiber	None
IEQc5: Indoor Chemical and Pollutant Source Control	ASHRAE 52.2-1999 for MERV ratings
IEQc6.1: Controllability of Systems: Lighting	None
IEQc6.2: Controllability of Systems: Thermal Comfort	• ASHRAE Standard 55-2004, w/errata: Thermal Environmental Conditions for Human Occupancy • Balance with ASHRAE Standard 62.1-2007 for Acceptable Indoor Air Quality
IEQc7.1: Thermal Comfort: Design	ASHRAE Standard 55-2004, w/errata: Thermal Environmental Conditions for Human Occupancy, Section 6.1.1
IEQc7.2: Thermal Comfort: Verification	ASHRAE Standard 55-2004, w/errata: Thermal Environmental Conditions for Human Occupancy
IEQc8.1: Daylighting and Views: Daylight	ASTM D1003-07e1: Std. Test Method for Haze and Luminous Transmittance of Transparent Plastics
EQc8.2: Daylighting and Views: Views	None (EP: Heschong Mahone Group Study: Windows and Offices)
IDc1: Innovation in Design	None or as listed previously for applicable EP point subcategories
IDc2: LEED Accredited Professional	Green Building Certification Institute (GBCI) accreditation
RP: Regional Priorities	None or as listed previously for applicable RP subcategories

Table 8.2.1 (*Continued*)

Summary of the Credits

Sustainable construction is composed of many diverse objectives, criteria, impacts, and design features. It is a systematic way to build that includes multiple disciplines and various requirements that sometimes have possibly conflicting outcomes. Due to the complexity of sustainability, it is difficult to keep track of all the objectives. The USGBC LEED rating system and other sustainability guides do an excellent job of organizing and ranking many of these complex system objectives. However, it is still difficult to remember all the various parts, particularly since most people cannot be experts in everything. To help overcome the enormity of it all, it is recommended to have as many project participants as possible familiar with the prerequisites and credits. One way to accomplish this is to have many project decision makers accredited in LEED or by other environmental certifications. To facilitate both studying for the LEED accreditation and using it as a tool for any interested party to become familiar with the prerequisites and credits, the author has developed a study table. Two examples of the template for study tables are presented in Tables 8.2.2 and 8.2.3 for Sustainable Sites and Water Efficiency, respectively.

The first thing one might notice is that most of the cells in Tables 8.2.2 and 8.2.3 are empty. This is so that individuals can complete the table by reviewing the Reference Guide or the chapters in this text and complete them in the manner most useful for that individual. A civil engineer will need different information in various cells to understand the items than a person trained in architecture or planning or mechanical engineering. Thus, these tables can be used both for reviewing the prerequisites and credits and for customizing the information to each individual. One of the most import columns to complete is the variable/concepts column. What fits here will vary tremendously from credit to credit, especially since many credit criteria are based on quantitative variables and other criteria on qualitative concepts. For instance, for SSc2, there are specific quantitative variables such as density radius and development density that are used, whereas for EQc3.1 the concepts include the five principles of SMACNA.

It is intended that the individual make similar tables for the other LEED-NC categories and complete them for a comprehensive review of the rating system and as a tool for future reference.

8.3 Additional Tools and Education

Energy Star, BEES, and More

Two national rating tools have been developed in recent years that can aid in many of the sustainable design decisions. One has been developed by the EPA and rates individual building products based on energy consumption. It is the Energy Star program. The Energy Star label is a common feature on many consumer products also. The second tool has been developed by the National Institute of Standards Technology (NIST). It is called Building for Environmental and Economic Sustainability (BEES) and can be used to aid in choosing building materials based on both environmental and economic considerations. The current version in 2007 was *BEES 4.0*. It can be downloaded for free from the NIST website.

The BEES model usually assesses products based on a 50-year model, and it looks into various stages of the product life. The main stages are

- Raw materials acquisition
- Product manufacture

Credit	Exemplary Performance Point Availability	Exemplary Performance Point Criterion	Options	Standards/Codes	Criterion	Submittals	Comments	Variables/Concepts
SSp1: Construction Activity Pollution Prevention	n.a.							
SSc1: Site Selection	No							
SSc2: Development Density and Community Connectivity	Yes	Double (Option 1)						
SSc3: Brownfield Redevelopment	No							
SSc4.1: Alternative Transportation—Public Transportation Access	Yes	Double plus Minimum Transit Rides/Day						
SSc4.1: Alternative Transportation: Public Transportation Access								
SSc4.2: Alternative Transportation: Bicycle Storage and Changing Rooms	Yes	Plan (not available if attain EP for SSc4.1)						
SSc4.3: Alternative Transportation: Low-Emitting and Fuel-Efficient Vehicles								
SSc4.4: Alternative Transportation: Parking Capacity								
SSc5.1: Site Development: Protect or Restore Habitat	Yes	75%						
SSc5.2: Site Development: Maximize Open Space	Yes	Double						
SSc6.1: Stormwater Management: Quantity Control	No	Plan to exceed both						
SSc6.2: Stormwater Management: Quality Control	No							
SSc7.1: Heat Island Effect: Non-Roof	Yes	100%						
SSc7.2: Heat Island Effect: Roof	Yes	100% Green						
SSc8: Light Pollution Reduction	No							

Table 8.2.2 SS Prerequisite and Credit Summary Table Study Guide—LEED 2009

Credit	Exemplary Performance Point Availability	Exemplary Performance Point Criterion	Options	Standards/Codes	Criterion	Submittals	Comments	Variables/Concepts
WEp1: Water Use Reduction Required	n.a.							
WEc1: Water Efficient Landscaping	No							
WEc2: Innovative Wastewater Technologies	Yes	100%						
WEc3: Water Use Reduction	Yes	45%						

TABLE 8.2.3 WE Credit Summary Table Study Guide—LEED 2009

- Transportation
- Installation
- Operation and maintenance
- Recycling and waste management

Various impacts are evaluated in the BEES model, and initially the main environmental items included were GWP, acid-rain impact, nitrification potential, natural resource depletion, solid waste impact, and indoor air quality considerations. As the model developed, additional factors were added. The various subcategories evaluated in version 4 are listed in Table 8.3.1.

Other decision tools for choosing sustainable materials and products are being developed by many organizations, both public and private. As the movement evolves, these will become more readily accessible and inclusive. Even trade organizations are developing sustainable design manuals and guides for their product and services customers. An example is the National Ready Mixed Concrete Association (NRMCA) which has published a guide to LEED-NC with respect to concrete applications. It is a good idea to ask the various trade organizations representing the building materials that one uses if they can supply you with their current updated sustainability guide.

Incorporation into CSI Formats

There has been much interest into how best to incorporate sustainable construction into the established building process methods in the United States. Even though there may

Environmental Categories	Economic Categories
Global warming	First Costs
Acidification	Future Costs
Eutrophication	
Fossil fuel depletion	
Indoor air quality	
Habitat alteration	
Water intake	
Criteria air pollutants	
Human health	
Smog	
Ozone depletion	
Ecological toxicity	

TABLE 8.3.1 Various Modules Assessed in BEES 4.0

be many methods tried and recommended, one of the most important things is to establish some consistency so that the lessons learned and experiences can be evaluated and handed down to the next generation as effectively as possible. With this in mind, it is important to establish some consistent reporting mechanisms. Since the Construction Specification Institute (CSI) has developed the CSI MasterFormat Division List, which is used extensively in the United States for construction management and data reporting, particularly with respect to materials and products, it makes sense to also follow this format with the management of data with respect to the LEED rating system. In response to this, the CSI recently developed the GreenFormat. This format requested data on products and was organized into 14 categories of questions:

- MasterFormat 2004 section number and name of material/product
- Manufacturer's information
- Product description
- Regulatory agency sustainability approvals
- Sustainable standards and certifications
- Sustainable performance criteria
- Sustainable compositions of product
- Material extraction and transportation
- Manufacturing phase
- Construction phase
- Facility operations phase
- Deconstruction and recycling phase

- Additional information
- Certification

Additional information about the use of GreenFormat can be found at www.greenformat.com.

Education

Recent research has shown that increased familiarity with the LEED system is one of the major factors in improving performance and the cost-effectiveness of green building. The two major components of this are experience with the system and education about the system. Several universities across the country have set the standard with green building, making commitments to using the LEED rating system for all new campus construction. At the University of South Carolina in Columbia, it was shown that this commitment had a significant positive impact on the familiarity of the various stakeholders in the area. A study also showed that regions of the state where universities and colleges had forged ahead with green building initiatives also became pockets of additional sustainable construction. Both experience and continual education are important.

The USGBC offers many workshops on the rating system and the various guidances. In addition, many facilities have been built that are becoming showcases for various sustainable alternatives. Some examples in South Carolina are the West Quad, Learning Center, at the University of South Carolina (see Fig. 2.7.3), the Sustainable Interiors Showcase at Fort Jackson in Columbia, and the Edisto Interpretive Center at Edisto Beach. The Southface Energy Institute in Atlanta, Georgia, and the Rocky Mountain Institute in Old Snowmass, Colorado, are both national showcases and offer a range of educational opportunities. On a federal level, in addition to the resources previously mentioned from the Department of Defense (DoD) there is the Office of the Federal Environmental Executive (OFFE) where additional resources have been collated and also documented in a report entitled *The Federal Commitment to Green Building: Experiences and Expectations* (http://www.ofee.gov/sb/fgb_report.html) and of course the EPA websites for the various environmental concerns, including a website devoted to green buildings (http://www.epa.gov/greenbuilding/).

As mentioned in Chap. 1, there is also a movement among higher education presidents and faculty to incorporate sustainability education into the curriculum. It is particularly important to add sustainability concepts to the tools that engineers use for future designs and innovations.

References*

Bilec, M., and R. Ries (2007), "Preliminary Study of Green Design and Project Delivery Methods in the Public Sector," *Journal of Green Building,* Spring, 2(2): 151–160.

CSI (2004), *The Project Resource Manual—CSI Manual of Practice,* 5th ed., Construction Specifications Institute, Alexandria, VA.

CSI (2006), "Supporting Sustainability with GreenFormat," *The Construction Specifier,* 59(12), accessed at http://www.csinet.org/s_csi/docs/13700/13644.pdf, July 26, 2007.

*Standards are as previously referenced in Chaps. 2 through 7.

Davidson, C. I., et al. (2007), "Adding Sustainability to the Engineer's Toolbox: A Challenge for Engineering Educators," *Environmental Science and Technology*, 41 (July 15): 4847–4850.

DuBose, J. R., S. J. Bosch, and A. R. Pearce (2007), "Analysis of State-Wide Green Building Policies," *Journal of Green Building*, Spring, 2(2): 161–177.

EPA (2007), http://www.epa.gov/greenbuilding/, U.S. Environmental Protection Agency webpage, accessed July 10, 2007.

GBCI (2010), http://www.gbci.org/, Green Building Certification Institute website, accessed February 6, 2010.

Hooper, P. A. (2007), "The Construction Specification Institute's Sustainable Reporting Data Format: GreenFormat," *Journal of Green Building*, Spring, 2(2): 3–13.

Horman, M. J., et al. (2006), "Delivering Green Buildings: Process Improvements for Sustainable Construction," *Journal of Green Building*, Winter, 1(1).

Huff, W. (2007), "Obstacles and Rewards of Enterprise Sustainability and the Need for a Director of Sustainability," *Journal of Green Building*, Spring, 2(2): 62–69.

Kibert, C. J. (2005), *Sustainable Construction; Green Building Design and Delivery*, John Wiley & Sons, Hoboken, N.J.

NIST (2007), *BEES 4.0*, National Institute of Standards Technology, Gaithersburg, MD., http://www.bfrl.nist.gov/oae/software/bees/ website accessed August 2, 2007.

NRMCA (2005), *Ready Mixed Concrete Industry LEED Reference Guide—Updated with LEED Version 2.2 Information*, National Ready Mixed Concrete Association, Silver Spring, MD.

OFEE (2003), *The Federal Commitment to Green Building: Experiences and Expectations*, Office of the Federal Environmental Executive, http://www.ofee.gov/sb/fgb_report.html accessed July 10, 2007.

Portney, K. E. (2002), "Taking Sustainable Cities Seriously: A Comparative Analysis of Twenty-Four US Cities," *Local Environment*, 7(4): 363–380.

RMI (2007), http://www.rmi.org/, Rocky Mountain Institute webpage, accessed July 10, 2007.

Schexnayder, C. J., and R. E. Mayo (2004), *Construction Management Fundamentals*, McGraw-Hill, New York.

Southface (2007), http://www.southface.org/, Southface Energy Institute webpage accessed July 10, 2007.

Thomas, D., and O. J. Furuseth (1997), "The Realities of Incorporating Sustainable Development into Local-Level Planning: A Case Study of Davidson, North Carolina," *Cities*, 14(4): 219–226.

USGBC (2003), *LEED-NC for New Construction, Reference Guide*, Version 2.1, 2d ed., May, U.S. Green Building Council, Washington, D.C.

USGBC (2005–2007), *LEED-NC for New Construction, Reference Guide*, Version 2.2, 1st ed., U.S. Green Building Council, Washington, D.C., October 2005 with errata posted through Spring 2007.

USGBC (2009), *LEED Reference Guide for Green Building Design and Construction*, 2009 Edition, U.S. Green Building Council, Washington, D.C., April 2009.

USGBC (2009), *LEED 2009 for New Construction and Major Renovations Rating System, For Public Display*, www.usgbc.org, accessed April 20, 2009 and February 20, 2010, U.S. Green Building Council, Washington, D.C.

Watson, C. H. (2006), *The University of South Carolina: A Catalyst for Sustainable Development?* MS thesis, University of South Carolina, Columbia.

Exercises

1. Put together a CPM schedule for sequencing the installation of absorbent materials related to SMACNA for EQc3.1 and EQc3.2 for materials stored outside. Two options that might be used are staggered procurement and on-site protected storage.

2. Put together a CPM schedule for sequencing the installation of absorbent materials related to SMACNA for EQc3.1 and EQc3.2 with respect to interior painting and finishing and VOCs.

3. Look up a paint product on the GreenFormat system and list the information.

4. Look up a wood furniture product on the GreenFormat system and list the information.

5. Look up a plumbing product on the GreenFormat system and list the information.

6. Download the BEES program and evaluate the performance of a particular product.

7. Go to the appliance store and compare the Energy Star labels on four different refrigeration units.

8. Compare the Energy Star labels on four different television sets.

9. Compare the Energy Star labels on four different driers.

10. Identify three different materials or products used to construct your classroom. Find out if green information is available on the product websites.

11. Go to the paint department in a store and compare the labels on similar paint products from three different manufacturers. Compile and compare any environmental data. Go to the manufacturers' websites and see if green standards are listed for these products. Summarize them.

12. Go to the carpet department in a store and find the names of three different manufacturers. Go to the manufacturers' websites and see if green standards are listed for these products. Summarize them.

13. Go to a furniture department and find three manufacturers of a specific type of furniture product (for example, wood chair, mattress, couch, dining table). Go to the manufacturers' websites and see if green standards are listed for these products. Summarize them.

14. Visit a vacant lot in your area and make a list of existing site features that you can see from outside the property line and other areal information such as direction, access, and nearby utilities.

15. Go online and see what additional information you can collect on the site in Exercise 14. This may include mapping programs, land record databases, MPO GIS websites, and state websites. There are also websites that list areal environmental items such as designated brownfields and floodplain information.

16. Develop an initial site assessment questionnaire that can be used for preliminary site assessments from a site visit.

17. Develop an expanded site assessment questionnaire that can be used to collect information from local land record offices, departments of public works, and other public record agencies that will be useful in initial site assessments.

18. Look up one of the Code of Federal Regulation (CFR) codes referenced in Table 8.2.1. Summarize your findings. These can usually be found online.

19. Look up one of the EPA referenced items in Table 8.2.1. Summarize your findings. These can usually be found online.

20. Go to a library (or other resource) and read one of the ASTM standards referenced in Table 8.2.1. Summarize your findings.

21. Go to a library (or other resource) and read one of the ASHRAE standards referenced in Table 8.2.1. Summarize your findings.

22. Look up EPAct 1992 as referenced in Table 8.2.1. Summarize your findings.

23. Look up one of the Green Seal standards as referenced in Table 8.2.1. Summarize your findings.

24. Update Table 8.2.1 based on any addenda to LEED-NC 2009 or for the current version of LEED-NC.

25. Complete Table 8.2.2.

26. Complete Table 8.2.3.

27. Make and complete a summary study guide similar to Table 8.2.2 for the EA category.

28. Make and complete a summary study guide similar to Table 8.2.2 for the MR category.

29. Make and complete a summary study guide similar to Table 8.2.2 for the IEQ category.

CHAPTER 9

Department of Defense (DoD) Sustainable Construction and Indoor Air Quality (IAQ)

Construction spending in the United States as of December 2006 totaled approximately $1200 billion annually. The federal government alone pumped over $19 billion a year into the economy for the construction of new facilities, and the U.S. government had an inventory of approximately 445,000 buildings occupying nearly 3.1 billion ft^2 of floor space. These facilities consumed approximately 0.4 percent of the nation's total energy usage and spewed out about 2 percent of all building-related greenhouse gases. The federal government is a major consumer of construction materials and a significant user of energy and natural resources. To minimize the impact that federal facilities have on the environment, the U.S. government has become a leader in implementing policy and objectives for energy reduction and green building.

9.1 Government Mandates

The President of the United States exercises authority over all the federal agencies within the executive branch of the U.S. government. One of the ways the President provides guidance is through written directives called Executive Orders (EOs). In the late 1990s, President Clinton began issuing a series of EOs aimed at "Greening the Government." Among other objectives, these have served to set the baseline for energy reduction and implementation of green building within all federal agencies. EO 13101, "Greening the Government Through Waste Prevention, Recycling, and Federal Acquisition," was the first of this series of EOs and was signed in September 1998. Some of the requirements of this order included promotion of waste prevention and recycling and the use of environmentally preferable products. Federal agencies are required to establish procurement programs and develop specifications to ensure all products meet or exceed Environmental Protection Agency (EPA) guidelines. This requires maximizing the use of materials that are reused, recovered, or recycled and have reduced or eliminated toxicity. Additionally, this order required each agency to establish goals for solid waste prevention and recycling. In June 1999, EO 13123, "Greening the Government Through Efficient Energy Management," was signed, setting aggressive goals for reducing the government's energy consumption. This order established the following targets:

- Reduce building-related greenhouse gases by 30 percent by 2010 compared to 1990 levels.
- Reduce energy consumption for facilities by 30 percent by 2005 and 35 percent by 2010 relative to 1985 energy usage.
- Install 2000 solar energy systems for federal facilities by the year 2000 and 20,000 by the year 2010.
- Reduce petroleum use by switching to natural gas or renewable fuels or by eliminating unnecessary use.
- Reduce water consumption to meet Department of Energy water conservation goals.

This order also established objectives for federal agencies to maximize use of Energy Star products, pursue Energy Star building criteria, and implement sustainable building design. Energy Star is a program of the U.S. Environmental Protection Agency and the U.S. Department of Energy which has the purpose of helping the United States save money and protect the environment through the use of energy-efficient products and practices. The sustainable design principles developed by the Department of Defense and General Services Administration were also specifically cited as benchmarks for other agencies.

Another federal publication, the Office of Management and Budget's Circular A-11, provided more details and updates to the goals behind EO 13123 and specifically encouraged agencies to evaluate the use of Energy Star or LEED standards and guidelines for incorporation in designs for new building construction or renovation in Sec. 55 in its 2002 version. A summary of how the different federal agencies have responded to this order as reported by the Office of the Environmental Executive can be found in the 2003 report *The Federal Commitment to Green Building: Experiences and Expectations.*

Pollution prevention (P2) and environmentally preferable purchasing (EPP) were programs set forth by EO 13148, "Greening the Government Through Leadership in Environmental Management," signed in April 2000. This order furthered the use of life-cycle considerations and implemented additional acquisition and procurement practices to select products and develop facilities that have reduced toxic chemicals, hazardous substances, and pollutants. This order also required the planning and development of environmentally and economically beneficial landscaping.

In January of 2007, President Bush signed Executive Order 13423, "Strengthening Federal Environmental, Energy, and Transportation Management." This order further identified sustainable strategies and goals for the agencies of the federal government. Each agency was ordered to meet certain energy, transportation, and environmental goals and to maintain an environmental management system (EMS) structure. The specific manners in which these were met are to be developed by the agencies. EO 13423 also revoked the three previously mentioned orders signed by President Clinton. Then in October of 2009, President Obama signed Executive Order 13514, "Federal Leadership in Environmental, Energy, and Economic Performance." EO 13514 enhanced the provisions in EO 13423. Some major requirements included the naming of a Senior Sustainability Officer (SSO) by November 5th, 2009 and the

Executive Order	Title	Federal Register Date	Status as of July 2007
EO 13101	Greening the Government through Waste Prevention, Recycling, and Federal Acquisition	September 16, 1998	Revoked Jan. 26, 2007
EO 13123	Greening the Government through Efficient Energy Management	June 8, 1999	Revoked Jan. 26, 2007
EO 13148	Greening the Government through Leadership in Environmental Management	April 26, 2000	Revoked Jan. 26, 2007
EO 13423	Strengthening Federal Environmental, Energy, and Transportation Management	January 26, 2007	Current and Updated by EO 13514
EO 13514	Federal Leadership in Environmental, Energy, and Economic Performance	October 5, 2009	Current

TABLE 9.1.1 Recent U.S. Sustainable Construction Related Executive Orders

setting of 2020 greenhouse gas (GHG) emission reduction targets by each Federal agency. A summary of these orders is presented in Table 9.1.1.

9.2 Department of Defense Facilities

In 2006, the Department of Defense (DoD) owned a total of over 330,000 buildings occupying more than 2.0 billion ft^2 of floor space. The DoD accounted for approximately two-thirds of the federal government's buildings by number and square footage and used 244 trillion Btu of energy annually at a cost of $2.6 billion. The DoD continues to grow and modernize through new facility construction. Its military construction (MILCON) budget has averaged $6 billion per year during the 5-year period from 2001 through 2005 and grew to $8 billion for fiscal year 2006 (FY06). Based on the volume of projects for the development of facilities for new and emerging requirements, renovation and modernization of aging facilities, and support for moving commands and service members due to base realignment and closure (BRAC), the DoD budget showed a MILCON volume of nearly $8 billion for FY07 and almost doubled to $15 billion per year for FY08 through FY11. The DoD has become a leader among government agencies in implementing sustainable principles, such as energy reduction, recycling and waste minimization, and lessening the impact on the environment, for planning, design, and construction through experience and by necessity. Its implementation involves not only buildings (vertical construction), but also sustainability of the ranges and other horizontal construction. The DoD has also adopted *low-impact development* (LID) practices at many facilities. After the signing of EO 13514 in late 2009, Aston B. Carter was named as the SSO for the DoD.

U.S. Army and the U.S. Army Corps of Engineers

Within the Department of the Army, installation management services are provided by the Office of the Assistant Chief of Staff for Installation Management (ACSIM).

This includes establishing policy and program management for MILCON. However, the actual coordination and management of MILCON projects are the responsibility of the U.S. Army Corps of Engineers (USACE). The Army is the largest federal facility owner of all agencies with over 160,000 buildings occupying 1.1 billion ft^2 of floor space. This is nearly one-third of all federal buildings and one-half of those owned by the DoD. The USACE also manages an equivalent percentage of the MILCON budget each year. Following the signing of EO 13123, the Office of the Deputy Assistant Secretary of the Army for Installations and Housing (DASA-I&H) issued a memo in April of 2000 addressing sustainable design and development. DASA-I&H directed ACSIM to implement policy and USACE to develop technical guidance for sustainable construction. In May 2000, ACSIM published the *sustainable design and development* (SDD) policy, defining *SDD* as "the systematic consideration of current and future impacts of an activity, product or decision on the environment, energy use, natural resources, the economy and quality of life" and mandating SDD considerations for Army installations. One year later, USACE developed and released the Sustainable Project Rating Tool (SPiRiT) and provided technical guidance for implementation in Technical Letter ETL 1110-3-491, Sustainable Design for Military Facilities. SPiRiT is a tool similar to and based on LEED, although it incorporates operations and maintenance issues, allows for flexibility in design for building modifications based on need changes, and is a self-rating tool that does not require third-party certification as with LEED. ACSIM issued a policy statement in May 2001 for the implementation of SPiRiT with the initial requirement of a bronze rating for all new construction projects. Over the next several years, the sustainable rating requirements were raised to silver and then to gold, and several version changes were made to further refine SPiRiT. In late 2005, the USACE's Construction Engineering Research Laboratory (CERL) released the results of a study comparing SPiRiT rated projects to LEED and recommended that the Army adopt LEED as the primary rating tool. On January 5, 2006, the Office of the Assistant Secretary of the Army for Installations and the Environment (OASA-I&E) released a memorandum mandating the change from SPiRiT to LEED effective with the FY08 MILCON program and establishing LEED silver as the standard. The U.S. Army has aggressively pursued sustainable construction in the MILCON program and seems poised to continue raising the standard as LEED grows as the industry benchmark.

The Army has been progressively establishing 25-year sustainability goals at many of its facilities. This effort is led out of the Sustainability Division ODEP (Office of Director of Environmental Programs) in the Pentagon. In response to this initiative, in 2007, a group at Fort Jackson in Columbia, S.C., developed a showcase for "Sustainable Interiors" as both a way to incorporate sustainability into the purchasing procedure on post and an educational outlet for the Army and the community. Figure 9.2.1 shows an educational cutaway in the showcase.

Naval Facilities Engineering Command

For the Department of the Navy, the Commander, Naval Installations Command (CNIC), owns and manages all the Navy's installations and facilities thereon. CNIC relies on Naval Facilities Engineering Command (NAVFAC) for all engineering support, including the establishment of technical guidance and policy, related to facility sustainability and execution of the MILCON program. The Department of the Navy, including the Marine Corps, owns about 30 percent of the DoD's buildings and has

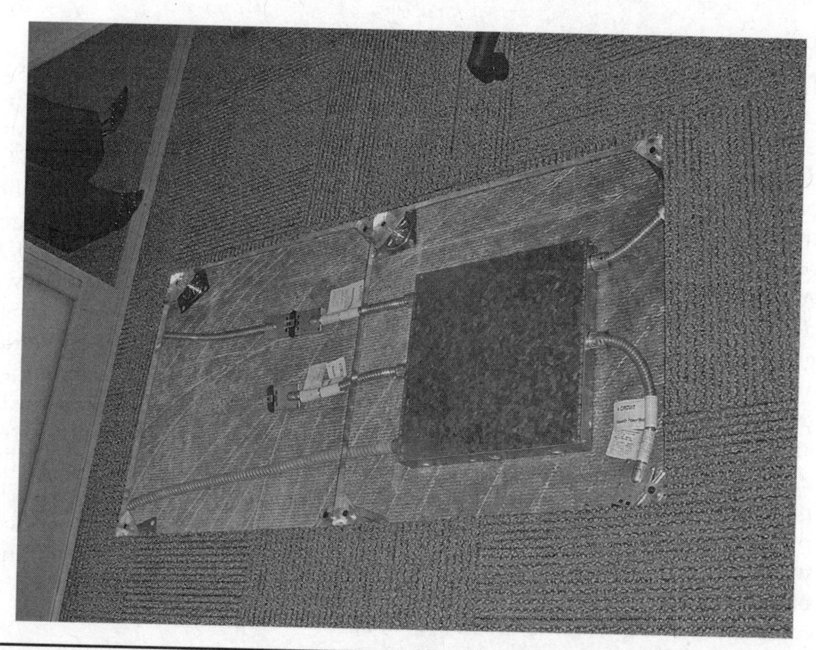

FIGURE 9.2.1 Army Sustainable Interiors demonstration and education site. The cutaway shows a TecCrete variable-height floor system by Haworth, Inc. Raised floor systems can help provide space to facilitate energy-efficient controls and air distribution in buildings as related to several EA credits. (*Photograph taken at the Sustainable Interiors ribbon cutting at the Strom Thurmond Building on Fort Jackson, Columbia, S.C., June 2007.*)

recently had a robust MILCON program, managing above normal construction volume mainly due to the naval installations on the Gulf Coast that sustained significant damage due to hurricanes Ivan in 2004 and Katrina, Rita, and Wilma in 2005. The Navy was the first federal agency to pursue implementation of sustainable design features into MILCON projects. The *Whole Building Design Guide* (www.wbdg.org) was launched by the Navy in 1997 to begin incorporating sustainable requirements into construction standards, specifications, and guidelines. The Construction Criteria Base (CCB) was built within the WBDG website as a data store for standardizing construction specifications for all the military services. The WBDG is now maintained by the National Institute for Building Sciences and is supported by eight federal agency partners.

NAVFAC completed DoD's first LEED certified facility with the construction of a 365,000 ft^2, $55 million Bachelor Enlisted Quarters at Great Lakes Naval Training Center in FY99. Since then several major renovation projects at Washington Navy Yard and the Pentagon have been showcases for applying LEED for existing buildings and the use of LID. In July 2002, NAVFAC's Chief of Engineering issued a memorandum adopting the LEED rating system as a tool and metric for MILCON. In June 2003, NAVFAC Instruction 9830.1, Sustainable Development Policy, was issued requiring all construction, renovation, and repair projects over $750,000 to attain the LEED certified level and Navy Family Housing to implement the EPA Energy Star Label Homes Program. The Department of the Navy was a pioneer in making sustainable design and development a standard for military construction and continues these efforts through utilization of LEED.

Air Force Center for Environmental Excellence

The U.S. Air Force programs and budgets for MILCON requirements go through the Office of the Civil Engineer. Field agencies within this office include the Air Force Civil Engineer Support Agency (AFCESA), which establishes construction standards and design criteria, and the Air Force Center for Environmental Excellence (AFCEE), which provides technical support and guidance. Although it is the smallest facility owner of the military services, the Air Force still maintains a sizable MILCON budget; however, execution of Air Force MILCON projects is divided between USACE and NAVFAC (the Air Force does not staff its own construction agent). The Air Force was the first department within the DoD to adopt the U.S. Green Building Council's (USGBC) LEED green building rating system as the standard for measuring sustainable development and construction. The Sustainable Development Policy was issued by a memorandum from the U.S. Air Force (USAF) Civil Engineer Deputy Chief of Staff for Installations and Logistics in December 2001. The goals set forth in this memo were to achieve 20 percent of Air Force MILCON projects selected as LEED pilots by FY04 and all projects to become LEED certified by FY09. The Air Force, through AFCEE, has also published a number of guides to assist project teams with developing specific action plans to incorporate LEED guidance into their design and construction planning. From the beginning, the Air Force has supported LEED and continues to develop and publish tools to aid in implementation of sustainable development.

9.3 LEED Application Guide for Lodging

Lodging facilities for soldiers, sailors, airmen, and marines are referred to as barracks, dormitories, or bachelor quarters on military installations. Barracks projects make up a large percentage of military construction projects, as the military forces grow and troops are relocated from one installation to another, and in recent years are part of major recapitalization programs to replace older facilities dating back to the World War II era or older. The quality and design of barracks have a vital role in the health, welfare, and morale of the service members who call them home; therefore, sustainable design contributes benefits not only to the global environment but also to the local living environment for men and women in uniform.

The most common design for barracks facilities is a three- or four-story building or grouping of these buildings in a "campus" area. Since LEED-NC 2.2 was primarily developed for new and renovated commercial buildings and general office facilities, different challenges are encountered, particularly in achieving energy efficiency and indoor environmental quality, for low-rise residential buildings (less than four stories) which are designed for continuous occupancy. To provide guidance and promote sustainable construction for barracks facilities, AFCEE worked with the USGBC to develop and publish *Application Guide for Lodging Using the LEED Green Building Rating System*. This publication was prepared by Paladino and Company, Inc., in 2001 and is more commonly referred to as the LEED Application Guide for Lodging. The intent of this guide is to provide interpretation of LEED credits in application to low-rise, lodging building projects and specific to Air Force dormitory construction, although it also allows for generic application on other similar facilities.

The LEED Application Guide for Lodging addresses each of the six LEED categories: Sustainable Sites (SS), Water Efficiency (WE), Energy and Atmosphere (EA), Materials and Resources (MR), Indoor Environmental Quality (IEQ), and Innovation in Design (ID).

The guide is laid out in a format that details the official intent and requirements of LEED for each credit; then it provides special considerations for lodging facilities and unique applications and requirements for USAF dormitories. Also, for each credit, a list of steps is provided to evaluate or achieve the credit, including a summary of the official LEED submittals required and applicable Air Force references. The intent of this format is to ensure the project planning and design teams have consolidated information for review and consideration as a tool to interpret LEED for lodging facilities.

The LEED Application Guide for Lodging is a useful tool for design and project teams involved in construction of low-rise lodging facilities and specifically MILCON for U.S. Air Force dormitories. This guide expands on the USGBC's LEED green building rating system by giving suggestions for interpreting the intent of many of the credits and how they can be applied to the unique considerations for facilities of this type and those located on Air Force installations. In addition to the information provided relative to each credit and the steps to evaluate and achieve the credit, the references listed in the guide provide a direct link to additional resources available to project teams engaged in USAF lodging designs. The following summary is provided as examples of the information included in the LEED Application Guide for Lodging for each of the LEED categories and specifically for USAF dormitory projects.

Sustainable Sites

There are eight credit subcategories in the sustainable sites category of LEED. The write-up in the USAF application guide has a significant amount of guidance for interpretation in this category, perhaps due to the unique considerations for development on federal government property and the planning and programming methodology used by the Air Force for construction projects.

The LEED Application Guide for Lodging addresses credit subcategories such as site selection, redevelopment, and alternative transportation, stating that achieving points for these may not be possible due to the existing installation location, relationship and layout of existing facilities, and installation general plan (which is the basis for site selection and style of construction). Also, since much of the project planning is done in advance to get funding approval for MILCON, the design and construction architect-engineer (AE) may not be able to have any impact on these issues once a project is approved. Credits for reducing site disturbance and providing for stormwater management many times rely on local zoning regulations and defined (subdivided) parcels of land. However, zoning and subdividing are usually not applicable on government property, although information is provided to interpret the intent as it would apply for an Air Force project.

In some cases, design features that would meet the intent of the credit cannot be achieved due to restrictions in military design. One example is construction of parking underneath a building to reduce the heat island effect. This is prohibited for government facilities due to Anti-Terrorism/Force Protection (AT/FP) concerns.

However, other useful sustainable features are promoted throughout the guide, such as use of pervious concrete and design of constructed wetlands. The primary references for this area include the USAF Landscape Design Guide, the Land Use Planning Bulletin, and general plan for the specific USAF installation.

Water Efficiency

There are only three credit subcategories in the area for Water Efficiency, each of which is well addressed in current Air Force guidances. Most of these guidances are available at the the Air Force Center for Engineering and the Environment which provides Air Force leaders with the expertise on facilities management and construction. The Air Force Water Conservation Program under the Air Force Civil Engineer Support Agency addresses wastewater technologies and water use reduction.

Energy and Atmosphere

The prerequisites in the Energy and Atmosphere category are covered by federal, DoD, and Air Force policy or MILCON specifications; however, some of the six credit subcategories may be difficult or costly to achieve. Energy reduction and the related LEED credit points are a high-priority goal for DoD, and targets are set by Air Force policy. Guides such as the USAF Passive Solar Handbook (available from the Air Force Center for Engineering and the Environment) encourage designs that promote use of natural light to help reduce energy use. Certain sustainable strategies, such as the use of many types of renewable energy and green power, are completely dependent on availability and are currently not as likely to be attained on military installations. However, many installations are looking into opportunities for future energy self-sufficiency.

Credits for additional commissioning, measurement and verification, and ozone depletion may be achievable, but first cost is an issue that must also be considered in the budget cycles. The Air Force has implemented a process of life-cycle assessment (LCA) for MILCON projects to evaluate issues such as these not only for first cost, but also with consideration of what the return would be over the life of the facility.

Materials and Resources

The LEED section for Materials and Resources contains one prerequisite and eight credit subcategories. In this area, some sustainable elements are standard practice in the military, but others that would otherwise seem achievable may be hampered by procurement regulations. The prerequisite for storage and collection of recyclables is a given, as it has become common practice on military installations. The building reuse credit depends on the project but is not common since many older facilities have exceeded their design life and do not meet current military design requirements; therefore, they would be too expensive to upgrade rather than replace. The credit for construction waste management is covered by the Construction and Demolition (C&D) Waste Management Guide (available from the Air Force Center for Engineering and the Environment), and the Defense Reutilization and Marketing Service (DRMS) is a practical avenue to incorporate resource reuse.

What is most interesting in this section is that credits for recycled content, local and regional materials, rapidly renewable materials, and certified wood all should be achievable with proper planning and product selection, although the LEED Application Guide for Lodging includes a disclaimer stating, "Government procurement regulations may prohibit achievement of this credit because there may only be a single supplier for the qualifying material." Recent efforts at many military installations have helped ease the procurement obstacles. An example is the Sustainable Interiors Showcase that opened in 2007 at Fort Jackson in South Carolina, where procurement and design teams collaborated to develop a grouping of interior finishes and furnishings which could be easily purchased and used to attain many credits.

Indoor Environmental Quality

This section includes two prerequisites and eight credit subcategories. The prerequisites are easily met since IAQ performance requirements are included in construction specifications, and environmental tobacco smoke is typically achieved by smoking bans and the establishment of outdoor designated smoking areas. Some credit points are fairly easy, such as daylight views and operable windows (for increased ventilation and controllability of systems), which are inherent in lodging design. Some credits are not as feasible, such as CO_2 monitoring, since it may be difficult and expensive due to the decentralized HVAC design common to lodging facilities. Other credits could be accomplished by simply making them a requirement in the design standards and construction specifications. Although the LEED Application Guide for Lodging cites the importance of IAQ due to the long lifetime of lodging facilities and the extended duration of exposure to contaminants by occupants, it also states that there are no standards or product selection criteria for construction IAQ management and low-emitting materials which could easily aid in mandating achievement of these credits. However, as noted in the Materials and Resources section, recent initiatives at many military facilities to form sustainable initiatives between procurement and design teams have resulted in these materials becoming more readily accessible for installation use. Indoor air quality (IAQ) is one of the main components of this category and is one of the most important sustainable features for lodging facilities and Air Force housing projects since it is of major importance for the health of occupants and has a direct impact on training and mission if service members become ill. Items related to IAQ in Air Force facilities are further detailed in Sec. 9.4.

Innovation in Design

No specific guidance is provided for these credits as they involve individually substantiated evidence that additional innovation beyond the other categories and credits was applied to the specific project. There is very little that could be added to help interpret these broadly defined credits to a lodging faculty or Air Force project.

Many of the items in the LEED Application Guide for Lodging can be applied in so many other industries in the United States. Many states are promoting "Green Lodging" programs in the hotel and tourism industry too. Considering that lodging is the fourthlargest energy consumer in the U.S. commercial sector, it does make both environmental and economic sense for this industry to also go green.

9.4 Importance of Indoor Air Quality

IAQ is one of the most vital design considerations for military lodging facilities since indoor pollution-related illnesses can create losses of personnel for critical training or mission requirements. IAQ also has a higher relative significance in residential or lodging facilities; due to the longer sustained time period residents occupy these facilities. The health and comfort of occupants of facilities and the impact this has on productivity are a recognized value of a sustainable facility. Poor IAQ may result in temporarily degrading an individual's physical well-being and can lead to serious long-term illnesses. Types of symptoms commonly linked to poor IAQ include headaches, fatigue, shortness of breath, sinus congestion, coughing, sneezing, irritation of the eye, nose, and throat, dizziness, and nausea. Immediate reactions may occur after a single exposure and can be further exacerbated by repeated exposures. These health effects could be a minor inconvenience or significantly debilitating. LEED-NC 2.2 and 2009 contain two prerequisites and nine credit areas pertaining specifically to IAQ.

Indoor air pollution results from a variety of contaminants from sources in the outside air, equipment used in the facility, building components and furnishings, and human activities conducted within the facility. Outside air contaminants include pollen, dust, industrial pollutants, vehicle exhaust, and soil contamination such as radon or even moisture, which may breed mold or other microbial growth. Facility equipment pollutants result from dust or microbes in HVAC components, use of office equipment and supplies, and operations of elevator motors or other mechanical systems. Building components and furnishings contribute to poor IAQ from dust collected within carpeting and hard-to-clean areas, microbiological growth from water damage or moisture, and volatile organic compounds (VOCs) released from furnishings and building products. Human activities produce indoor pollution from cleaning solvents, cooking, cosmetics, pesticides, and VOCs from maintenance products (e.g., paint, caulk, adhesives). Additionally, inadequate ventilation may increase the indoor pollutant levels by not diluting the indoor sources and taking the pollutants out.

Although the primary function of HVAC systems is to control the temperature and humidity of supply air, they are increasingly being designed for maintaining IAQ as well. According to the EPA IAQ Guide, "a properly designed and functioning HVAC system provides thermal comfort, distributes adequate amounts of outdoor air to meet ventilation needs of all building occupants, and isolates and removes odors and contaminants through pressure control, filtration, and exhaust fans." Ultimately, there are three primary strategies for controlling IAQ: source control, ventilation, and air cleaning.

Source Control

Source control can have an immediate and ongoing impact on IAQ by limiting contaminant-producing materials used in the construction and furnishing of the facility and implementing controls to reduce cleaning and maintenance products which produce indoor pollutants. According to the EPA IAQ Guide, "source control is generally the most cost effective approach to mitigating IAQ problems in which point sources of contaminants can be identified." Source control is directly addressed by one of the prerequisites and three of the subcategories in the Indoor Environmental Quality category of LEED-NC 2.2 and 2009. They are as follows:

- IEQ Prerequisite 2: Environmental Tobacco Smoke (ETS) Control prohibits or restricts smoking within or near the facility.
- IEQ credits 3.1 and 3.2 require construction IAQ management plans to minimize the potential of contaminants from the construction phase having an impact after occupancy.
- IEQ credits 4.1 through 4.4 require selection of low-emitting materials for construction products.
- IEQ credit 5 addresses entryway and other source area designs to minimize the intrusion of outdoor contaminants.

There are also other design issues not specifically credited through LEED that help prevent poor IAQ. The siting of the facility and the location of ventilation intakes are related to source control in that these design considerations can help maximize separation from outdoor sources of contaminants. Additionally, it is important that safeguards, policies, or practices be put in place with sufficient related enforcement to ensure product selection for cleaning and maintenance activities conducted throughout the life of the facility that utilizes low-emitting solutions and compounds. It is evident from the number of credits available within LEED that source control is an important part of IAQ management, although there are additional measures that need to be considered for both the design and the operation of facilities to maintain a healthy IAQ.

Ventilation

The common catch phrase "dilution is the solution to pollution" has actual application as a control mechanism for IAQ. Increasing the amount of outdoor air brought in through the HVAC system is another approach to lowering the concentrations of indoor air contaminants. According to the EPA IAQ Guide, some of the main design considerations for HVAC systems to dilute indoor air contaminants are to "increase the total quantity of supply air (including outdoor air), increase the proportion of outdoor air to total air, and improve air distribution." To measure this effectiveness, "the term 'ventilation efficiency' is used to describe the ability of the ventilation system to distribute supply air and remove internally generated pollutants." LEED-NC 2009 stresses proper ventilation through IEQ Prerequisite 1: Minimum IAQ Performance which establishes ASHRAE 62.1-2007, Ventilation for Acceptable Indoor Air Quality, as the standard ventilation system design. Additionally, IEQ credit 1 provides for monitoring CO_2 concentrations as a means to ensure ventilation is maintained as designed, and IEQ credit 2 recognizes efforts to increase ventilation above the ASHRAE 62.1-2007 standard. Special ventilation system considerations for establishing negative pressure in specific spaces are addressed within IEQ prerequisite 1 for ETS control and IEQ credit 5 for indoor chemical and pollutant source control. The EPA IAQ Guide also specifies that although introduction of sufficient outdoor air for proper dilution of contaminants is key, "inadequate distribution of ventilation air can also produce IAQ problems." This comment emphasizes the importance of commissioning and proper testing and balancing of HVAC systems for enhancing ventilation efficiency. EA prerequisite 1 and EA credit 3 relate to the commissioning requirements, and EA credit 5 addresses proper measurement and verification. Although these credits are within the Energy and Atmosphere category of LEED-NC 2009, they have direct contributions to IAQ as well.

Lodging facilities have other unique considerations when it comes to proper ventilation for good IAQ. For example, the use of decentralized HVAC systems is a fairly common practice; however, design solutions need to be evaluated for other cost-effective alternatives that still allow some degree of occupant control while maintaining minimum ventilation requirements. Another concern for lodging is proper ventilation of cooking equipment and fireplaces, if installed. Since not all sources of indoor air pollution can be eliminated, proper ventilation is critical to maintaining good IAQ within a facility. LEED contains several direct and related prerequisites and credits that emphasize the importance of ventilation for IAQ management.

Air Cleaning

Air cleaning, or filtration, should not be considered a primary control mechanism for IAQ management since it is most effective when used in conjunction with either source control or ventilation. Filtration has two main functions—removing contaminants from the airstream that are circulated within the facility and preventing buildup of contaminants within the HVAC system which can reduce equipment efficiency, lead to reentrainment of contaminants in the airstream, and serve as a breeding ground for other microbes. Some technologies that remove contaminants from air are particulate filtration, electrostatic precipitation, negative ion generation, and gas sorption. The first three technologies are designed to remove particulates, and the fourth is designed to remove gaseous contaminants.

Particulate filtration involves the use of mechanical devices placed in the airstream often using tightly woven fibers to trap the particulates as air flows through the filter medium. Filters are classified by their minimum efficiency reporting value (MERV) which relates to the efficiency the filter has for removing particles from 0.3 to 10 μm. High-efficiency particulate air (HEPA) filters and ultra-low penetration air (ULPA) filters are also available for the removal of even smaller particulates. The disadvantage of particulate filtration systems is that the higher the efficiency of the filter, the more it will increase the pressure drop within the air distribution system and reduce total airflow. This in turn can overwork fan motors causing increase in energy usage and can reduce the air exchange rate critical to dilution of indoor air pollutants. Also, particulate filtration systems may require frequent maintenance for replacement or cleaning of filters.

Electrostatic precipitation and negative ion generation are similar in principle because they apply a charge to particles to enable removal from the airstream. With electrostatic precipitators (ESPs), the charged particles become attracted to an oppositely charged surface on which they are collected. These systems provide relatively high-efficiency filtration of small respirable particles at low pressure losses, although ESPs must be serviced regularly and produce ozone. Negative ion generators (ionizers) apply a static charge to remove particles from the indoor air. The charged particles then become attracted to surfaces such as walls, floors, draperies, or even occupants. This is not always an ideal method since it requires deliberate cleaning of the space to actually remove the particles, although more effective designs include collectors to attract particles back into the unit. Like ESPs, ionizers require frequent maintenance and may produce ozone.

Gas sorption is a process for the control of gaseous contaminants, such as formaldehyde, sulfur dioxide, ozone, or VOCs. It involves the use of a sorption material (e.g., activated carbon, chemically treated clays) and processing of the gaseous contaminant, such as chemical reaction with the sorbent, binding of the contaminant with the sorbent, or diffusion of the contaminant from areas of higher concentration to areas of lower concentration. There are currently few standards for rating the performance of these systems, which makes design and evaluation problematic. Also, operating expenses of gas sorption systems can be quite high, and sorbent filters can reach breakthrough conditions at higher temperatures and relative humidity which causes them to desorb VOCs back into the airstream. Recent system development and testing for gaseous contaminant removal have been conducted with photocatalytic oxidation (PCO). PCO is a process involving a photocatalytic filter and UV light for the elimination of VOCs through an oxidation reaction producing carbon dioxide and water.

Air cleaning through filtration systems can be a critical link to long-term maintenance of good IAQ. It is not likely that sources of contaminants can be completely avoided through product selection, and dilution through increased ventilation does not fully eliminate contaminants; therefore, filtration can drastically improve the effectiveness of removing indoor pollutants as the third leg of the IAQ strategic triad. Currently, LEED-NC 2009 provides no specific credits for application of air cleaning or filtration systems, although if a unique system design is selected with appropriate calculations demonstrating removal efficiency, credits could be applied for under the Innovation and Design Process (ID) category. However, LEED 2009 does address the filters in mechanical HVAC systems and requires certain minimum MERV ratings for them, or replacement of these filters prior to occupancy, for attaining IEQc3.1, IEQc3.2 or IEQc5.

Future Combined Technologies

Several other methods are being developed for improvement of IAQ. Many combine both source control and air cleaning in the HVAC system. One mentioned previously related to the UV disinfection and VOC removal methods. Another related to microbes and mold production is the introduction of antimicrobial surfaces such as copper and other antimicrobial products into the HVAC systems. Due to the importance of improved IAQ within military facilities, particularly in housing, medical facilities, and other 24-h occupied spaces such as submarines and ships, the DoD is interested in research into many of these areas. Current research on the opportunities for copper as an antimicrobial in HVAC systems is being led by the Copper Development Association (CDA) and funded through a Congressional Special Interest (CSI) Medical Program overseen by TATRC, the Telemedicine and Advanced Technology Research Center of the U.S. Army Medical Research & Materiel Command (USAMRMC). Copper was recently approved by the U.S. EPA as a solid antimicrobial material for touch surfaces.

The occupants of a facility are perhaps the most significant portion of the "environment" that should be considered when designing for sustainability. Reduction or elimination of indoor air contaminants is vital to enabling healthy working or living conditions. Ultimately, the most effective means of maintaining good IAQ involves effective facility design and proper application of ventilation systems considering all strategies for IAQ management.

9.5 Summary

As the largest facility manager within the federal government, the DoD bears a great responsibility for leadership in adopting strategies and developing guidance for reducing energy consumption and minimizing environmental impact. Although all the military services started from different approaches, they have merged in recent years by incorporating sustainable requirements into standardized construction specifications and uniformly establishing the LEED green building rating system as the primary metric for sustainable military construction. Since the construction of military facilities, particularly low-rise lodging facilities such as dormitories or barracks, often entails unique requirements, it has been recognized that additional guidance and interpretation of the intent of certain USGBC LEED green building rating system credits are necessary. The U.S. Air Force, through AFCEE, has developed one such guide, the LEED Application Guide for Lodging. One of the greatest impacts LEED has on the performance of military personnel comes from the Indoor Environmental Quality (IEQ) category, specifically related to the prerequisites and credits for IAQ. Source control, ventilation, and air cleaning are the three primary strategies for controlling IAQ. LEED-NC 2009 provides specific credits for the first two, and some minimum filter requirements for air cleaning. Regardless, it is important for designers and engineers to consider all possible strategies and technologies available to maximize the facility's IAQ and provide a healthy working and living environment for occupants.

References

AFCESA (2007), http://www.afcesa.af.mil/ces/cesc/water/cesc_watercons.asp, Air Force Water Conservation Program of the Air Force Civil Engineer Support Agency website, accessed July 23, 2007.

Brzozowski, C. (2007), "Sleeping Green," *Distributed Energy, The Journal for Onsite Power Solutions,* July/August, 5(4): 24–33.

Bush, G. W. (2007), "Executive Order 13423—Strengthening Federal Environmental, Energy, and Transportation Management," *Federal Register*, 72 FR 3919, January 26.

CDA (2007), http://www.copper.org/health/, Copper Development Association webpage, accessed July 10, 2007.

CDA (2010), "Copper Touch Surfaces," Copper Development Association webpage, http://coppertouchsurfaces.org/antimicrobial/index.html, accessed February 4, 2010.

Clinton, W. J. (1998), "Executive Order 13101 Greening the Government Through Waste Prevention, Recycling, and Federal Acquisition," *Federal Register,* 63 FR 49643, September 16.

Clinton, W. J. (1999), "Executive Order 13123 Greening the Government through Efficient Energy Management," *Federal Register,* 64 FR 30851, June 8.

Clinton, W. J. (1999), "Executive Order 13148 Greening the Government through Leadership in Environmental Management," *Federal Register* 65 FR 24595, April 26.

DoD (2004), *Unified Facilities Criteria (UFC) Design: Low Impact Development Manual,* Department of Defense, October 24, 2004, accessed July 10, 2007, http://www.lowimpactdevelopment.org/lid%20articles/ufc_3_210_10.pdf.

DRMS (2007), http://www.drms.dla.mil/, Defense Reutilization and Marketing Services website, accessed July 23, 2007.

EPA (2007), http://www.energystar.gov/, Energy Star website, U.S. Environmental Protection Agency, U.S. Department of Energy, Washington, D.C., accessed July 21, 2007.

EPA OAR (1991), *Building Air Quality: A Guide for Building Owners and Facility Managers,* EPA Document 402-F-91-102, December, U.S. Environmental Protection Agency, Office of Air and Radiation, Washington, D.C.

EPA OAR (1998), *Building Air Quality Action Plan,* EPA Document 402-K-98-001, June, U.S. Environmental Protection Agency, Office of Air and Radiation, Washington, D.C.

EPA and U.S. Consumer Product Safety Commission Office of Radiation and Indoor Air (1995), *The Inside Story: A Guide to Indoor Air Quality*, EPA Document # 402-K-93-007, April 1995, U.S. Environmental Protection Agency, Washington, D.C.

Ingersoll, W., and L. M. Haselbach (2007), "RS-VSP: A New Approach to Support Military Range Sustainability," *Environmental Engineering Science,* May, 24(4): 525–534.

Michels, H. T. (2006), "Anti-Microbial Characteristics of Copper," *ASTM Standardization News,* October, 34(10): 3–6.

Obama, B. M. (2009), "Executive Order 13514—Federal Leadership in Environmental, Energy, and Economic Performance," *Federal Register*, 74(194), accessed March 3, 2010.

OFEE (2003), *The Federal Commitment to Green Building: Experiences and Expectations*, Office of the Federal Environmental Executive, http://www.ofee.gov/sb/fgb_report.html, accessed July 10, 2007.

OFEE (2010), *List of Agency Senior Sustainability Officers under E.O. 13514*, Office of the Federal Environmental Executive, http://www.ofee.gov, accessed March 3, 2010.

OMB (2002), Circular A-11, Section 55, Office of Management and Budget, Washington, D.C.

OMB (2007), *Budget of the United States Government Fiscal Year 2008*, Office of Management and Budget, http://www.whitehouse.gov/omb/budget/fy2008/, accessed April 3, 2007.

USAF (2006), *The Air Force Center for Environmental Excellence Environmental Quality and Compliance Resources, November 20, 2006*, Air Force Center for Environmental Excellence, http://www.afcee.brooks.af.mil/eq/sustain/ExecOrderCrossReference.doc, accessed April 3, 2007.

USAF, USGBC, and Paladino and Company, Inc. (2001), *Application Guide for Lodging Using the LEED Green Building Rating System*, U.S. Air Force Center for Environmental Excellence, U.S. Green Building Council, Washington, D.C.

USAMRMC (2007), https://mrmc-www.army.mil/crpindex.asp, website on the Congressional Research Programs managed by the U.S. Army Medical Research and Materiel Command accessed July 18, 2007.

U.S. Census Bureau (2007), *Construction Spending*, U.S. Census Bureau Manufacturing, Mining, and Construction Statistics, http://www.census.gov/const/www/c30index.html, accessed April 2007.

USGBC (2004–2005), *LEED-EB Green Building Rating System for Existing Buildings, Upgrades, Operations and Maintenance*, Version 2, October 2004, updated July 2005, U.S. Green Building Council, Washington, D.C.

USGBC (2005), *LEED-NC Application Guide for Multiple Buildings and On-Campus Building Projects, for Use with the LEED-NC Green Building Rating System Versions 2.1 and 2.2*, October, U.S. Green Building Council, Washington, D.C.

USGBC (2005–2007), *LEED-NC for New Construction, Reference Guide*, Version 2.2, 1st ed., U.S. Green Building Council, Washington, D.C., October 2005 with errata posted through Spring 2007.

USGBC (2007), *LEED for Schools for New Construction and Major Renovations*, approved 2007 version, April, U.S. Green Building Council, Washington, D.C.

U.S. GPO (2007), *Access Budget of the United States Government, January 30, 2007*, U.S. Government Printing Office, http://www.gpoaccess.gov/usbudget/browse.html, accessed April 3, 2007.

Yu, K.-P.,et al. (2006), "Effectiveness of Photocatalytic Filter for Removing Volatile Organic Compounds in the Heating, Ventilation, and Air Conditioning System," *Journal of the Air & Waste Management Association*, May, 56: 666–674.

Exercises

1. What is the current status of EO 13514? Has it been superseded or enhanced by any other Executive Orders, and if so, what are they?

2. What is the current legislation in your state with respect to green building of state facilities?

3. Visit the Army sustainability website. What are some current projects and programs in sustainability in the Army?

4. Many Army bases have developed 25-year sustainability plans. Which ones have adopted these plans?

5. Many Army bases are developing 25-year sustainability plans. What is the status of the plans for Army facilities in your region of the country?

6. The website http://www.FedCenter.gov is a website for Federal Facilities Environmental Stewardship and Compliance Assistance. What current regulations does it list with respect to green building?

7. The Office of Management and Budget's Circular A-11, which specifically encouraged agencies to evaluate the use of Energy Star or LEED standards and guidelines for incorporation in designs for new building construction or renovation in Section 55 in its 2002 version, was still valid when President Bush issued EO 13423. What is its status now?

8. What are some current practices with respect to the sustainability of live firing ranges at DoD facilities?

CHAPTER 10

Low-Impact Development and Stormwater Issues

Stormwater management is an important design and construction practice. Without proper management, the resulting runoff can cause many problems including flooding and poor water quality. Traditionally, the design of stormwater facilities is taught in hydrology and watershed management classes. These types of courses are important to stormwater management, but focus more on the big picture, not individual construction projects. The purpose of this chapter is to look at stormwater management from the construction and individual site perspective so that it can be better integrated into the LEED or green building process. However, the actual design of any stormwater management system should be done by practitioners experienced in civil engineering systems, hydrology, and watershed management in addition to these newer concepts.

The concepts that are of most interest for stormwater management are the rate of stormwater runoff from a site or area, the amount of stormwater runoff from an area, the pollutant loadings from the runoff, and the impact on erosion and subsequent pollution that runoff has on the downstream areas. Over the decades many methods have been developed to estimate runoff. They are used for calculating rates, volumes, and a concept referred to as the time of concentration. A simplified definition of the *time of concentration* is the time it takes for various runoff contributions from a defined site or area to come together to a point producing a peak rate at that point. Akan and Houghtalen (2003) list three methods for determining the time of concentration: the SCS (Soil Conservation Service) time-of-concentration method, kinematic time-of-concentration formulas, and the Kirpich formula. They also list the following runoff rate calculation methods: unit hydrograph method, TR-55 graphical peak discharge method and tabular hydrograph method, the Santa Barbara urban hydrograph method, U.S. Geological Survey (USGS) regressions equations, the rational method, and the kinematic rational methods.

The concept of percent imperviousness as used by the USGBC, and as discussed in Chap. 2, is based on the rational method. The rational method is a very old (nearly a century in use) and fairly simplified method. It is applicable to smaller watersheds, usually much smaller than 100 acres. Most project areas associated with the LEED certification process are relatively small (10 acres or less) and therefore fall into this category. It is mainly a method to estimate peak runoff based on a 2- to 5-year storm. The calculations for both peak runoff rates and volumes usually require some stormwater path information, and therefore the other hydrological models are frequently used, expecially the TR-55 method for larger watersheds. Regardless, the following discussions concern methods to control runoff from a project, and proper hydrological modeling should be used to determine the rates and volumes. This text is not intended to teach hydrology; there are many excellent references that already do that. Instead, it is intended to guide the designers to approach stormwater management in a new evolving and inclusive way for more sustainable results.

The field of stormwater management is rapidly growing for a number of reasons, of which two important ones come to mind. The first is the ever-increasing development of the land in both the United States and the rest of the world, as populations grow and gain higher standards of living. This development usually translates to an increase in impervious surfaces and changes to site hydrologics and thus usually increases stormwater runoff. The second reason is based on the concern about global climate change, which may result in alterations to precipitation and weather patterns in many areas of the world and then could impact the hydrology and subsequent runoff, even from existing developments.

10.1 Nonpoint Source Pollution, BMPs, and LID

Most of the focus in the last century has been on collecting stormwater runoff and discharging it somewhere else to prevent flooding in the immediate area caused by development. However, in the last few decades it has been recognized that this general design may not be the best alternative in all cases. Collecting stormwater can cause powerful flows off-site that may result in property and ecological damage in addition to poor water quality. This pollution from stormwater runoff is commonly referred to as *nonpoint source* (NPS) pollution, as it has areal sources.

In response to the concerns about stormwater runoff and NPS pollution, the United States has adopted a permitting program, with the intent that it will help decrease off-site impacts of stormwater runoff. It is called the National Pollutant Discharge Elimination System (NPDES). Phases of NPDES have slowly been implemented in the 1990s and early 2000s, and the practices and policies resulting from their implementation are considered to be important for stormwater management sustainability, both in the construction phase and during the operational life of a facility.

There is no standard set of solutions to manage stormwater and NPS pollution. Each site and each watershed has its own needs and special conditions. Therefore, stormwater management design in the 1990s developed into a concept called Best Management Practices (BMPs). BMPs are a compilation of various methods and policies to choose from that are intended to serve as stormwater quantity controls and stormwater quality control measures. Several BMPs together may "best" serve to reduce stormwater impacts of construction and development.

BMPs are frequently subdivided into several broad categories. They can be grouped by project phase, by the form they take, and by function. Differentiation by project phase is whether they serve to reduce NPS pollution during the demolition and construction phase, or whether they are used to manage stormwater and reduce water quality impacts during the operational phase of a facility, as depicted in Fig. 10.1.1.

BMPs can also be differentiated by form, that is, whether they are structural or nonstructural, as noted in Fig. 10.1.2. A structural BMP is one that is physically put in place, such as a detention pond, swale, or a catch basin. A nonstructural BMP is a policy or practice. (Note that the LEED-NC 2.2 Reference Guide tends to put the more natural BMPs such as swales also in the nonstructural category.) An example of a BMP that is a policy might be a regulation that all facilities which change motor oil also have facilities

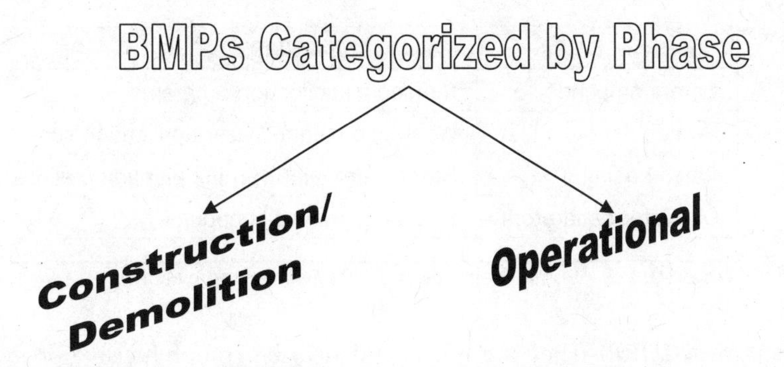

FIGURE 10.1.1 BMPs can be categorized by the life-cycle phase of the facility.

for accepting used motor oil for recycle at no charge. This helps reduce the potential for people to put used motor oil on the ground or into stormwater devices and lessens the chance for oil pollution downstream of a site. Another example of a practice is a change in fertilizer schedules based on weather patterns so that excess fertilizer does not get carried off-site during a storm. A nonstructural practice common in the Bay Area of California is to label stormwater inlets with pictures of fish to warn people that the inlet collects runoff that discharges into the Bay and anything put into the inlet might harm life in the Bay. Table 10.1.1 lists some categorizations of BMPs by form.

Finally, BMPs can be broadly categorized by function. However, there are many different ways that BMPs are organized by function. Sometimes BMPs are segregated as quality or quantity control measures. Sometimes they are separated into groupings based on whether they control erosion or control sedimentation. When they are categorized by the latter, they are typically referred to as *erosion and sedimentation control* (ESC) measures. Figure 10.1.3 gives some different categorizations by function.

Specific structural BMPs are listed under various names and functions that have been adopted regionally or locally or by various types of practitioners. Since they are too numerous to list or organize, an attempt will be made in the following sections to give a few examples of BMP terminology and categorization. These examples come from entities such as a watershed-related agency in the Chesapeake Bay area, a southeastern state department of transportation (DOT) as related to ESC, and research into BMPs by the Transportation Research Board (TRB) National Cooperative Highway Research

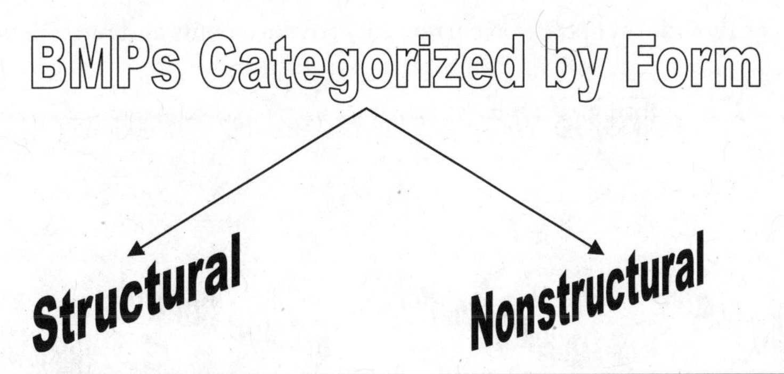

FIGURE 10.1.2 BMPs can be categorized by form.

Structural BMPs	Nonstructural BMPs
Detention ponds	Reused oil collection programs
Swales	Weather-based fertilizer application schedules
Catch basins	Stormwater inlet warning painting practices
Oil/water separators	Seasonal grading schedules

TABLE 10.1.1 Some Examples of BMPs Segregated by Form

Program (NCHRP). They are not intended to be comprehensive or inclusive for the United States, but together give a good introduction to some of the terminology used.

The use of BMPs is starting to achieve wide acceptance in the United States. It is a method to increase the sustainability of stormwater management systems. However, like the older, more traditional stormwater management systems, it again focuses on what is discharged from a site. So, to also start addressing what remains on a site or in a development, an even more focused concept is being introduced. It is commonly referred to as *low-impact development* (LID), although some refer to a similar concept as IMPs (Integrated Management Practices). These LID practices use many of the concepts from traditional hydrologic models and many BMPs, but they also focus on how the stormwater impacts the site and other "preventive" land development techniques to keep the site hydrology as close to the natural state as possible. Traditionally, hydrology management deals with the larger storms. LID focuses more on the frequently occurring storms, which are of a short duration and high frequency. These smaller, more frequent storms commonly carry an abundance of pollutants from a site, and even though they may not be individually as destructive as a larger storm, since they are more frequent, if not controlled, collectively they may cause substantial erosion and other deleterious impacts.

The following sections of this chapter will

- Outline some of the reasons why BMPs were developed
- Outline many structural BMPs as categorized by one agency in the Chesapeake area
- Outline many BMPs as categorized by a state DOT for ESC
- Outline many structural BMPs as evaluated by the NCHRP
- Introduce LID and IMPs concepts
- Provide an update on current LID advancements and references

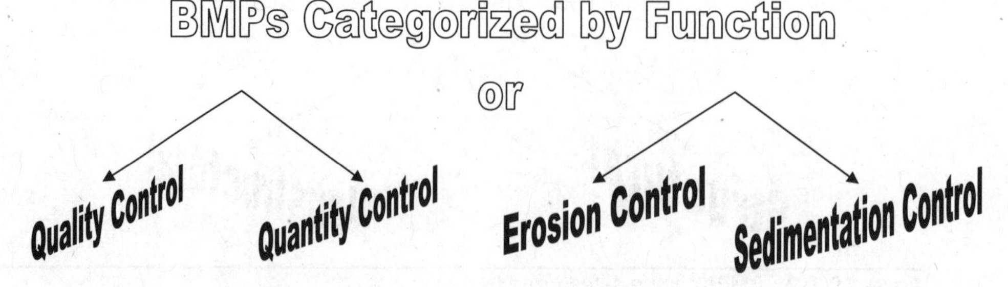

FIGURE 10.1.3 BMPs can be broadly categorized by function.

Percent Impervious Cover	Effect on Downstream Receiving Waters
<10	Sensitive
10–25	Impacted
>25	Nonsupporting

Table 10.1.2 Downstream Effect Rankings on Receiving Waters from Nonmitigated Impervious Cover for Some Example Watersheds (CWP, 2000)

Why BMPs Were Developed: Stormwater Impacts of Urbanization

The Center for Watershed Protection (CWP) in Ellicott City, MD, has compiled, reviewed, and synthesized extensive research and data on the impacts of urbanization on stormwater runoff and watershed protection methods. It has been shown that one of the main parameters that increase stormwater runoff and downstream impacts is an increase in impervious areas caused by development. These increases can be both on the ground (pavements) and from roofs. Each watershed is different and responds uniquely to the amount and location of impervious cover, so there is no single simple model based only on impervious cover. Table 10.1.2 shows a range of impacts that was found to occur for a number of watersheds studied.

In these, the impacts of urbanization are usually categorized into four areas—hydrological, geomorphological, water quality, and habitat-related. Some of the effects and impacts in each of these categories are listed in Table 10.1.3.

The Center for Watershed Protection also summarized some of the methods or tools which can be used to protect the watershed and puts them into eight approximate groupings. These are summarized in Table 10.1.4.

Hydrological	Geomorphological	Water Quality	Habitat/Wildlife
Increased frequency of flooding	Widened streams	Increased temperature	Decreased habitat value
Increased volumes of flooding	Modified access for fish passage	Increased beach closures	Decreased large woody debris
Higher peak flows	Loss of riffle pools	Increased pollutants - SS (suspended solids) - Nutrients (N and P) - Metals (Cu, Zn, Pb, Cd) - Oil and greases - Bacteria - Various toxics (pesticides/ herbicides)	Increased fish barriers
Increased variation in storage needs	Fragmentation of tree cover		Shift in food chain
Modified channelization	Silt embedded in banks and bottom		Decreased speciation
Lower dry weather flows	Degraded bottom substrate quality		Decreased weather impact buffers
	Eroded stream banks		Increased algal problems

Table 10.1.3 Some Impacts of Urbanization and Increased Impervious Cover (CWP, 2000)

Methodologies	Examples
Land use planning	Stormwater "hot spots" away from water bodies
Increased land conservation	Conservation easements, cluster housing
Increase aquatic buffers	Setbacks from banks for development
Better site design	Less impervious cover
Erosion and sediment control	Geofabrics, planting schedules
BMPs	Sedimentation basins, infiltration swales
Control of nonstormwater discharges	Point source pollutant control measures
Watershed stewardship	Cleanup days, educational programs

TABLE 10.1.4 Various Methodologies for Watershed Protection (CWP, 2000)

The remainder of this chapter focuses on many of the methodologies in the ESC and BMP grouping, which are typically part of the engineering design of a project.

BMP Terminology

There are many different names for similar BMPs in the United States and many variations of BMPs that are regionally or locally used. There are even variations on the term *BMP*, such as calling some of them STPs (Stormwater Treatment Practices). Sometimes the variations are more random, but at other times the variations are based on adaptations to local conditions such as soil types or topography. Very frequently the variations are due to the focus of the group or profession that has categorized them. For instance Tables 10.1.5 and 10.1.6, list some typical categorizations of BMPs as given by a watershed group and a highway-related research organization. In the ESC section, there will be a table referring to some categorization of these methods typical of many state DOTs. Table 10.1.5 gives a listing of many BMPs that have been categorized by the CWP.

Pond-Type BMPs	Wetlands	Infiltration	Filtering	Open Channels
Micropool ED* pond	Shallow marsh	Infiltration trench	Surfaces and filter	Dry swale
Wet pond	ED shallow wetlands	Infiltration basin	Underground sand filter	Wet swale
Multiple pond system	Pond/wetland system	Porous pavement	Perimeter sand filter	Grass channel
Pocket ponds	Pocket marsh		Organic filter	
	Submerged gravel wetland		Pocket sand filter	
			Bioretention	

*Extended detention.

TABLE 10.1.5 Some Typical Categorizations of Structural BMPs with a Watershed Focus (CWP, 2000)

Runoff Flow and Volume Control	Runoff Volume Control	Physical, Chemical, or Biological Sorption, Ion Exchange	Physical Screening	Force-Related Physical Screening	Aerating, Flocculating, Volatilizing, Coagulating	Disinfection	Microbial-Mediated Processes
ED* basins	Infiltration basins	Engineered media	Screens, racks	ED basins	Sprinklers	Chlorine	Wetlands
Retention ponds	Infiltration trenches	Granular AC,† compost or peat filters	Biofilters	Retention basins	Aerators	Ultraviolet systems	Bioretention
Detention ponds	Permeable pavements	Sand/gravel filters	Permeable pavements	Wetlands	Coagulant systems	Ozone	Biofilters
Tanks, vaults	Dry swales	Subsurface wetlands	Infiltration basins	Settling basins	Flocculent systems	Advanced systems	Retention ponds
Wetlands	Dry wells	Zeolites or complexation media	Infiltration trenches	Tanks, vaults			Engineered, sand, compost filters
Equalization basins	ED basins	Bioretention	Bioretention systems	Check dams			
		Infiltration trenches/ basins	Engineered media	Oil/grit separators			
		Wetlands	Sand/gravel filters	Hydrodynamic catch basins			
		Biofilters	Catch basin inserts				

*Extended detention.
†Activated carbon.

TABLE 10.1.6 Functional Unit Operations of Many BMPs (NCHRP, 2006)

There are also components to a BMP that provide additional pretreatment or water quality protection. Some of these are referred to as water quality (WQ) inlets, and they may take many forms and names. Some common pretreatment units are forebays or sedimentation basins before a pond, or specialty premanufactured items such as oil/grit separators or cyclonic catch basin inserts. There are also structural additions to the BMP systems whose primary purpose is to decrease the flow rate of runoff, which is important for erosion and damage protection in the receiving structure or water body. Many of these are referred to as level spreaders or energy dissipators, although sometimes they are just called by the name of the material they are made of such as riprap. Figure 10.1.4 depicts a detention pond where additional riprap has been used for erosion control.

Table 10.1.6 lists many of the structural BMPs that have been segregated by functional unit operations by the NCHRP (2006). Many of the BMPs serve multifunctional purposes. Figure 10.1.5 depicts a detention pond with added retention which aids in stormwater quality control. Figure 10.1.6 is of a retention pond. The main differences between detention and retention ponds are that the former *detain* most of the stormwater, releasing it at controlled rates to downstream structures or waterbodies, while the latter *retain* most of the stormwater, allowing it to eventually infiltrate into the ground or evaporate. Each also usually provides overflows for larger storm events.

In addition to the BMPs already listed in the tables, there are many new or combined variations on all these themes. For instance, there is a BMP called an *ecology ditch*. It can best be described as a combination of a swale and a bioretention cell.

FIGURE 10.1.4 Detention pond with riprap erosion control around the inlets and banks in Irmo, S.C. (*Photograph taken July 2007.*)

FIGURE 10.1.5 Detention pond with some retention for additional quality control in Irmo, S.C. (*Photograph taken July 2007.*)

FIGURE 10.1.6 Retention pond in Greenville, S.C. (*Photograph taken February 2003.*)

There are also classifications of BMPs with respect to adaptation for climate change. In these cases, other BMPs and functions are added to the mix. Some of the additional functions considered are

- Urban heat island reduction
- Urban forestry support
- Carbon sequestration
- Micro hydropower potential
- Combined sewer overflow (CSO) reduction

ESC

Erosion and sediment control measures are important for both the construction phase and the operational (built) phase, but are particularly important during the construction phase of a project. Construction sites are sometimes referred to as nonpoint source pollution "hot spots" as the pollution in the runoff, if not controlled and treated, is frequently several times higher than from a developed site.

ESCs are varied in function, form, and applicability just as other BMPs are. The best practices are usually site-specific and sometimes season-specific. Therefore, there are many options to choose from and many manufacturers. Since there are such a broad range of products, there are various testing facilities which are analyzing the options throughout the country. Some of the testing facilities include

- The Hydraulics, Sedimentation, and Erosion Control Laboratory operated by the Texas Transportation Institute and the Texas Department of Transportation
- The San Diego State University Soil Erosion Research Laboratory
- The St. Anthony Falls Laboratory operated by the University of Minnesota
- Alden Research Laboratory in Holden, MA

Information from these and other testing facilities, from other research reports, and from the various manufacturers can be used to determine which ESC measures may be applicable for a site, but there is an immense amount of information and with the different options, a need developed to establish some form of organized inspection and organization of the methods. From this has sprung the programs of the Certified Professional in Erosion Prevention and Sediment Control (CPESC) and the Certified Erosion Prevention and Sedimentation Control Inspector (CEPSCI) which are being recognized by many state agencies as a means to facilitate erosion and sediment control on project sites, particularly in departments of transportation, where there are always many horizontal maintenance and construction projects underway. Table 10.1.7 lists the various categories of erosion and sedimentation control practices that are taught in the CEPSCI certification classes in South Carolina.

The most common ESC measures used during construction are the silt fence around the perimeter of the disturbed areas so that sediments do not wash out from the site, construction entrances at the openings to the disturbed areas so that vehicles can come in and out, and some form of inlet protection on any stormwater device installed prior to or during construction so that it does not become a conduit for sediment flow off the site during the construction phase. Based on weather and soil conditions, some type of dust control may also be needed to prevent soil from escaping the site by wind mechanisms.

Erosion Prevention Practices	Sediment Control Practices
Surface roughening	Sediment basin
Mulching	Multipurpose basin
Erosion control blankets (ECBs)	Sediment dam (trap)
Turf reinforcement mats (TRMs)	Silt fence
Dust control	Ditch check • Rock • Sediment tube
Polyacrylamide (PAM)	Stabilized construction entrance
Stabilization • Temporary • Riprap for channel • Final	Inlet protection • Filter fabric • Block and gravel • Hardware fabric and stone • Block and gravel curb • Prefabricated
Outlet protection	Rock sediment dikes

TABLE 10.1.7 Typical Erosion Prevention and Sediment Control Practices Taught in ESC Certification Classes (CEPSCI, 2005)

When the construction is nearing completion, additional measures are permanently installed on the site so that erosion does not occur after construction. These range from vegetative measures to using riprap or constructed erosion prevention devices.

LID

The main concepts of low-impact development (LID) are to minimize the impacts of development, mainly the stormwater-related impacts. This is done usually by mimicking the natural hydrologic cycle at the site or project as much as possible and mimicking the natural hydrologic runoff cycle. So how does one implement the design goal of mimicking the former site hydrology with LID features? There are essentially four main parameters of site hydrology which can be used to do this:

- Storage
- Infiltration and the recharge of groundwater
- Evaporation and evapotranspiration
- Detention

And there are a series of many concepts that can help do this. The main ones as listed in the Prince George's County (1999) LID manual and as amended are as follows:

- Reduce added imperviousness.
- Conserve natural resources.
- Maintain natural drainage courses.
- Reduce use of pipes, use open-channel sections.

- Minimize clearing and grading.
- Provide dispersed areas for stormwater storage uniformly throughout the site.
- Disconnect flow paths.
- Keep predevelopment time of concentration.
- Employ public education and operational procedures.
- Maintain sustainability of LID practices and BMPs.

These can be summarized as follows: reduce new development, conserve existing and natural systems, and disperse and disconnect the site stormwater.

New development introduces many conveyance mechanisms for stormwater which need to be disconnected, reduced, or staged and planned. If these are not reduced or mediated, then they can facilitate a rapid transport of runoff off-site and the resulting environmental and flooding problems. Keeping water on-site and disconnecting flow paths can also reduce pollutant transport off-site.

Typical built items that can become flow paths include

- Impervious roadways
- Roofs
- Gutters
- Downspouts
- Impervious drives
- Curbs
- Pipes
- Swales
- Impervious parking areas
- Grading

Reducing these items or designing them in alternative fashions can aid in lowering the impacts of development and can also mean a reduction in construction costs. However, development means we are building or engineering something, so for the site to be accessible and functional, many of the surface features that may also become flow paths will be installed. Some measures will need to be taken to minimize the impacts these have on the site hydrology. The main focus of LID is to think micromanagement and integrated management practices.

These practices can be multifunctional. In fact, many of the LID and IMP concepts can be implemented in otherwise "nonfunctional" spaces. Dr. Allen Davis gives the example of a grass swale or bioretention landscape strip being located in the area needed for the overhang of fronts or backs of cars in parking spaces.

In response to the economic benefit of multifunctional best management practices as promoted in the LID concepts, many agencies have expanded their stormwater guides to include in detail some of the IMPs. On July 2, 2007, the North Carolina Department of the Environment and Natural Resources released its new BMP manual. There are chapters in this manual devoted to bioretention, permeable pavements, and other BMPs that aid in attaining the LID concepts.

Then, in 2007, the Environmental and Water Resources Institute (EWRI) of the American Society of Civil Engineers (ASCE) formed an LID committee in its Urban

Water Resources Research Council (UWRRC). It has been recognized that there is a great need for research and development of manuals, best practices, and standards with respect to the LID concepts. The first three initiatives started by this committee are to look into the technology associated with green roofs, porous and permeable pavements, and rainwater harvesting. In 2008 through 2010, additional initiatives were added including modeling, ultra urban issues, and green streets or highways.

Another recent action that is helping to foster the LID trend in design and development is a new credit added to the USGBC LEED for Schools rating system in 2007 which was further updated in LEED 2009. In the Sustainable Sites category an additional credit was added called *site master plan*. Basically, this credit requires a master plan to be prepared for a school location covering current and planned development, and in this master plan the future plans should include adhering to at least four out of seven specified SS credits. The specified credits are SSc1: Site Selection; SSc5.1: Site Development: Protect or Restore Habitat; SSc5.2: Site Development: Maximize Open Space; SSc6.1: Stormwater Management: Quantity Control; SSc6.2: Stormwater Management: Quality Control; SSc7.1: Heat Island Effect: Non-Roof; and SSc8: Light Pollution Reduction. If one thinks about these credits, most together promote many of the LID concepts.

Combining a number of credits is very similar to the concept of integrating various management practices. For instance, in most cases one needs landscape areas for aesthetics, but why not also use these for stormwater management? Shading is needed for urban heat island effect reduction, but if trees are used, the tree islands can also be used for stormwater management.

There are a number of issues that must be addressed when adopting LID practices for good design. Some of these include the following:

- Make sure that flood control is adequately considered.
- Be careful that water is not infiltrated where it might cause structural damages such as under some impervious pavements (hydrostatic forces), under buildings (water damage), and into unstable slopes.
- Some of the quasi-structural BMPs may not be sustainable since they may not appear to have readily noticeable site functionality. They need some way of fostering permanence. Some methods to do this may be structural such as adding rock walls; others may be nonstructural such as establishing conservation easements for infiltration areas.
- The decrease in paved areas must take into consideration safety and access concerns.

LID and Land Development Design

This chapter in no way attempts to be a comprehensive resource for LID. It would take an entire book or manual to even start any detailed design. However, a few of the main concepts have been reviewed, and this chapter gives some site design examples and an example method to perform some calculations on one of the newer stormwater management technologies that may be used for LID designs. Four other commonly referenced resources for information on how to incorporate LID into land development design are

- Prince George's County, Department of Environmental Resources (1999), *Low-Impact Development Design Strategies: An Integrated Design Approach*, Prince George's County, MD.

- CWP (1998), *Better Site Design*, Center for Watershed Protection, Ellicott City, MD.
- DoD (2004), *Unified Facilities Criteria (UFC) Design: Low Impact Development Manual*, Department of Defense, October 24.
- Hinman, C. (2005), *Low Impact Development: Technical Guidance Manual for Puget Sound*, Puget Sound Action Team, Seattle, WA.

When one is developing a site, many local land development and zoning regulations may also impact LID designs. Some common ones are listed in Table 10.1.8.

In addition to considering the items listed in Table 10.1.8, it may be helpful to keep a list of other concepts in the engineering toolbox for site design to aid in attaining LID goals. Many of these are listed or reiterated in Table 10.1.9.

There are several important variables used in LID calculations which are many times also used for local zoning calculations. Table 10.1.10 summarizes several of the common basic ones.

Zoning Requirement	Meaning	Applicability to LID
Minimum landscape areas	A minimum percentage of the lot must be landscaped	These areas will reduce impervious areas and may also be used for IMPs for stormwater management such as bioretention cells
Required landscape buffers	A minimum amount of setback or buffer must be landscaped between the development on the site and neighbors	These areas will reduce impervious areas and may also be used for IMPs for stormwater management such as bioretention cells
Minimum trees in parking areas	A minimum amount of tree shading is required in parking areas	These areas will reduce impervious areas and may also be used for IMPs for stormwater management such as bioretention cells
Maximum slopes on drives/paved areas	For vehicular safety and access reasons there are maximum allowed slopes in paved areas	Minimizing slopes on paved surfaces decreases the flow rate of runoff
Maximum ADA (Americans with Disabilities Act) slopes	For ADA access reasons there are maximum allowed slopes along the accessibility route	Minimizing slopes on surfaces decreases the flow rate of runoff
FAR	Minimum or maximum floor area ratios are sometimes established to limit either the use or the land disturbance impact on the site, respectively	Balancing FARs with lowered impervious cover can help maintain more areas on the lot which mimic the natural hydrology
Minimum and maximum parking	Minimum parking requirements establish adequate parking for the use and maximum limits site impacts	Minimizing parking areas will decrease impervious surfaces
Road/drive width requirements	There are usually minimum and maximum road and drive widths for safety and access reasons	Minimizing paved areas will decrease impervious surfaces

Table 10.1.8 Example Zoning or Land Development Requirements That Impact LID Design Features

Tool	Meaning	Applicability to LID
Limit development envelope and minimize grading	Only disturb during and after construction the minimum amount of land necessary	Limited disturbance will reduce impervious areas and maintain natural areas for stormwater management
Remember "out of sight" may be "out of mind"	Do not install LID features that will not be known to users	Making LID features a prominent part of the site will aid in proper use for stormwater management
Choose alternatives to lawns	Consider alternative landscapes to lawns	Alternative landscapes may reduce runoff and aid in evapotranspiration
Spread concentrated flows	Use level spreaders or energy dissipators to decrease flow energy	Concentrated flows from gutters and leaders can cause increased erosion and runoff
Modify curb types	Use no curbing or modified curbing to not concentrate flow and allow infiltration on-site	Concentrated flows from curb gutters can cause increased erosion and runoff
Reduce imperviousness	Limit impervious surfaces and seek alternative hardscapes that are more pervious	Reduced imperviousness helps limit runoff and aids in on-site infiltration
Conserve natural resources	Conserve natural areas that may aid in LID such as wetlands and good infiltrating soil areas	This will help maintain on-site infiltration and limit runoff
Conserve natural drainage	Maintain natural drainage courses	Preexisting drainage courses usually have already stabilized
Reduce pipe use	Use open channels where possible	This increases awareness and may also allow for more infiltration and evaporation
Disperse	Provide smaller stormwater storage and infiltration features throughout the site	This decreases flow rates at individual areas and helps in infiltrating and watering throughout the site
Disconnect	Disconnect flow paths such as from leaders to drives	This limits the increase of spot runoff rates by superposition, and disruptive flows have more time to infiltrate
Maintain time of concentration	Maintain predevelopment time of concentration t_c where possible	Altering the t_c may mean that runoff will combine with runoff from other areas in a different way, which could increase flows
Educate	Consider options for educating designers and users from the activities on the site	Site users should understand the importance of the LID features and be familiar with maintenance requirements. Also consider using the site for education of others, whether with signage, publications, or visits

TABLE 10.1.9 Other Miscellaneous Tools for LID Designs (*Continued*)

Tool	Meaning	Applicability to LID
Sustainability of LID and BMPs	Consider if the practice will remain effective over time	BMPs/LID practices may require maintenance for continued performance. This must be incorporated into the site operations. Also, some practices are removed because of the lack of knowledge with respect to their importance by future users or designers. Making the features more structurally evident or using educational tools will improve their sustainability
Soil Amendments or Scarification	Scarify or add organic or other amendments to soils previously compacted or compacted during construction	Reestablishes or improves infiltration into soils

TABLE 10.1.9 Other Miscellaneous Tools for LID Designs (*Continued*)

Variable	Title	Chapter Introduced	Use
C	Rational runoff coefficient	2	To estimate the runoff from a site simply based on site surface types
CN	Curve number	10	To estimate the runoff from a site using more advanced models
DF	Development footprint	2	To determine total building and hardscape areas. Useful to estimate the impact on the land with respect to hydrology, vegetation, and heat island issues
FAR	Floor area ratio (also refer to building density in App. B)	10	To estimate the intensity of use of the site based on built floor area. Useful to estimate the impact on areal utilities, transportation, and services
GFA	Gross floor area	2, 10	Usually the built floor area used for the FAR and development density
%imp	Percent impervious	2	To estimate the percent of the site with impervious hardscape. Useful to estimate the impact on the land with respect to hydrology, vegetation, and heat island issues
%impness	Percent imperviousness	2	To estimate overall imperviousness of the site. Useful to estimate the impact on the land with respect to hydrology. Can correlate with the rational runoff coefficient

TABLE 10.1.10 Some Common Variables Used for LID Determinations

The variables C, DF, %imp, and %impness are explained in Chap. 2 and will not be repeated here. However, there is a need for a closer look at CN, FAR, and GFA.

CN is similar to the simpler rational runoff coefficient, but also includes other site factors such as soil types. It is used in very well-established hydrology modeling, and more information about its use and application can be found in most hydrology texts and from the Natural Resources Conservation Service. How to modify the curve number to adequately take into account LID design modifications is under development. Some mechanisms for doing this have been introduced by Prince George's County Department of Environmental Resources (1999) and the DoD (2004). In addition, methods for incorporating LID practices into other established hydrology models can be found in the Puget Sound Technical Guidance as previously listed.

The gross floor area of a building (GFA_i) can have many definitions depending on the local ordinances used. Usually it includes all the floors of a building, taken from the outer dimensions, except for unfinished attic and basement spaces. Chapter 2 discusses some modifications which can be used to also include parking garages for the purposes of the LEED calculations. For every case, the actual areas used as such should be summarized.

The *floor area ratio* (FAR) is usually defined as the gross floor area of all the buildings on a site divided by the total site area, as in Eq. (10.1.1).

$$\text{FAR} = \frac{\sum \text{GFA}_i}{\text{A}_T} \quad \text{for all buildings } i \text{ on the site} \tag{10.1.1}$$

Why is it necessary to have so many variables? Well, each is useful in aiding in the determination of different impacts of construction. Several examples are also listed in Table 10.1.10. One simple example to explain the importance of determining many of these variables for sustainable construction purposes can be seen if the floor area ratio is examined. A metamodel called Sustainable Futures 2100 has been recently developed to examine many of the complex environmental and natural resource impacts of construction, both on-site and regional. It was used to compare a single-story house and a two-story house of similar gross floor area, which in turn means similar FARs. It was determined that the single-story home had less of an energy impact during construction, but the two-story home had a higher operational energy demand. Hydrologic impacts are different too and were not compared in that analysis, but obviously the single-story home will use more of the land and therefore leave less for keeping the natural hydrologic cycle of the land without additional LID measures. So, knowledge of only the FAR is not sufficient for determining the complex impacts of construction, although it is a good indicator of many items such as utility and regional transportation usage related to the site.

Green Streets and Urban LID Practices

As LID practices are incorporated into the landscape, more and more references and details for applications under various different scenarios are being developed. Some of these are focusing on applications to linear projects such as highways and more urban streetscapes. Special considerations may also be given to other development parameters in conjunction with the LID practice such as vehicular flow, traffic calming, pedestrian access, bicycle paths, landscape needs and wildlife passage, in addition to other context sensitive solutions. Figure 10.1.7 is a photograph of a traffic circle combined with a rain

Figure 10.1.7 Bainbridge, WA, High School traffic circle rain garden. (*Photograph courtesy Coughlin Porter Lundeen taken October 2009.*)

garden in the Puget Sound area of Washington. Figure 10.1.8 is of a pervious concrete parking lot in Oregon which aided in infiltrating runoff to protect a nearby wetland. These two LID practices—rain gardens and permeable pavements—are becoming more and more popular in urban settings for their multiple uses and flexibility in design. Table 10.1.11 lists some useful green streets and urban LID manuals and references.

Figure 10.1.8 Lincoln City Oregon pervious concrete parking lot. (*Photograph courtesy Scott Erickson of Evolution Paving taken November 2009.*)

Design Aid	Agency	Website
2008 Stormwater Management Facility Monitoring Report: Sustainable Stormwater Management Program	Portland Bureau of Environmental Services, Portland, OR	http://www.portlandonline.com/bes/index.cfm?c=36055&a=232644
Street Design Manual	New York City Department of Transportation	http://www.nyc.gov/html/dot/downloads/pdf/sdm_lores.pdf
Phytotechnology Technical and Regulatory Guidance and Decision Trees (2009)	The Interstate Technology & Regulatory Council, Washington, D.C.	http://www.itrcweb.org/Documents/PHYTO-3.pdf
San Mateo County Sustainable Green Streets and Parking Lots Design Guidebook	San Mateo County, CA	http://www.flowstobay.org/ms_sustainable_guidebook.php
Seattle Right-of-Way Improvement Manual	City of Seattle, WA	http://www.seattle.gov/transportation/rowmanual/
Maryland Stormwater Design Manual, Volumes I & II (2000)	Maryland Department of the Environment	http://www.mde.state.md.us/Programs/WaterPrograms/SedimentandStormwater/stormwater_design/index.asp
New Hampshire Stormwater Manual (2008)	New Hampshire Department of Environmental Services	http://des.nh.gov/organization/divisions/water/stormwater/manual.htm
LID Center – Green Streets Website	Low Impact Development Center, Inc., Beltsville, MD	http://www.lowimpactdevelopment.org/greenstreets/background.htm
Low Impact Development (LID) Urban Design Tools Website	Low Impact Development Center, Inc., Beltsville, MD	http://www.lid-stormwater.net/index.html
Mix Design Development for Pervious Concrete in Cold Weather Climates	Iowa State University Center for Transportation Research and Education	http://www.intrans.iastate.edu/reports/mix_design_pervious.pdf
Pervious Concrete	National Ready Mix Concrete Association	http://www.perviouspavement.org/
Specifiers Guide to Pervious Concrete Pavements in the Greater Kansas City Area	The Concrete Specifiers Group of Greater Kansas City	http://www.concretepromotion.com/

TABLE 10.1.11 Green Streets and Urban Stormwater References

Rain Gardens

Rain gardens or bioretention cells or swales are becoming common landscape features which serve the dual purposes of landscaping and low-impact development stormwater management. They are usually specially designed landscaped beds which have some similar characteristics. They usually have a concave shape, instead of the convex or mounded landscape beds which were popular in the past. This concave shape serves to aid in water storage to allow for stormwater infiltration into the ground. There are usually many separate rain gardens distributed throughout a site. Figures 10.1.9 and 10.1.10 are photographs of some of the rain gardens distributed throughout a middle school yard in Tacoma, WA.

The design of rain gardens usually includes several layers of material to allow for enhanced water storage, infiltration, and contaminant removal. The plantings are chosen for the appropriate saturation levels found in the various portions of the rain garden and the local climate. Some designs include nonvegetative media also for aesthetics, especially in arid regions or when flows would damage vegetation. Rain gardens can be very small or much larger, but generally are not as large as the typical stormwater ponds. They can have a more centralized ponding area or be arranged along channels or within sidewalk and other linear shoulders. Figure 10.1.11 is a photograph of an open conveyance feature in a low-impact development project in Central Washington that has little or no vegetation, while Figs. 10.1.12 and 10.1.13 are photographs of a rain garden along a walkway in its first year, and its third year as the planted vegetation becomes established.

FIGURE 10.1.9 A Gray Middle School rain garden. (*Photograph courtesy Coughlin Porter Lundeen taken November 2009.*)

FIGURE 10.1.10 Another Gray Middle School rain garden. (*Photograph courtesy Coughlin Porter Lundeen taken November 2009.*)

FIGURE 10.1.11 Decorative open conveyance landscape feature at Central Washington University. (*Photograph courtesy Coughlin Porter Lundeen taken August 2008.*)

Figure 10.1.12 Recently planted rain garden at Ben Franklin Elementary school in Kirkland, WA. (*Photograph courtesy Coughlin Porter Lundeen taken October 2007.*)

Figure 10.1.13 Established rain garden at Ben Franklin Elementary school in Kirkland, WA. (*Photograph courtesy Coughlin Porter Lundeen taken June 2009.*)

Permeable Pavements

Another novel group of LID practices includes the various permeable pavements. Permeable pavements serve as a structured surface for vehicles, bicycles, or pedestrians, while also aiding in stormwater management. There are various different types of permeable pavements, but the four main categories are permeable pavers, permeable grids, porous asphalt, and pervious concrete. Permeable paver placements are typically made with traditional impermeable pavers such as brick or concrete precast pavers, although there are some pervious pavers being marketed. With the traditional pavers, the water infiltrates to the ground below through the interstitial spaces between the pavers.

Permeable grid systems have a grid made up of various materials such as concrete or plastic with open spaces filled with usually gravel or grass in between. The grass-filled grid systems are usually placed where grass is desired, but occasional traffic will be in that location such as for fire lanes around schools or for occasion parking for large events at campuses or arenas. Both porous asphalt and pervious concrete are similar to traditional asphalt or concrete placements, except that the materials are designed and carefully placed with an interconnected void structure which allows water to pass through the pavement. Both of these systems usually allow water to pass at a very rapid rate and are frequently part of a layered system with an aggregate sub-layer for additional below-ground water storage. These additional layers are used when placed over poorly draining soils to allow time for the water to infiltrate and also for additional loading capacity for the pavement system. Porous asphalt and pervious concrete are becoming popular pavement alternatives in parking lots, for trails and for low volume streets. Figure 10.1.14 is a photograph of a pervious concrete parking lot in Oregon.

It is important that standards, specifications, and testing methods are developed for these novel pavements for municipalities to use in determining acceptance and also for

FIGURE 10.1.14 Pervious concrete parking lot in Salem, OR. (*Photograph courtesy Scott Erickson of Evolution Paving taken November 2009.*)

monitoring for maintenance needs. The ASTM C09/49 pervious concrete technical committee has recently released two standard testing methods specific for pervious concrete. It is also developing additional standards with respect to durability, hardened density, and strength. The two ASTM standards already available are ASTM C1688/C1688M-08 Standard Test Method for Density and Void Content of Freshly Mixed Pervious Concrete and ASTM C1701/C1701M-09 Standard Test Method for Infiltration Rate of In Place Pervious Concrete. The American Concrete Institute technical committee on pervious concrete (ACI 522) has also released a specification for pervious concrete: ACI 522.1-08: Specification for Pervious Concrete Pavement.

10.2 Modeling LID BMPs

There are many commercial stormwater models that are commonly used for design and assessment. Many of these have opportunities for including some of the LID scenarios for stormwater management into the assessment. In the future, more individual LID practices and technologies will be included in them. However, given the many options for LID and the site-specific needs and variations for each, it is helpful to have some simple methods to perform preliminary design and performance estimates of the practices. The following section gives some simple mass balance methods for preliminary design and assessment of many LID technologies.

Overall Mass Balances

In stormwater management there are two types of mass balances that are of interest: a mass balance on the water and a mass balance on the pollutant in question. Regardless of the type of substance for which its mass is being balanced, the balances can be summarized with Eqs. (10.2.1) and (10.2.2). Given the "system" or unit operation that is being evaluated, Eq. (10.2.1) represents a dynamic situation where the mass of the substance in question within the system changes with time. Equation (10.2.2) represents a steady-state situation where the mass of the substance in question within the system is constant.

$$\text{Rate of accumulation} = \Sigma\,\text{Rates in} - \Sigma\,\text{Rates out} \pm \Sigma\,\text{Internal reactions/processes} \quad \text{dynamic mass balance} \tag{10.2.1}$$

$$0 = \Sigma\,\text{Rates in} - \Sigma\,\text{Rates out} \pm \Sigma\,\text{Internal reactions/processes} \quad \text{steady-state mass balance} \tag{10.2.2}$$

Stormwater Mass Balances

The first mass balance explained here is the total stormwater mass balance around a system such as a BMP or other LID practice. Let the BMP be represented by a box, and let the typical flows in and out be as shown in Fig. 10.2.1.

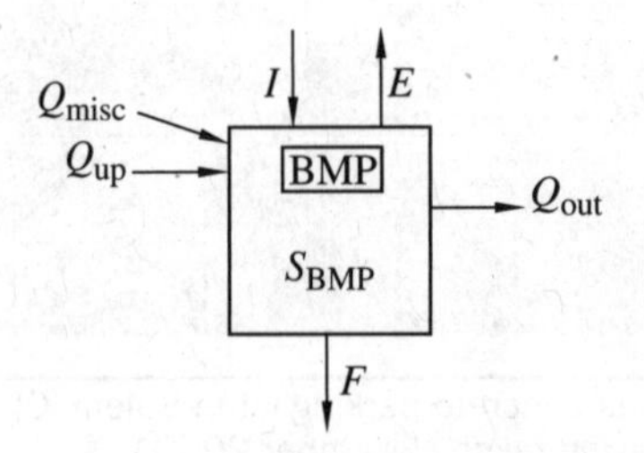

FIGURE 10.2.1 Simple box model of a stormwater mass balance on a BMP.

In this simple box model of a stormwater mass balance around a BMP, the following variables are used:

A_{BMP} Surface area of BMP (length squared, usually acres or ft^2)
E Evaporation rate (length/time, usually in/h)
F Infiltration rate (length/time, usually in/h)
I Rainfall rate (length/time, usually in/h)
Q_{misc} Miscellaneous flows into a BMP ($length^3$/time, usually ft^3/s or cfs)
Q_{out} Total runoff out of a BMP ($length^3$/time, usually ft^3/s)
Q_{up} Runoff from upslope areas into a BMP ($length^3$/time, usually ft^3/s)
S_{BMP} Storage volume in a BMP ($length^3$, usually ft^3)
t_{fill} Time to fill

Assuming that there are no internal reactions within the BMP that remove or add water, these variables can be substituted into Eqs. (10.2.1) and (10.2.2) and the following water mass balance equations result:

$$\text{Accumulation rate} = Q_{misc} + Q_{up} + I(A_{BMP}) - E(A_{BMP}) - F(A_{BMP}) - Q_{out}$$
stormwater dynamic mass balance (10.2.3)

$$0 = Q_{misc} + Q_{up} + I(A_{BMP}) - E(A_{BMP}) - F(A_{BMP}) - Q_{out}$$
stormwater steady-state mass balance (10.2.4)

In addition to these equations, the initial state of the BMP with respect to the mass of water within it must be known. For the initial condition where the storage volume within the BMP is not full, Eq. (10.2.3) represents the condition when the storage volume within the BMP is filling with water if the flows in are greater than the flows out; or if there is any water within the BMP, then losing water when the flows out are greater than the flows in. Equation (10.2.4) represents the condition when the amount of stormwater that comes in equals what goes out. This usually means the storage volume is full, or that the storage volume is not "filling" since the potential for flow out is greater than what comes in.

What is needed now is some way to estimate the flows. The simplest model to use for many of the flows in the stormwater mass balance is based on the rational runoff method, and this is presented here. (However, flows from other models can be substituted into the overall mass balances if they are used.) In the *rational runoff method,* the governing generic equation relates the total runoff Q from a site to the average rational runoff coefficient C, the rainfall rate I, and the land area A_T in the following manner:

$$Q = CIA_T \quad \text{generic rational method equation} \qquad (10.2.5)$$

[If the units used are runoff in cubic feet per second, rainfall rate in inches per hour, and area in acres, then it just so happens that the conversion factor from acre-inches per hour (acre·in/h) to cubic feet per second is approximately 1, and no unit conversions are needed. See Sec. 2.6 for additional information about the rational method and C.]

In the case of most BMPs, we may have upslope overland flow into the system, rainfall directly on the top surface of the BMP, some additional water sources (such as pipe flows from elsewhere), evaporation from the top of the BMP, infiltration into the ground/subsurface underneath, and then the flow out of the BMP, which is the main flow rate we are interested in determining. The direct rainfall flow into the BMP has been estimated in Eqs. (10.2.3) and (10.2.4) as the rainfall rate times the surface area of the BMP, and both the soil infiltration flow out (exfiltration) and the evaporation loss have been estimated by some sort of areal soil/subsurface infiltration rate and some sort of areal evaporation rate (see also Chap. 3 for evapotranspiration rates). The flow from the upslope areas can be estimated by Eq. (10.2.5), using the following variables:

A_{up} Land area upslope of BMP that drains into it (length squared, usually acres)

C_{up} Average stormwater runoff coefficient of land surface upslope of BMP

This results in Eq. (10.2.6).

$$Q_{up} = C_{up} I A_{up} \quad \text{upslope rational method runoff estimate contribution} \quad (10.2.6)$$

By using these equations, many different situations can be modeled. For the first scenario, assume that it is raining at a constant rainfall rate I, that the only flows in are from upslope runoff and the rain, that there is negligible evaporation, and that there was no water initially stored in the BMP. The equations can be used to estimate the time it might take to start runoff out of the BMP (which is essentially the time to fill the storage volume during which the runoff out of the BMP is zero) if the flows in are greater than the infiltration rate.

$$t_{fill} = S_{BMP}/[C_{up} I A_{up} + I(A_{BMP}) - F(A_{BMP})] \quad \text{time to fill in first scenario} \quad (10.2.7)$$

Now, assume that the storage volume S_{BMP} has filled and the rain continues. The equations can also be used to estimate the steady-state runoff out of this BMP. In this case, the outlet for the flow out is above the storage volume height, a common condition in many cases. If there are multiple outlets, then the equations will need to be solved for each stage of storage within the BMP.

$$Q_{out} = C_{up} I A_{up} + I(A_{BMP}) - F(A_{BMP}) \quad \text{steady-state flow out in second scenario} \quad (10.2.8)$$

The main way to model the stormwater mass balance around any BMP is to first make a simplified sketch of the system, labeling potential flows in and out and internal items such as storage as in Fig. 10.2.1. Then the initial conditions are used along with Eqs. (10.2.3) and (10.2.4) to model the system. For solutions, the equations are further simplified by allowing negligible flows to be neglected for the model and by using accepted models and parameters such as the rational method or soil infiltration rates to substitute for many of the variables.

Pollutant Mass Balances

The mass balance of pollutants around a BMP can be modeled by using the box model in the stormwater mass balance model and including the concentration of the pollutant in each flow, or within the BMP, using the following definitions:

C_{PBMP} Average concentration of a pollutant in a BMP

C_{PF} Average concentration of a pollutant infiltrating into ground

C_{PI} Average concentration of a pollutant in rainfall

C_{Pmisc} Average concentration of a pollutant coming from miscellaneous flows into a BMP

C_{Pout} Average concentration of a pollutant flowing out of BMP

C_{Pup} Average concentration of a pollutant in upslope runoff

Note that the concentration of a pollutant has been ignored in the evaporation portion, but this may need to be included as a pollutant evaporation rate if the contaminant in question is volatile. Also, many pollutants may have reactions internal to the BMP, such as microbial degradation of organics to inorganic compounds, and the various internal reaction rates would need to be added. However, for cases where neither internal reactions nor pollutant evaporation is important, the stormwater equations can be rewritten as pollutant mass balances as in Eqs. (10.2.9) and (10.2.10).

$$\text{Accumulation rate of pollutants at time } t = Q_{\text{misc}}C_{P\text{misc}} + Q_{\text{up}}C_{P\text{up}} + IA_{\text{BMP}}C_{PI} - FA_{\text{BMP}}C_{PF} - Q_{\text{out}}C_{P\text{out}}$$

pollutant dynamic mass balance, no internal reactions or evaporation (10.2.9)

$$0 = Q_{\text{misc}}C_{P\text{misc}} + Q_{\text{up}}C_{P\text{up}} + IA_{\text{BMP}}C_{PI} - FA_{\text{BMP}}C_{PF} - Q_{\text{out}}C_{P\text{out}}$$

pollutant steady-state mass balance, no internal reactions or evaporation (10.2.10)

The storage volume is then used along with the time and the equations to calculate the concentration within the BMP (C_{PBMP}).

There are additional models that are needed to aid in this estimation, and these are models of how the pollutant is distributed within the BMP so that it can be determined what the pollutant concentrations might be of the flows out of the system. The two extremes are usually the completely mixed assumption (the concentration is equal throughout the BMP) and the plug flow assumption. Plug flow means that there is no mixing in the direction of flow within the BMP so that there is a pollutant concentration gradient in this direction. If it is assumed that the pollutant is evenly distributed within the BMP (completely mixed model), then it can be assumed that C_{PBMP} is equal to both C_{PF} and C_{Pout}. Plug flow and other models are beyond the scope of this text and are better presented in environmental modeling or chemical operation texts, but may be important in some BMP modeling, particularly in systems such as bioretention cells where there may be nutrient uptake within some layers of the cell.

If the completely mixed model is assumed for the pollutant mass balances, then Eqs. (10.2.9) and (10.2.10) simplify to

$$\text{Accumulation rate of pollutants at time } t = Q_{\text{misc}}C_{P\text{misc}} + Q_{\text{up}}C_{P\text{up}} + IA_{\text{BMP}}C_{PI} - (FA_{\text{BMP}} + Q_{\text{out}})C_{PBMP}$$

pollutant dynamic completely mixed model, no internal reactions/evaporation (10.2.11)

$$0 = Q_{\text{misc}}C_{P\text{misc}} + Q_{\text{up}}C_{P\text{up}} + IA_{\text{BMP}}C_{PI} - (FA_{\text{BMP}} + Q_{\text{out}})C_{PBMP}$$

pollutant steady-state completely mixed, no internal reactions/evaporation (10.2.12)

Pervious Concrete

Pervious concrete is an alternative paving surface that allows infiltration into the subbase and ground below. It is a novel paving material that is being developed to aid in maintaining preexisting hydrologic conditions while still giving a structural base for vehicular and other uses. There is a need to be able to model the various pervious concrete systems and their impacts on hydrology and pollutant transport.

Many studies show that the infiltration rate through the pervious concrete itself is usually much greater than that through any natural soil type. The infiltration rate may therefore be limited by the subbase and subgrade below. Typically, pervious concrete can be placed directly over well-drained soils such as sands, but should be placed over additional storage subbases such as gravel, when used over poorly drained soils such as clays. This allows for additional time for the stormwater to infiltrate into the ground instead of running off. Usually overflow pipes are also included for stormwater management purposes for large storms.

Based on the LEED-NC 2.2 and 2009 Reference Guides, pervious concrete areas can be considered to be pervious. However, there are also local or state interpretations of how to count the highly pervious concrete and asphalt pavements with respect to imperviousness, and these must be taken into consideration for local requirements and interpretations. LEED usually defers to the local code if it is more stringent than the LEED credit criteria. For example, in North Carolina, pervious concrete and porous asphalt may count toward 40 to 60 percent as managed grass with respect to percent imperviousness for a site.

A pervious concrete application is a good method by which to use the simple box model to estimate runoff from a site. Consider a system consisting of pervious concrete with a gravel subbase and then soil below and with the pavement system essentially flat. This system receives upslope runoff and direct rainfall. The condition being modeled is a storm with no antecedent precipitation (storage volume initially empty) and during the time when evaporation can be assumed negligible (during the precipitation event). There are two questions being asked:

- How long until the onset of runoff?
- After the storage volume is full, what is the steady-state rate of runoff?

The system is depicted as a two-layer simple box model as shown in Fig. 10.2.2. Since there are two different layers to the pavement system, the total storage can be estimated as the sum of the pervious concrete and the gravel layer storage volumes. Given:

$A_{\text{BMP-PC}}$ Surface area of pervious concrete BMP
S_{Gravel} Storage volume in gravel layer (length^3, usually ft^3)
S_{Pervious} Storage volume in pervious concrete layer (length^3, usually ft^3)

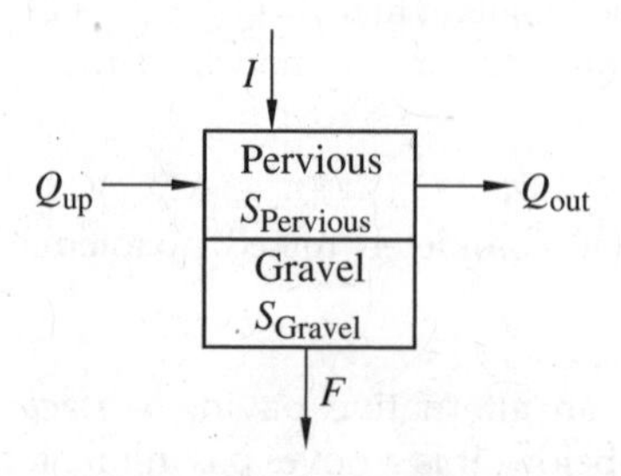

FIGURE 10.2.2 Simple box model of pervious concrete over gravel system.

Equations (10.2.7) and (10.2.8) can thus be modified for the pervious concrete system model as follows:

$$t_{\text{fill}} = (S_{\text{Gravel}} + S_{\text{Pervious}})/[C_{\text{up}} I A_{\text{up}} + (I - F)(A_{\text{BMP-PC}})] \quad \text{time to fill} \qquad (10.2.13)$$

$$Q_{\text{out}} = C_{\text{up}} I A_{\text{up}} + (I - F)(A_{\text{BMP-PC}}) \quad \text{steady-state flow out} \qquad (10.2.14)$$

Many other initial conditions and configurations can be modeled for pervious concrete systems. In like manner, other BMP systems can be modeled using the basic concepts of a box model and mass balances. First determine the stormwater flows in, then how and where stormwater might flow out, and internally determine if there are separate layers or cells within the BMP. For instance, rain gardens collect water that falls on them to decrease runoff from these areas and may also collect and infiltrate stormwater from neighboring areas. Figures 10.2.3 and 10.2.4 show how rainwater that falls on a roof may be directly collected into a roof top garden (green roof) or diverted to a rain garden on the site. Note how the green roof has low-water needs-succulent plant species.) Figure 10.2.5 is a photograph of a storage facility in Oregon where the pervious concrete collects rainwater that falls directly on it and also stormwater from the roof. Then perform water and pollutant mass balances around each layer, and for the overall LID practice.

FIGURE 10.2.3 Roof top garden at Rachel Carson Elementary School in Sammamish, WA. (*Photograph courtesy Coughlin Porter Lundeen taken September 2008.*)

Figure 10.2.4 Roof stormwater leaders directed into a rain garden under construction at Bothell High School in Washington. (*Photograph courtesy Coughlin Porter Lundeen taken September 2008.*)

Figure 10.2.5 Self storage facility in Florence, OR, with pervious concrete hardscape. (*Photograph courtesy Dan Clem of Evolution Paving taken February 2010.*)

References

ACI (2008), *ACI 522.1-08: Specification for Pervious Concrete Pavement*, American Concrete Institute, Farmington Hills, MI.

Akan, A. O., and R. J. Houghtalen (2003), *Urban Hydrology, Hydraulics and Stormwater Quality, Engineering Applications and Computer Modeling*, John Wiley & Sons, Hoboken, NJ.

ASCE (2010), 20101 International Low Impact Development Conference, American Society of Civil Engineers, Website http://content.asce.org/conference/lid10/program.html, accessed March 13, 2010.

ASTM (2008), *ASTM C1688/C1688M-08 Standard Test Method for Density and Void Content of Freshly Mixed Pervious Concrete*, ASTM, West Conshohocken, PA.

ASTM (2009), *ASTM C1701/C1701M-09 Standard Test Method for Infiltration Rate of In Place Pervious Concrete*, ASTM, West Conshohocken, PA.

Atlanta Regional Commission and Georgia Department of Natural Resources (2001), *Georgia Stormwater Management Manual*, vol. 2: *Technical Handbook*.

Barber, M. E., S. King, D. Yonge, and W. Hathhorn (2003), "Ecology Ditch: A Best Management Practice for Storm Water Runoff Mitigation," *ASCE Journal of Hydrologic Engineering*, May/June, American Society of Civil Engineers, Reston, VA.

Bean, E. Z. (2005), *A Field Study to Evaluate Permeable Pavement Surface Infiltration Rates Runoff Quantity, Runoff Quality and Exfiltrate Quality*. MS thesis, North Carolina State University, Raleigh.

Berke, P. R., et al. (2003), "Greening Development to Protect Watersheds, Does New Urbanism Make a Difference?" *Journal of American Planning Association*, 69(4).

Brady, N. C., and R. Weil (2007), *The Nature and Property of Soils*, 14th ed. Prentice-Hall, Upper Saddle River, NJ.

CPESC (2007), http://www.cpesc.net/, Certified Professional in Erosion and Sediment Control website, accessed May 14, 2007.

CEPSCI (2005), *Erosion Prevention and Sediment Control Inspector Training Handout*, CEPSCI South Carolina, Clemson University.

CEPSCI (2007), https://www.clemson.edu/t3s/cepsci/index.htm, CEPSCI South Carolina, Clemson University website, accessed July 27, 2007.

CWP (1998), *Better Site Design*, CD, Center for Watershed Protection, Ellicott City, MD.

CWP (2000), *8 Tools of Watershed Protection*, CD, Center for Watershed Protection, Ellicott City, MD.

CWP (2000), *Impacts of Urbanization*, CD, Center for Watershed Protection, Ellicott City, MD.

CWP (2000), *The Practice of Watershed Protection*, Center for Watershed Protection, Ellicott City, MD.

CWP (2000), *A Review of Stormwater Treatment Practices*, CD, Center for Watershed Protection, Ellicott City, MD.

Davis, A. P. (2005), "Green Engineering Principles, Promote Low Impact Development," *ES&T*, August 15, pp. 338A–344A.

Davis, A. P., and R. H. McCuen (2005), *Stormwater Management for Smart Growth*, Springer, New York.

Debo, T., and A. J. Reese (2003), *Municipal Stormwater Management*, 2d ed., Forester Press, Santa Barbara, CA.

Deletic, A., and T. D. Fletcher (2006), "Performance of Grass Filters Used for Stormwater Treatment—A Field and Modelling Study," *Journal of Hydrology*, 317: 261–275.

DoD (2004), *Unified Facilities Criteria (UFC) Design: Low Impact Development Manual,* Department of Defense, October 24, 2004, accessed July 10, 2007, http://www.lowimpactdevelopment.org/lid%20articles/ufc_3_210_10.pdf.

Driscoll, E. (1986), "Methodology for Analysis of Detention Basins for Control of Urban Runoff Quality," EPA440/5-87-001, U.S. Environmental Protection Agency, Office of Water, Nonpoint Source Branch, Washington, D.C.

Elliot, A. H., and S. A. Trowsdale (2007), "A Review of Models for Low Impact Urban Stormwater Drainage," *Environmental Modelling & Software,* 22(3): 394–405.

EPA (1986), *Preliminary Data Summary of Urban Storm Water Best Management Practices,* EPA-821-R-99-012, U.S. Environmental Protection Agency, Washington, D.C.

EPA (2000), *Introduction to Phytoremediation,* EPA-600-R-99-107, U.S. Environmental Protection Agency, Cincinnati, OH.

EPA (2007), http://cfpub.epa.gov/npdes/, Environmental Protection Agency NPDES publication website, accessed May 24, 2007.

EPA OW (2007), http://www.epa.gov/watersense/, U.S. Environmental Protection Agency Office of Water, Water Sense; Efficiency Made Easy website, accessed July 10, 2007.

Ferguson, B. (2005), *Porous Pavements,* Taylor & Francis CRC Press, Boca Raton, FL.

FHWA (1971), *Hydraulic Engineering,* Circular No. 9 (HEC-9), U.S. Department of Transportation, Federal Highway Administration.

FHWA (1989), *Hydraulic Engineering,* Circular No. 11 (HEC-11), U.S. Department of Transportation, Federal Highway Administration.

Fifield, J. S. (2002), *Field Manual on Sediment and Erosion Control Best Management Practices for Contractors and Inspectors,* Forester Press, Santa Barbara, CA.

Funkhouser, L. (2007), "Stormwater Management as Adaptation to Climate Change," *Stormwater, The Journal for Surface Water Quality Professionals,* July/August, 8(5): 50–71, Forester Communications, Santa Barbara, CA.

Haselbach, L. M., and R. M. Freeman (2006), "Vertical Porosity Distributions in Pervious Concrete Pavement," American Concrete Institute *ACI Materials Journal,* November–December, 103(6): 452–458.

Haselbach, L. M., S. R. Loew, and M. E. Meadows (2005), "Compliance Rates for Storm Water Detention Facility Installation,"*ASCE Journal of Infrastructure Systems,* March, 11(1): 61–63.

Haselbach, L. M., S. Valavala, and F. Montes (2006), "Permeability Predictions for Sand Clogged Portland Cement Pervious Concrete Pavement Systems," *Journal of Environmental Management,* October, 81(1): 42–49.

Hinman, C. (2005), *Low Impact Development: Technical Guidance Manual for Puget Sound,* Puget Sound Action Team, Washington State University, Pierce County Extension, Olympia, WA.

Holman-Dodds, J., A. Bradley, and K. Potter (2003), "Evaluation of Hydrologic Benefits of Infiltration Based Urban Storm Water Management," *Journal of American Water Resources Association,* February, 39(1): 205–215.

Hunt, W. F., and B. A. Doll (2000), "Designing Stormwater Wetlands for Small Watersheds," *Urban Waterways Series,* AG-588-2, North Carolina State University, Raleigh.

Hunt, W. F., and W. G. Lord (2006), "Bioretention Performance, Design, Construction and Maintenance," *Urban Waterways Series,* AGW-588-05, North Carolina State University, Raleigh.

Hunt, W. F., and N. M. White (2001), "Designing Rain Gardens/Bioretention Areas," *North Carolina Cooperative Extension Service Bulletin, Urban Waterfronts Series,* AG-588-3, North Carolina State University, Raleigh.

Kadlec, R. H., and R. L. Knight (1996), *Treatment Wetlands,* Lewis Publishers, Boca Raton, FL.

Kim, H., E. A. Seagren, and A. P. Davis (2003), "Engineered Bioretention for Removal of Nitrate from Stormwater Runoff," *Water Environment Research,* 75: 355–367.

Li, K., et al. (2007), "Development of a Framework for Quantifying the Environmental Impacts of Urban Development and Construction Practices," *Environmental Science and Technology,* 41(14): 5130–5136.

Loew, S. R., L. M. Haselbach, and M. E. Meadows (2004), "Life Cycle Analysis Factors for Construction Phase BMPs in Residential Subdivisions," *Proceedings of the ASCE-EWRI 2004 World Water and Environmental Resources Congress,* 6/27–7/1/04, Salt Lake City, Utah, Best Management Practices (BMP) Technology Symposium: Current and Future Directions.

Malcom, H. R. (1989), *Elements of Urban Stormwater Design,* North Carolina State University, Raleigh.

Millen, J. A., A. R. Jarrett, and J. W. Faircloth (1996), "Reducing Sediment Discharge from Sediment Basins with Barriers and a Skimmer," ASAE Microfiche 96-2056.

Montes, F., and L. M. Haselbach (2006), "Measuring Hydraulic Conductivity in Pervious Concrete," *Environmental Engineering Science,* November, 23(6): 956–965.

Montes, F., S. Valavala, and L. M. Haselbach (2005), "A New Test Method for Porosity Measurements of Portland Cement Pervious Concrete," *Journal of ASTM International,* January, 2(1).

Moran, A. C., W. F. Hunt, and G. D. Jennings (2004), "A North Carolina Field Study to Evaluate Green Roof Quantity, Runoff Quality, and Plant Growth," in *Proceedings of Green Roofs for Healthy Cities Conference,* Portland, OR.

NCDENR, DLR (2006), *Erosion and Sediment Control Planning and Design Manual,* North Carolina Department of Environment and Natural Resources, Division of Land Resources, Raleigh.

NCDENR, DWQ (1999), *Neuse River Basin: Model Stormwater Program for Nitrogen Control,* July, North Carolina Department of Environment and Natural Resources, Division of Water Quality, Raleigh, http://h2o.enr.state.nc.us/su/Neuse_SWProgram_Documents.htm.

NCDENR, DWQ (2003), *Tar-Pamlico River Basin: Model Stormwater Program for Nutrient Control,* North Carolina Department of Environment and Natural Resources, Division of Water Quality, Raleigh, http://h2o.enr.state.nc.us/nps/Model-FINAL9-12-03.doc.

NCDENR, DWQ (2004), *Memorandum:Updates to Stormwater BMP Efficiencies.* September 8, North Carolina Department of Environment and Natural Resources, Division of Water Quality, Raleigh.

NCDENR, DWQ (2007), *Stormwater Best Management Practices Manual,* July 2, 2007, North Carolina Department of Environment and Natural Resources, Division of Water Quality, Raleigh, http://h2o.enr.state.nc.us/su/bmp_manual.htm, website accessed July 21, 2007.

NCDOT (2003), *Best Management Practices for Construction and Maintenance Activities,* August, North Carolina Department of Transportation, Raleigh.

NCHRP (2006), *Evaluation of Best Management Practices for Highway Runoff Control;* National Cooperative Highway Research Program (NCHRP) Report 565, Transportation Research Board (TRB), Washington, D.C.

NIPC (1993), *Urban Stormwater Best Management Practices for Northeastern Illinois*, Northeastern Illinois Planning Commission.

NRCS (2007), http://www.nrcs.usda.gov/, Natural Resources Conservation Service website, accessed July 27, 2007.

Pitt, R., S. Clark, and D. Lake (2006), *Construction Site Erosion and Sediment Controls: Planning, Design and Performance*, Forester Press, Santa Barbara, CA.

Prince George's County, Department of Environmental Resources (1999), *Low-Impact Development Design Strategies: An Integrated Design Approach*, Prince George's County, MD.

Prince George's County, Department of Environmental Resource (2002), *Bioremediation Manual*, Prince George's County, MD., accessed May 10, 2007, http://www.goprincegeorgescounty.com/Government/AgencyIndex/DER/ESD/Bioretention/bioretention.asp.

Rafter, D. (2007), "The Art of Testing: Putting Stormwater, Erosion, and Sediment Control BMPs through Their Paces," *Stormwater, The Journal for Surface Water Quality Professionals*, July/August, 8(5): 24–32, Forester Communications, Santa Barbara, CA.

SCDHEC (2003), *South Carolina Stormwater Management and Sediment Control Handbook for Land Disturbance Activities*, South Carolina Department of Health and Environmental Control; Bureau of Water, Office of Ocean and Coastal Resource Management http://www.scdhec.gov/environment/ocrm/pubs/docs/swmanual.pdf, accessed July 6, 2007.

Schnoor, J. L. (1996), *Environmental Modeling: Fate and Transport of Pollutants in Water, Air, and Soil*, John Wiley & Sons, New York.

Schueler, T. R. (1987), *Controlling Urban Runoff: A Practical Manual for Planning and Designing Urban BMPs*, Department of Environmental Programs, Metropolitan Washington Council of Governments.

Schueler, T. R. (1992), *Design of Stormwater Wetland Systems*, Department of Environmental Programs, Metropolitan Washington Council of Governments.

Schueler, T., and R. Claytor (1996), *Design of Stormwater Filtering Systems*, Chesapeake Research Consortium, Center for Watershed Protection, Ellicott City, MD.

USGBC (2009), *LEED Reference Guide for Green Building Design and Construction*, 2009 Edition, U.S. Green Building Council, Washington, D.C., April 2009.

U.S. Soil Conservation Service (1986), *Urban Hydrology for Small Watersheds*, Technical Release 55, Washington, D.C.

Valavala, S., F. Montes, and L. M. Haselbach (2006), "Area Rated Rational Runoff Coefficient Values for Portland Cement Pervious Concrete Pavement," *ASCE Journal of Hydrologic Engineering*, May–June, 11(3): 257–260.

Waschbusch, R. J., R. Selbig, and R. T. Bannerman (1999), *Sources of Phosphorus in Stormwater and Street Dirt from Two Urban Residential Basins in Madison, Wisconsin, 1994–95*, Water Resources Investigations Report 99-4021, U.S. Geological Survey.

Wossink, A., and W. Hunt (2003), *An Evaluation of Cost and Benefits of Structural Stormwater Best Management Practices in North Carolina*, North Carolina Cooperative Extension Service, Raleigh.

Zhang, Y.-K., and K. E. Schilling (2006), "Effects of Land Cover on Water Table, Soil Moisture, Evapotranspiration, and Groundwater Recharge: A Field Observation and Analysis," *Journal of Hydrology*, 319: 328–338.

Exercises

1. You are going to pave a parking lot on 0.5 acre of land. You are required to maintain a minimum of 10 percent of the area as landscaped or green space. You decide to incorporate bioretention cells into three landscape areas in the parking lot.

A. Sketch an example parking lot and show the stormwater mass balances around the parking lot.

B. Sketch one of the bioretention cells and make a box model around it showing the overall water mass balance. Define each variable and incorporate them into Eqs. (10.2.3) and (10.2.4). Assume that there is no overflow pipe in the bioretention cell.

2. Most bioretention cells have overflow pipes that are in the top of the gravel layer or somewhere else in the interior. This changes the amount of the storage volume filled until there is flow out from the cell. Using the site in Exercise 1, modify the equations to incorporate this additional outflow option and show a sketch of the new box model.

3. You are going to develop a 1-acre site with a building, parking areas, and a drive. You opt to use a detention pond for the site.

A. Sketch the site and depict the stormwater flows on the plan.

B. Sketch the detention pond and make a box model around it showing the overall water mass balance. Define each variable and incorporate them into Eqs. (10.2.3) and (10.2.4).

4. You are going to develop a 1-acre site with a building, parking areas, and a drive. You opt to use a retention pond for the site.

A. Sketch the site and depict the stormwater flows on the plan.

B. Sketch the retention pond and make a box model around it showing the overall water mass balance. Define each variable and incorporate them into Eqs. (10.2.3) and (10.2.4).

5. You are going to develop a 1-acre site with a building, parking areas, and a drive. You opt to use swales which will connect to an existing stormwater pipe in the street for drainage.

A. Sketch the site and depict the stormwater flows on the plan.

B. Sketch one of the swales and make a box model around it showing the overall water mass balance. Define each variable and incorporate them into Eqs. (10.2.3) and (10.2.4).

6. You are going to develop a 1-acre site with a building, parking areas, and a drive. You opt to use a small manmade wetland for the site.

A. Sketch the site and depict the stormwater flows on the plan.

B. Sketch the wetland and make a box model around it showing the overall water mass balance. Define each variable and incorporate them into Eqs. (10.2.3) and (10.2.4).

7. You are asked about phosphorus removal in the wetland from Exercise 6. The reeds in the wetland are known to uptake phosphorus. Sketch the wetland and make a box model around it showing the overall phosphorus mass balance. Assume that the concentration of phosphorus in the water in the wetland is completely mixed. Define each variable and incorporate them into Eqs. (10.2.11) and (10.2.12).

8. You are asked about phosphorus removal in the bioretention cell from Exercise 2. The organic layer in the cell has been shown to uptake phosphorus. Sketch the bioretention cell with the various layers and make a box model around and within it showing the overall phosphorus mass balance. Assume that the concentration of phosphorus in the water in each layer is completely mixed. Define each variable and incorporate them into Eqs. (10.2.11) and (10.2.12). (You will probably need a series of mass balance equations.)

9. You are asked about phosphorus removal in the swale from Exercise 5. The grass along the swale has been shown to uptake phosphorus at a specified rate based linearly on concentration. In this case, a completely mixed model is not appropriate, since phosphorus is being removed along the length of flow through the swale. Sketch the swale and make a box model around and within it showing the overall phosphorus mass balance. Define each variable and incorporate them into Eqs. (10.2.9) and (10.2.10).

10. You are building a home on a lot in Columbia, S.C., that is 160 ft^2 and currently undeveloped (wooded). Your local municipality has initiated a stormwater utility, and you will be taxed for any impervious surfaces that are developed. Please compare building a split-level (two-story) home to a traditional single-story home as to impervious area, percent impervious on the lot, and floor area ratio (FAR). You should make a rough sketch of your site options with all pertinent dimensions shown and labeled. The design should meet the following criteria:

- You should have 2400 ft^2 of gross floor area in the single-story home, which includes an attached two-car garage. To compensate for the stairwell, the split-level home would have 2500 ft^2 of gross floor area, and the garage is a side-access drive-under to the right of the house. The split-level home is the same dimension on both floors including the garage.
- To protect the exterior siding, you plan to put a 1-ft overhang/gutter setup around the entire roof (both types of homes).
- The front wall of the house must be at least 20 ft set back from the front property line (both types of homes).
- Minimum backout area from drive-under side-access garages is 30 ft (split-level home).
- The driveway will be paved and so will a 3-ft-wide walkway from the front door to the driveway (don't forget a landing). Your spouse insists that the front door be centered on the house (both types of homes).

11. The South Carolina DOT recommends using the rational method ($Q = CIA$) for runoff volume calculations for sites under 100 acres. For the site in Exercise 10 calculate the estimated runoff for each option (single-story versus split-level). You may assume that the overhangs are included in the impervious area. You may assume that the runoff coefficient is 0.9 for all paved and roofed surfaces and 0.30 for all other surfaces. Because of the small size of the lot, the time of concentration t_c can be assumed to be 5 min. Rainfall intensities are available from the SCDOT rainfall intensity charts (available on the SCDOT website). The municipality wants to know the value for both a 2-year and a 10-year storm.

12. Many MPOs have local requirements regulating the minimum volume in special BMPs for pollutant removal to handle what is believed to be the most frequent volumes of the most polluted runoff. According to many researchers, this is the first 0.5 in of rain to fall on impervious areas during a storm. You are building on a site with approximately 19,000 ft^2 of impervious surface area. You are installing a grass swale IMP, which is expected to have a removal efficiency of 50 to 60 percent for total phosphorus and is designed to handle this water quality volume. Typically, runoff from impervious surfaces in this area tends to have P concentrations of about 60 mg/L. Calculate the estimated uptake of phosphorus by the grass in pounds for a storm event.

13. We are interested in rainwater harvesting, and our uncle at the pickle factory is donating several 55-gal rain barrels for our new ranch house. The house has a roofed surface area of 2800 ft^2. If we have 4 in of rain, how many rain barrels would we need to capture the full rainfall without having to drain them? Please note that there is some storage, even on a roof. The roof is a typical 4V:12H asphalt shingle roof. Please use and reference a typical runoff coefficient.

14. You are placing a pervious concrete parking area over a natural sand subbase. The pervious has an average porosity of 25 percent, and the sand has an infiltration rate of 0.022 cm/s. The parking area will be about 0.27 acre and is essentially flat. It will also accept the drainage from an upslope area of about 5.38 acres. This upslope area is mixed use and has about one-half with a runoff coefficient of 0.80 and the other half around 0.30 (rational method). If you get a constant rainfall of 2.9 in/h, what would you expect the steady-state (after the storage in the pervious concrete is filled) runoff rate to be (volumetric flow) from the parking area from the entire drainage area? Draw a rough site sketch. Note that the rate of infiltration through the pervious concrete is much greater than the rate through the sand.

15. The parking lot in Exercise 10.14 is experiencing a lot of sand drifting and may occasionally become fully covered with blowing sand, causing a reduction in the effective permeability of the pervious system. It has been found by researchers that this reduces the system infiltration rate to a fraction of the sand infiltration rate, and this fraction is the same as the fraction of the porosity in the pervious, since this more accurately represents the flow paths to the sand. What is the worst volumetric steady-state runoff rate that might be expected at steady state for the same conditions in Exercise 14 if the pervious is fully clogged (covered) with this sand?

16. In Exercise 14, the engineer decides to alter the design and have this top 6-in layer of pervious concrete of 25 percent porosity over a 6-in layer of gravel with a porosity of 45 percent. This will be placed over the same native sand of 0.022 cm/s permeability. Assume the same upslope active runoff conditions and the same steady rainfall of 2.9 in/h. What is the expected time to the start of runoff from the pervious parking lot if the full upslope active runoff and the direct rainfall start at the same time? Assume the lot to be fairly flat. (Note that there is a time delay to maximum runoff from other areas based on the time of concentrations, but we are ignoring that here.)

17. Many engineers are worried that if placed on slopes, water will just pour out of pervious pavement, especially if it is placed over poorly draining soils. Typically soils have the following infiltration rate ranges:

Soil	Infiltration Rate Order of Magnitudes
Gravel	>1 cm/s
Sand	10^{-2} cm/s
Silt	10^{-6} cm/s
Clay	$<10^{-8}$ cm/s

If the rainfall rate is greater than the soil percolation rate, then when the "available" storage in the pervious and subbase layers is full, there will be runoff. In previous exercises we assumed that the available storage was all the voids in the pavement and gravel layers, but in this case the available storage can be reduced when built on a slope. You are proposing a 5-ft-wide sidewalk, 100 ft long with the length sloping with the grade (width is approximately horizontal). The sidewalk is made with a 4-in-thick top layer of pervious concrete of 25 percent porosity and a 4-in-thick subbase of gravel of 40 percent porosity. Calculate, for a rainfall rate of 1 in/h over silt, an estimate of the time until runoff starts if the sidewalk is on a 2 percent grade. Make the following assumptions to simplify the analysis. Assume that friction through the pervious and gravel is negligible, so the water fills the storage volumes right away. (Typically the flow rate through unclogged pervious concrete is much greater than 0.3 cm/s.) Draw a rough sketch to explain your estimates. Is it realistic to assume that the rainfall falls at this rate for the time estimated?

18. Repeat Exercise 17 if the sidewalk is on a 10 percent grade.

19. Repeat Exercise 17 if the sidewalk is built over sand.

20. Repeat Exercise 18 if the sidewalk is built over sand.

APPENDIX A
Notation

Acronyms and Chemical Symbols

The following is a list of commonly used acronyms and notations for chemical compounds as found throughout this text.

ACEEE	American Council for an Energy Efficient Economy
ACI	American Concrete Institute
ACSIM	Office of the Assistant Chief of Staff for Installation Management, U.S. Army
ADA	Americans with Disabilities Act
AE	Architect-engineer
AFCEE	Air Force Center for Environmental Excellence
AFCESA	Air Force Civil Engineer Support Agency
AIA	The American Institute of Architects
AISC	American Institute of Steel Construction
ANSI	American National Standards Institute
AP	Accredited Professional (LEED)
ARI	Air Conditioning and Refrigeration Institute
ASAE	American Society of Agricultural Engineers
ASCE	American Society of Civil Engineers
ASLA	American Society of Landscape Architects
ASHRAE	American Society of Heating, Refrigerating, and Air-Conditioning Engineers
ASTM	ASTM International, a.k.a. the American Society for Testing and Materials
AT/FP	Antiterrorism/Force Protection
ATM	Automated teller machine
B20	Mixture of 20 percent biodiesel and 80 percent petroleum diesel
B100	100 percent biodiesel
BD&C	Building Design & Construction

BEES	Building for Environmental and Economic Sustainability
BHT	Butylated hydroxytoluene
BMPs	Best Management Practices for stormwater control
BOD	Basis of design
BOD_5	5-Day Biological oxygen demand
BRAC	Base realignment and closure
BREEAM	Building Research Establishment's Environmental Assessment Method
BSC	California Building Standards Commission
CALGreen	California Green Building Standards developed by the California Building Standards Commission (BSC)
CARB	California Air Resources Board
CBECS	Commercial Buildings Energy Consumption Survey (U.S. DOE)
CCB	Construction criteria base
Cd	Cadmium
CDA	Copper Development Association
CEPSCI	Certified Erosion Prevention and Sedimentation Control Inspector
CEC	California Energy Commission
CERCLA	Comprehensive Environmental Response, Compensation, and Liability Act
CERL	USACE's Construction Engineering Research Laboratory
CFC	Chlorofluorocarbon
CFC11	Trichlorofluoromethane (a.k.a. R-11 or CCl3F)
CF_2BrCl	Bromochlorodifluoromethane (halon 1211)
CF_3Br	Bromotrifluoromethane (halon 1301)
CFR	U.S. Code of Federal Regulations
CGP	Construction general permit
CHP	Combined Heat and Power
CHPS	Collaborative for High Performance Schools
CIBSE	Chartered Institution of Building Services Engineers
CIR	Credit Interpretation Request (Credit Interpretation Ruling)
CN	Curve number
CNIC	Department of the Navy, Commander, Naval Installations Command
CO	Carbon monoxide
CO_2	Carbon dioxide
CON_2H_4	Urea [see also $(NH_2)_2CO$]
CoC	Forest Stewardship Council (FSC) Chain-of-Custody certification
CofO	Certificate of Occupancy

CPESC	Certified Professional in Erosion and Sediment Control
CPM	Critical path method (a scheduling method)
CRS	Center for Resource Solutions
CSAECS	Commercial Sector Average Energy Costs by State (U.S. DOE)
CSI	Congressional Special Interest medical program
CSO	Combined sewer overflow
Cu	Copper
Cx	Commissioning
CxA	Commissioning authority
C&D	Construction and demolition (usually refers to debris)
DASA-I&H	Office of the Deputy Assistant Secretary of the Army for Installations and Housing
DCV	Demand-controlled ventilation
DoD	U.S. Department of Defense
DOE	U.S. Department of Energy
DRMS	Defense Reutilization and Marketing Service
EA	Energy and Atmosphere
EB	Existing building
ECB	Energy cost budget (method)
or	
ECB	Erosion control blanket
ECM	Energy conservation measure (EAc5)
or	
ECM	Exceptional calculation method (EAp2 and EAc1)
EGB	Emerging Green Builders
EIA	Energy Information Administration, U.S. Department of Energy
EMS	Emergency medical services
EO	Executive Order
EPA	U.S. Environmental Protection Agency
EPAct	Energy Policy Act of 1992
EPDM	Ethylene propylene diene monomer, rubber roofing material
EPP	Environmentally preferable purchasing
EQ	Indoor Environmental Quality
ESC	Erosion and sedimentation control
ESP	Electrostatic precipitator
ETS	Environmental tobacco smoke
EVO	Efficiency valuation organization

EWRI	Environmental and Water Resources Institute of ASCE
FAR	Floor area ratio
FEMA	Federal Emergency Management Agency
FF&E	Fixtures, Furniture & Equipment
FHWA	Federal Highway Administration
FIRM	Flood insurance rate map
FPT	Functional performance testing
FSC	Forest Stewardship Council
FYxx	Fiscal Year 20xx or 19xx
GBCI	Green Building Certification Institute
GBI	Green Building Initiative
GFA	Gross floor area
GPO	Government Printing Office
GWP	Global warming potential
HCFC	Hydrochlorofluorocarbon
HEPA	High-efficiency particulate air (filters)
HET	High-efficiency toilet
HEU	High-efficiency urinal
HFC	Hydrofluorocarbon
HOV	High-occupancy vehicle
HOV2	High-occupancy vehicle with two or more people
HOV3	High-occupancy vehicle with three or more people
HUD	Department of Housing and Urban Development
HVAC	Heating, ventilating, and air conditioning
HVAC&R	Heating, ventilating, air conditioning, and refrigeration
H_2CO	Formaldehyde
H_2O	Water
IA	Irrigation Association
IAPMO	International Association of Plumbing and Mechanical Officials
IAQ	Indoor Air Quality
ICC	International Code Council
ID	Innovation and Design Process
IDA	International Dark-Sky Association
ID&C	Interior Design & Construction
IESNA	Illuminating Engineering Society of North America
IMP	Integrated Management Practice
ImpEE	Improving Engineering Education project, University of Cambridge

IPC	International Plumbing Code
IPCC	International Panel on Climate Change
IPMVP	International Performance Measurement and Verification Protocol
ISO	International Organization for Standardization
ITE	Institute of Transportation Engineers
LBNL	Lawrence Berkeley National Laboratory
LCA	Life-cycle assessment or analysis
LCC	Life-cycle cost
LCI	Life-cycle inventory
LEED	Leadership in Energy and Environmental Design
LID	Low-impact development
LPD	Lighting power density (W/ft^2 or W/lin ft, or W/item)
LZ1	Light zone 1: "dark" ambient illumination
LZ2	Light zone 2: "low" ambient illumination
LZ3	Light zone 3: "medium" ambient illumination
LZ4	Light zone 4: "high" ambient illumination
MDF	Medium-density fiberboard
MEP	Mechanical, electrical, plumbing
MERV	Minimum efficiency reporting value, for particulate removal by filters
MET	Metabolic rate
MILCON	Military construction
MPO	Municipal planning organization
MPR	Minimum Program Requirements
MR	Materials and Resources
MSDS	Material Safety Data Sheet
MTBE	Methyl-tertiary-butyl ether
M&V	Measurement and Verification
N	Nitrogen
NAAQS	National Ambient Air Quality Standard
NAHB	National Association of Home Builders
NAVFAC	Naval Facilities Engineering Command
NBI	New Buildings Institute
NC	New construction or major renovation
NCHRP	National Cooperative Highway Research Program
NH_3	Ammonia
$(NH_2)_2CO$	Urea (see also CON_2H_4)
NIST	National Institute of Standards and Technology

NO_x	Nitrogen oxides
NPDES	National Pollutant Discharge Elimination System
NPS	Nonpoint source (pollution)
NRCS	Natural Resources Conservation Service [formerly the Soil Conservation Service (SCS)]
NREL	National Renewable Energy Laboratory, DOE
NRMCA	National Ready Mixed Concrete Association
O_2	Oxygen molecule
O_3	Ozone
OASA-I&E	Office of the Assistant Secretary of the Army for Installations and the Environment
ODEP	Office of Director of Environmental Programs (Army)
ODP	Ozone-depleting potential
OFEE	Office of the Federal Environmental Executive
O&M	Operations and Maintenance
OMB	Office of Management and Budget
OPR	Owner's project requirements
OSB	Oriented strand board
OSHA	Occupational Safety and Health Administration
P	Phosphorus
P2	Pollution prevention
PAM	Polyacrylamide
Pb	Lead
4-PCH	4-Phenylcyclohexene
PCO	Photocatalytic oxidation
PEL	Permissible exposure limit
PM	Particulate matter
PM10	Particulate matter of 10-μm diameter or less
PM2.5	Particulate matter of 2.5-mm diameter or less
ppb	Parts per billion
ppm	Parts per million
PVC	Polyvinyl chloride
P&Z	Planning & zoning
QA/QC	Quality Assurance and Quality Control
RCRA	Resource Conservation and Recovery Act
REC	Renewable energy certificate (EAc6)
REC	Renewable energy cost (EAp2, EAc1, and EAc2)

RFCI	Resilient Floor Covering Institute
RH	Relative humidity
RMI	Rocky Mountain Institute
RP	Regional Priority
SBR	Styrene butadiene rubber
SCM	Supplementary Cementitious Materials
SCS	Soil Conservation Service [now the Natural Resources Conservation Service (NRCS)]
SDD	Sustainable design and development
SEC	Soil and erosion control
SMACNA	Sheet Metal and Air Conditioning National Contractors Association
SOV	Single-occupancy (motor) vehicle
SPiRiT	Sustainable project rating tool
SS	Suspended solids; Sustainable Sites
SSI	Sustainable Sites Initiative
SSO	Senior Sustainability Officer
STEL	Short-term exposure limit
STP	Stormwater Treatment Practice
SWAT	Smart Water Applications Technologies
TARP	Technology Acceptance Reciprocity Partnership, Washington State Department of Ecology
TATRC	Telemedicine and Advanced Technology Research Center of the U.S. Army Medical Research & Materiel Command
T&DI	Transportation and Development Institute of ASCE
TEAP	Technology and Assessment Panel (Montreal Protocol)
TES	Thermal energy storage
TRB	Transportation Research Board
TRC	Tradable renewable certificate (LEED 2.2 EAc6)
TRM	Turf reinforcement mat
TSS	Total suspended solids
TVOC	Total volatile organic compounds/carbons (see also VOC)
TWA	Time-weighted average
ULPA	Ultra-low-penetration air (filters)
UPC	Uniform Plumbing Code
USACE	U.S. Army Corps of Engineers
USAF	U.S. Air Force
USAMRMC	U.S. Army Medical Research & Materiel Command

USBG	U.S. Botanic Garden
USGBC	U.S. Green Building Council
USGS	U.S. Geological Survey
UST	Underground storage tank (usually petroleum)
UV	Ultraviolet
UWRRC	Urban Water Resources Research Council (ASCE-EWRI)
VAV	Variable Air Volume (HVAC devices)
VOC	Volatile organic carbon or compound (see also TVOC)
WBDG	Whole Building Design Guide
WE	Water Efficiency
WMO	World Meteorological Association
ZEV	Zero emission vehicle
Zn	Zinc

Variables and Parameters

The following is a list of symbols for variables and parameters developed for use in this text. In some cases units are given. These are ones typically used in this text, but they can vary depending on the preference of the user.

Symbol	*Chapter*	*Definition (typical units as applicable)*
A_{BMP}	10	Surface area of BMP
$A_{BMP\text{-}PC}$	10	Surface area of pervious concrete BMP
A_i	2, 3	Land area of land or landscape surface *i* at a site
A_{imp}	2	Total land area of the site covered with impervious surfaces
A_T	2, 10	Total land area of the project lot or site
A_{Ti}	2	Total land area of a lot *i*
A_{up}	10	Land area upslope of BMP that drains into it
AECS	4	Annual energy cost savings (percent)
AFV	2	Minimum number of alternative-fuel vehicles that can be adequately serviced with fuel
BAEC	4	Building annual energy cost (energy $/yr) (Option 2)
BAEE	4	Building annual electrical energy (kWh/yr) (Option 2)
BBP	4	Baseline building performance (energy $/yr)
BBPP	4	Baseline building performance process load (energy $/yr)
BBPR	4	Baseline building performance regulated load (energy $/yr)

Symbol	*Chapter*	*Definition (typical units as applicable)*
BF	2	Building footprint—planar projection of built structures onto land
BR	2	Minimum number of bicycle rack spaces (commercial/institutional uses)
BSF	2	Minimum number of bicycles for which secure and covered bicycle storage facilities are provided for residential uses
BWB	3	Blackwater baseline generation rate (gal/yr)
BWD	3	Blackwater design generation rate (gal/yr)
BZA_i	6	Breathing zone horizontal area for zone *i* (ft^2)
BZAR	6	Breathing zone minimum area rate [$ft^3/(min \cdot ft^2)$ or (cfm/sf)]; usually 0.06 cfm/sf as given in the standard
$BZOD_i$	6	Breathing zone occupant density for zone *i* (people/ft^2)
$BZPR_j$	6	Breathing zone minimum people rate for use *j* [ft^3/min per person (cfm per person)] as given in the standard (for example, this is 5 ft^3/min per person for general office spaces and conference or meeting rooms, and it is 7.5 ft^3/min per person for lecture rooms and classrooms)
C	2, 10	Overall stormwater runoff coefficient of the total site area (see %Impness)
C_i	2	Stormwater runoff coefficient of land surface *i*
C_{PBMP}	10	Average concentration of a pollutant in a BMP
C_{PF}	10	Average concentration of a pollutant infiltrating into ground
C_{PI}	10	Average concentration of a pollutant in rainfall
C_{Pmisc}	10	Average concentration of a pollutant coming from miscellaneous flows into a BMP
C_{Pout}	10	Average concentration of a pollutant flowing out of BMP
C_{Pup}	10	Average concentration of a pollutant in upslope runoff
C_{up}	10	Average stormwater runoff coefficient of land surface upslope of BMP
$CBECSE_j$	4	Commercial buildings energy consumption survey median electrical use for building use *j* [$kWh/(ft^2 \cdot yr)$]
$CBECSF_j$	4	Commercial buildings energy consumption survey median fuel use for building use *j* [$kBtu/(ft^2 \cdot yr)$]
CDF_i	6	Combined daylight factor for each window type *i*
CE	3	Controller efficiency factor
CE_i	3	Controller efficiency factor for area *i*

Symbol	*Chapter*	*Definition (typical units as applicable)*
CP	2	Number of covered parking (CP) spaces for which top roofing material above them has an SRI of at least 29. Under cover includes underground, under the building, and under shade structures such as canopies
$CSAECSE_i$	4	Local electrical energy cost rate or the commercial sector average "electrical" energy costs for state/district *i* ($/kWh)
$CSAECSF_i$	4	Local fuel cost rate or the commercial sector average "fuel" energy costs for state/district *i* ($/kBtu)
CVPP1	2	Minimum number of preferred parking spaces or passes for carpools or vanpools for SSc4.4 Option 1
CVPP2	2	Minimum number of preferred parking spaces or passes for carpools or vanpools for SSc4.4 Option 2
DB	2	Density boundary
DC	6	Daylighting compliance (percent)
DD	2	Development density (ft^2/acre)
DD_{area}	2	Development density of the area within the density boundary as determined by the density radius for the density boundary
DD_{site}	2	Development density of site (ft^2/acre)
DDD_{area}	2	Development density of double the surrounding area as determined by the density radius for double (DRD) the density boundary area
DeBLDGREUSE	5	Items reused in building as defined in MRc1.1 if MRc.1.1 is not applied for, and the items reused in the building as defined in MRc1.3 if MRc.1.3 is not applied, and any other building materials salvaged and reused on-site if they are not included in credit 3 calculations
DeCARD	5	Cardboard recycled off-site
DeDIVERT	5	C&D debris diverted
DeGYPSUM	5	Gypsum-type wallboard recycled off-site
DeLANDFILL	5	All C&D debris excluding hazardous wastes, land-clearing debris, and soils which are disposed of off-site in a landfill or other waste facility without recycling or reuse
DeMETAL	5	Metals recycled off-site
DeMISC	5	Miscellaneous items recycled off-site
DePAVEREUSE	5	Existing pavement debris reused on-site
DeRUBBLE	5	Any rubble (bricks, concrete, asphalt, masonry, etc.) recycled off-site

Symbol	*Chapter*	*Definition (typical units as applicable)*
DeWOOD	5	Any wood recycled off-site
DF	2	Development footprint: total building and hardscape areas
DO	2, 3	Design occupancy, residential
DR	2	Density radius for density boundary area (ft)
DRD	2	Density radius for double the density boundary area (ft)
E	10	Evaporation rate (length/time, usually in/h)
EQR	2	Total roof areas covered by equipment and/or solar appurtenances and other appurtenances
ET	3	Evapotranspiration rate (in/time)
ET_0	3	The reference evapotranspiration rate based on local climate and a reference plant (grass or alfalfa). It is usually also based on conditions in that area for the month of July (in/time)
ETB_{Li}	3	Baseline evapotranspiration rate for landscape area i
ETD_{Li}	3	Design evapotranspiration rate for landscape area i
EV	6	Overall system ventilation effectiveness
EZ_i	6	Zone distribution effectiveness. This is typically 1.0 for overhead distribution systems and 1.2 for under-floor distribution systems
F	10	Infiltration rate (length/time, usually in/h)
FAR	10	Floor area ratio
FTE	2	Full-time equivalent occupant during the busiest time of day
FTE_j	2, 3	Full-time equivalent building occupant during shift j (people per shift)
$FTE_{j,i}$	2, 3	Full-time equivalent building occupancy of employee i during shift j (persons per shift)
$FTEF_j$	3	Female full-time employee equivalent for shift j (people per shift)
$FTEM_j$	3	Male full-time employee equivalent for shift j (people per shift)
GE	4	Green electricity purchased or "traded" from off-site (kWh/yr)
GF_j	6	Glazing factor of regularly occupied room area j
GFA_i	2	Gross floor area of building associated with lot i (ft^2)
GFA_P	2, 5	Gross floor area of the subject project (ft^2)
GFA_{PEX}	2, 5	Gross floor area of existing building portion of the subject project to be added onto (ft^2)

Symbol	*Chapter*	*Definition (typical units as applicable)*
GW	3	Annual graywater or other alternative nonpotable water source usage in wastewater fixtures
GWP_r	4	Global warming potential of refrigerant *r* (lb_{CO2}/lb_r)
HS	2	Hardscape: all nonbuilding manmade hard surface on the lot such as parking areas, walks, drives, and patios. Hardpacked areas such as gravel parking areas are included. Pools, fountains, and similar water-covered areas are included if they are on impervious surfaces which can be exposed when the water is emptied
I	10	Rainfall intensity or rate (length/time, usually in/h)
ICCL	6	Minimum number of individual comfort control locations
IE	3	Irrigation efficiency factor
IEB_i	3	Baseline irrigation efficiency for area *i*
IED_i	3	Design irrigation efficiency for area *i*
ILCL	6	Minimum number of individual lighting control locations
$INTFIN_{RE}$	5	The total of all the surfaces included in $INTFIN_T$ which have been reused from the existing building, even those which may have been moved
$INTFIN_T$	5	Total of all interior nonstructural surfaces in the *proposed design* (including the additions if applicable). This includes all finished interior ceilings, flooring surfaces, doors, windows, casework, walls (count both sides), and visible casework surfaces. It is easy to understand if one pictures the finished interior surfaces that can be seen in all the rooms, but with doors and windows only counting once
IWUB	3	Baseline indoor water usage (gal/yr)
$IWUB_{auto}$	3	Baseline indoor water usage (metered with autocontrols on lavatory faucets) (gal/yr)
$IWUB_{no\text{-}auto}$	3	Baseline indoor water usage (without autocontrols on lavatory faucets) (gal/yr)
IWUD	3	Design indoor water usage (gal/yr)
$IWUD_{auto}$	3	Design indoor water usage (metered with autocontrols on lavatory faucets) (gal/yr)
$IWUD_{no\text{-}auto}$	3	Design indoor water usage (without autocontrols on lavatory faucets) (gal/yr)
k_d	3	Irrigation vegetation density factor
k_{dB}	3	Irrigation vegetation density factor–Baseline
k_{dD}	3	Irrigation vegetation density factor–Design
k_{mc}	3	Irrigation site-specific microclimate factor

Symbol	*Chapter*	*Definition (typical units as applicable)*
k_{mcB}	3	Irrigation site-specific microclimate factor–Baseline
k_{mcD}	3	Irrigation site-specific microclimate factor–Design
k_s	3	Irrigation vegetation species factor
k_{sB}	3	Irrigation vegetation species factor–Baseline
k_{sD}	3	Irrigation vegetation species factor–Design
KSR_i	3	Kitchen sink water use rate for fixture type *i* (usually gal/min or gpm)
KSU_i	3	Number of daily kitchen sink uses for occupant type *i*
$LCGWP_i$	4	Life-cycle-based global warming potential for system *i* in pounds of CO_2 per ton of cooling capacity of system *i* per year $[lb_{CO2}/(ton_{ac} \cdot yr)]$
$LCODP_i$	4	Life-cycle-based ozone-depleting potential for system *i* in pounds of CFC11 per ton of cooling capacity of system *i* per year $[lb_{CFC11}/(ton_{ac} \cdot yr)]$
LEFEVP1	2	Minimum number of low-emitting and fuel-efficient vehicles and preferred parking spaces for SSc4.3 Option 1
LEFEVP3	2	Minimum number of low-emitting and fuel-efficient vehicles and preferred parking spaces or passes for SSc4.3 Option 3
LEFEVP4	2	The number of low-emitting and fuel-efficient shared vehicles and preferred parking spaces for SSc4.3 Option 4
LFR_i	3	Lavatory faucet water use rate for fixture type *i* (usually gal/min or gpm)
LFU_i	3	Number of daily lavatory uses for occupant type *i*
$Life_i$	4	Equipment life of system *i* (years)
Lr_i	4	Annual refrigerant leakage rate of system *i* (lb refrigerant leaked per lb of refrigerant in the unit per year with a default of 0.02)
LSMAN	2	Manmade or graded landscaped area totals
LSNAP	2	All previously disturbed areas on ground that are replanted with native and adapted plants
LSNAT	2	Natural landscaped area totals
lx	6	Abbreviation of unit of measure lux
$MASS_{FSCi}$	5	Mass (weight) of the FSC *new* wood portion of unit *i* (volume or value can be substituted for mass as appropriate for an item)
$MASS_{WOODi}$	5	Total mass (weight) of all *new* wood in unit *i* (volume or value can be substituted for mass as appropriate for an item)

Symbol	*Chapter*	*Definition (typical units as applicable)*
$MASS_{RECPOCi}$	5	Mass (weight) of postconsumer recycled portion of unit *i*
$MASS_{RECPRCi}$	5	Mass (weight) of preconsumer recycled portion of unit *i*
$MASS_{REGi}$	5	Mass (weight) of regional portion of unit *i*
$MASS_{RRMi}$	5	Mass (weight) of rapidly renewable portion of unit *i*
$MASS_{UNITi}$	5	Total mass (weight) of unit *i* (volume or value can be substituted for mass as appropriate for an item in certified wood calculations)
$MATL\$_{FSCi}$	5	Value of FSC *new* wood portions of unit *i*
$MATL\$_{RECPOC}$	5	Total value of postconsumer recycled portion of materials (CSI 2-10) plus any value of postconsumer recycled portions of furniture or furnishing if included consistently in MR subcategories 2 to 7. Value determinations are based on postconsumer recycled weight percent of value of each of individual items
$MATL\$_{RECPOCi}$	5	Value of postconsumer recycled portions of unit *i*
$MATL\$_{RECPRC}$	5	Total value of preconsumer recycled portion of materials (CSI 2-10) plus any value of preconsumer recycled portions of furniture or furnishing if included consistently in MR subcategories 2 to 7. Value determinations are based on preconsumer recycled weight percent of value of each of the individual items.
$MATL\$_{RECPRCi}$	5	Value of preconsumer recycled portions of unit *i*
$MATL\$_{REG}$	5	Total value of regionally extracted, processed, and manufactured portions of materials (CSI 2-10) plus any similar portions of furniture or furnishing if included consistently in MR subcategories 2 to 7. Value determinations are based on the weight percent of the regional portion of each of the individual item values
$MATL\$_{REGi}$	5	Value of the regional portions of unit *i*
$MATL\$_{RRM}$	5	Total value of rapidly renewable materials (CSI 2-10) plus any similar portions of furniture or furnishing if included consistently in MR subcategories 2 to 7. Value determinations are based on the weight percent of the rapidly renewable portion of each of the individual item values
$MATL\$_{RRMi}$	5	Value of the rapidly renewable portions of unit *i*
$MATL\$_{SAL}$	5	Total cost of salvaged materials plus any salvaged furniture/furnishing costs if included consistently in MR subcategories 2 to 7 (salvaged items are valued at greater of market value or actual cost)

Symbol	*Chapter*	*Definition (typical units as applicable)*
$MATL\$_T$	5	Total cost of all project materials in CSI categories 2 to 10 plus any furniture/furnishing costs included consistently in MR subcategories 2 to 7 (salvaged items are valued at greater of market value or actual salvage cost; default is 45% of total construction costs plus any furniture or furnishing costs included consistently in MR subcategories 2 through 7)
$MATL\$_{UNITi}$	5	Total cost of unit *i*
$MATL\$_{WOODi}$	5	Value of all the *new* wood portions of unit *i*
MIDAL	2	Maximum allowed percent of total initial designed fixture lumens emitted at an angle of 90 degrees or higher from nadir
Mr_i	4	End-of-life refrigerant loss of system *i* (in pounds refrigerant lost per pound of refrigerant in the unit with a default of 0.10)
n	6	Number of moles of a gas
ND_{FTE}	3	Number of days in a year that FTE-type occupants use the building
ND_{TOW}	3	Number of days in a year that transient-type occupants use the building
Non-EPActB	3	Annual baseline water use for applicable non-EPAct fixtures
Non-EPActD	3	Annual design water use for applicable non-EPAct fixtures
O	2	O is defined as the area of the hardscape which is covered in open-grid pavement systems which are at least 50% pervious with vegetation in the open cells
O_{mod}	2	Sum of the areas with open grid pavement systems that are not shaded or have high SRI values
$OAIZ_i$	6	Outdoor air intake to zone *i* (ft^3/min or cfm)
ODP_r	4	Ozone depletion potential of refrigerant *r* (lb_{CFC11}/lb_r)
OS	2	Open space: all natural areas and landscaped areas, and wet areas with vegetated and low slopes
OSZONE	2	The percent of the total area required to be open space by local code, if such a requirement exists
P	6	Absolute pressure
PBE	4	Proposed building electrical performance (kWh/yr) (the PDEE less on-site renewable electrical energy credits-REE)
PBP	4	Proposed building performance (energy $/yr) (the PDEM less on-site renewable energy cost credits-REC)

Symbol	*Chapter*	*Definition (typical units as applicable)*
PBU_j	2	Peak building users during shift *j* (people per shift)
PD	2	Previously developed: all areas that have been built upon, landscaped, or graded (this does not include areas that may have been farmed in centuries past and have returned to natural vegetation)
PDEE	4	Proposed design electrical energy from the PDEM model (kWh/yr)
PDEM	4	Proposed design energy model (energy $/yr)
PDEMP	4	Proposed design energy model process load (energy $/yr)
PDEMR	4	Proposed design energy model regulated load (energy $/yr)
PED25LS	2	The special pedestrian areas with a minimum of 25% landscaping
PGEBAE	4	Percent green electricity (from off-site) based on the building annual energy (BAE) model if EAc1 Option 1 is not sought
PGEPB	4	Percent green electricity (from off-site) based on the proposed building performance (PBP) model if EAc1 Option 1 is sought
PREBAEC	4	Percent renewable energy based on the building annual energy cost (Option 2)
PREPDEM	4	Percent renewable energy based on the proposed design energy model (Option 1)
PRR_i	3	Prerinse spray water use rate for fixture type *i* (usually gal/min or gpm)
Q	10	Runoff rate (length^3/time, usually ft^3/s or cfs)
Q_{misc}	10	Miscellaneous flows into a BMP (length^3/time, usually ft^3/s or cfs)
Q_{out}	10	Total runoff out of a BMP (length^3/time, usually ft^3/s or cfs)
Q_{total}	4	The gross ARI rated total cooling capacity of all HVAC or refrigeration (ton_{ac})
$Q_{unit i}$	4	The gross ARI rated cooling capacity of an individual HVAC or refrigeration unit *i* in ton_{ac}. (A ton_{ac} of cooling capacity is based on an older system of rating refrigeration when tons of ice were delivered for iceboxes. The tonnage given is actually in tons of ice melted in 24 h. One ton_{ac} of cooling can be directly converted to 12,000 Btu/h)
Q_{up}	10	Runoff from upslope areas into a BMP (length^3/time, usually ft^3/s or cfs)

Symbol	*Chapter*	*Definition (typical units as applicable)*
R	2	Defined as the area of the hardscape that has an SRI of at least 29
R_{Ideal}	6	Ideal gas law constant. One value is 0.0821 L·atm/(mol · K)
R_{mod}	2	Sum of the areas with high SRI excluding those areas that are shaded
RA_j	6	Floor area of regularly occupied room area j
RAT	6	Total floor area of all regularly occupied room areas
Rc_i	4	Refrigerant charge of system i (lb of refrigerant/ton_{ac} of cooling capacity)
REC	4	Renewable energy cost credit for on-site renewable energy (energy $/yr)
REE	4	Renewable electrical energy credit for on-site renewable energy (kWh/yr)
RH	3	Relative humidity (%)
RRL	2	Sum of the low-slope roofed areas with an *areal weighted average* SRI greater than 78
RRS	2	Sum of the steep-slope roofed areas with an areal weighted average SRI greater than 29 (RRS)
RW	3	Reuse water used for irrigation
S	2	Effective area of the hardscape that is shaded
S_{BMP}	10	Storage volume in a BMP (length^3, usually ft^3)
S_{Gravel}	10	Storage volume in the gravel layer (length^3, usually ft^3)
S_{Pervious}	10	Storage volume in the pervious concrete layer (length^3, usually ft^3)
SD	5	Standard deviation
SHOWERS	2	Number of showers required
SR_i	3	Shower water use rate for fixture type i (usually gal/min or gpm)
SRI	2	Solar reflectance index
$STRDECK_{RE}$	5	Total square footage of structural roof decking, interior floor decking, and the foundation with decking or slab on grade which are intended to be reused from the existing structure, excluding any areas that are considered to be structurally unsound or hazardous materials
$STRDECK_T$	5	Total square footage of structural roof decking, interior floor decking, and the foundation with decking or slab on grade, excluding any areas that are considered to be structurally unsound or hazardous materials

Symbol	*Chapter*	*Definition (typical units as applicable)*
$STREXWALL_{RE}$	5	Total square footage of exterior wall areas excluding windows and doors which are intended to be reused from the existing structure. Exclude any areas that are considered to be structurally unsound or hazardous materials
$STREXWALL_{T}$	5	Total square footage of exterior wall areas excluding windows and doors. Exclude any areas that are considered to be structurally unsound or hazardous materials
$STRINWALL_{RE}$	5	Total square footage of interior *structural* wall area excluding openings and counting both sides of the wall elements which are intended to be reused from the existing structure, excluding any areas that are considered to be structurally unsound or hazardous materials
$STRINWALL_{T}$	5	Total square footage of interior *structural* wall area excluding openings and counting both sides of the wall elements, excluding any areas that are considered to be structurally unsound or hazardous materials
SU_i	3	Number of daily shower uses for occupant type *i*
t	10	Time
t_c	10	Time of concentration
t_{fill}	10	Time to fill
T	6	Absolute temperature
T_{vis}	6	Visible transmittance
TBW	3	Annual volume of blackwater treated and infiltrated and/or reused on-site
TIDAL	2	Total initial designed angled lumens for all exterior lighting that breaks the horizontal plane
TIDL	2	Sum of the total initial designed lumens for all exterior lighting
TKSU	3	Total number of annual kitchen sink uses in a building
TLFU	3	Total number of annual lavatory faucet uses in a building
TO_j	2	Transient occupancy during shift *j* (people per shift)
TOW	3	Total daily transient occupancy for water usage calculations
TOWF	3	Total daily female transient occupancy for water usage calculations
TOWM	3	Total daily male transient occupancy for water usage calculations
TP	2	Total number of parking spaces

Symbol	*Chapter*	*Definition (typical units as applicable)*
TPRU	3	Total number of prerinse spray uses in a building
TPWA	3	Total potable water applied for irrigation for the project (design)
TR	2	Total roof area which is defined as the sum of the horizontal projections of the roofs on the site plan less the sum of the roofed areas that hold equipment, solar energy panels, or other appurtenances (BF-EQR)
TSU	3	Total number of annual shower uses in a building
TUU	3	Total number of annual urinal uses in the facility
TWA	3	Total water applied (for irrigation)
TWAB	3	Baseline total water applied for irrigation for the project
$TWAB_i$	3	Baseline total water applied for landscape area *i*
TWAD	3	Design total water applied for irrigation for the project
$TWAD_i$	3	Design total water applied for landscape area *i*
TWCU	3	Total number of annual water closet uses in the facility
UR_i	3	Urinal water use per flush for fixture type *i*
UUF_i	3	Number of daily urinal uses by a female for occupant type *i*
UUM_i	3	Number of daily urinal uses by a male for occupant type *i*
V	6	Volume of a gas
VNAPR	2	All the roofed areas that are planted with native and adapted plants
VR	2	Sum of the vegetated roof areas
WA_{ij}	6	The sum of the areas of window type *i* in room area *j*
WCR_i	3	Water closet water use per flush for fixture type *i*
$WCUF_i$	3	Number of daily water closet uses by a female for occupant type *i*
$WCUM_i$	3	Number of daily water closet uses by a male for occupant type *i*
WETMAN	2	Total manmade wetlands, ponds, or pool areas
WETMANLS	2	Total manmade wetlands, ponds, or pools with vegetated low slopes (<25%) leading to its edges. Does not include pools and fountains with impervious bottoms as these are part of the hardscape
WETMANSS	2	Total manmade wetland, ponds, or pools with any steep ($\geq$25%) or unvegetated slopes leading to its edges. Does not include pools and fountains with impervious bottoms as these are part of the hardscape

Symbol	*Chapter*	*Definition (typical units as applicable)*
WETNAT	2	Total natural wetlands, ponds, stream, or pool areas
$WOOD\$_{FSC}$	5	Total value of FSC *new* wood portions of all items installed in the project
$WOOD\$_{T}$	5	Total value of *new* wood portions of all items installed in the project
WQ_{VA}	2	Water quality volume for arid climates which must be infiltrated on-site, used on-site, or treated prior to discharge off-site
WQ_{VH}	2	Water quality volume for humid climates which must be infiltrated on-site, used on-site, or treated prior to discharge off-site
WQ_{VS}	2	Water quality volume for semiarid climates which must be infiltrated on-site, used on-site, or treated prior to discharge off-site
%Imp	2	Percent of the land area that is covered with "impervious" surfaces
%Impness	2	Relative percent imperviousness of all the surfaces at a site (see runoff coefficient)
2V:12H	2	Slope such that for every 2 units raised in the vertical direction, 12 units are moved in the horizontal direction

APPENDIX B

Definitions

The following definitions as quoted come with permission from the LEED-NC 2.2 Reference Guide with additional modifications or explanations as noted. [USGBC (2005–2007), *LEED-NC for New Construction, Reference Guide*, Version 2.2, 1st ed., U.S. Green Building Council, Washington, D.C., October 2005 with errata posted through Spring 2007.] Several new definitions from LEED 2009 have also been added with addenda through December 2, 2009.

Acid Rain 'The precipitation of dilute solutions of strong mineral acids, formed by the mixing in the atmosphere of various industrial pollutants (primarily sulfur dioxide and nitrogen oxides) with naturally occurring oxygen and water vapor.'

Adapted (or Introduced) Plants 'Plants that reliably grow well in a given habitat with minimal attention from humans in the form of winter protection, pest protection, water irrigation, or fertilization once root systems are established in the soil. Adapted plants are considered to be low-maintenance but not invasive.'

Adaptive Reuse 'The renovation of a building or site to include elements that allow a particular use or uses to occupy a space that originally was intended for a different use.'

Adhesive 'Any substance that is used to bond one surface to another surface by attachment. Adhesives include adhesive bonding primers, adhesive primers, adhesive primers for plastics, and any other primer.'

Aerosol Adhesive 'An adhesive packaged as an aerosol product in which the spray mechanism is permanently housed in a nonrefillable can designed for handheld application without the need for ancillary hoses or spray equipment. Aerosol adhesives include special-purpose spray adhesives, mist spray adhesives, and web spray adhesives.'

Agrifiber Board 'A composite panel product derived from recovered agricultural waste fiber from sources including, but not limited to, cereal straw, sugarcane bagasse, sunflower husk, walnut shells, coconut husks, and agricultural prunings. The raw fibers are processed and mixed with resins to produce panel products with characteristics similar to those derived from wood fiber. The products must comply with the following requirements:

1. The product is inside of the building's waterproofing system.
2. Composite components used in assemblies are to be included (e.g., door cores, panel substrates).
3. The product is part of the base building systems.'

Air Changes per Hour (ACH) 'The number of times per hour a volume of air, equivalent to the volume of space, enters that space.'

Air Conditioning Refer to ASHRAE 62.1-2004. [ASHRAE (2004), ASHRAE 62.1-2004: *Ventilation for Acceptable Indoor Air Quality*, American Society of Heating, Refrigerating and Air-Conditioning Engineers, Atlanta, GA.]

Albedo 'Synonymous with solar reflectance.'

Alternative-Fuel Vehicles 'Vehicles that use low-polluting, nongasoline fuels such as electricity, hydrogen, propane or compressed natural gas, liquid natural gas, methanol, and ethanol. Efficient gas-electric hybrid vehicles are included in this group for LEED purposes.' Please note the additional comments about this in SS credit 4.3.

Angle of Maximum Candela 'The direction in which the luminaire emits the greatest luminous intensity.'

Anti-Corrosive Paints 'Coatings formulated and recommended for use in preventing the corrosion of ferrous metal substrates.'

Aquatic Systems 'Ecologically designed treatment systems that utilize a diverse community of biological organisms (e.g., bacteria, plants, and fish) to treat wastewater to advanced levels.'

Aquifer 'An underground water-bearing rock formation or group of formations, which supplies groundwater, wells, or springs.'

Assembled Recycled Content This 'includes the percentages of postconsumer and preconsumer content. The determination is made by dividing the weight of the recycled content by the overall weight of the assembly.'

Automatic Fixture Sensors 'Motion sensors that automatically turn on and off lavatories, sinks, water closets, and urinals. Sensors may be hardwired or battery-operated.'

Baseline Building Performance 'The annual energy cost for a building design intended for use as a baseline for rating above standard design, as defined in ASHRAE 90.1-2204 Informative App. G.'

Basis of Design (BOD) 'Design information necessary to accomplish the owner's project requirements, including system descriptions, indoor environmental quality criteria, other pertinent design assumptions (such as weather data), and references to applicable codes, standards, regulations, and guidelines.'

Bay 'LEED 2009 gives the definition of a bay for SSc8 as a component of a standard, rectilinear building design. It is the open area defined by columns or windows and there are typically multiple identical bays in succession.'

Biodiversity 'The variety of life in all forms, levels, and combinations, including ecosystem diversity, species diversity, and genetic diversity.'

Biomass 'Plant material such as trees, grasses and crops that can be converted to heat energy to produce energy.'

Bioremediation 'Involves the use of microorganisms and vegetation to remove contaminants from water and soils. Bioremediation is generally a form of in situ remediation, and it can be a viable alternative to landfilling or incineration.'

Blackwater 'Blackwater does not have a single definition that is accepted nationwide. Wastewater from toilets and urinals is, however, always considered blackwater. Wastewater from kitchen sinks (perhaps differentiated by the use of a garbage disposal), showers, or bathtubs may be considered blackwater by state or local codes. Project teams should comply with the blackwater definitions as established by the authority having jurisdiction in their areas.'

Breathing Zone 'The region within an occupied space between planes 3 and 6 ft above the floor and more than 2 ft from the walls or fixed air-conditioning equipment.'

Building Density 'The floor area of the building divided by the total area of the site (square feet per acre).' See FAR in Chap. 10.

Building Envelope 'The exterior surface of a building's construction—the walls, windows, roof, and floor. Also referred to as the "building shell."'

Building Footprint 'The area on a project site that is used by the building structure and is defined by the perimeter of the building plan. Parking lots, landscapes, and other nonbuilding facilities are not included in the building footprint.'

Carpool 'An arrangement in which two or more people share a vehicle for transportation.' Many state transportation agencies also provide special commuting lanes for carpools or vanpools. These are referred to as HOV (high-occupancy vehicle) lanes. Sometimes the definition of an HOV is for two or more people, at other times it is for three or more. Many times these are therefore further designated HOV2 or HOV3, respectively.

Car Sharing 'A system under which multiple households share a pool of automobiles, either through cooperative ownership or through some other mechanism.'

CERCLA 'Comprehensive Environmental Response, Compensation, and Liability Act (CERCLA), commonly known as Superfund. CERCLA addresses abandoned or historical waste sites and contamination. It was enacted in 1980 to create a tax on the chemical and petroleum industries and provided federal authority to respond to releases of hazardous substances.'

Chain-of-Custody 'A document that tracks the movement of a wood product from the forest to a vendor and is used to verify compliance with FSC guidelines. A *vendor* is defined as the company that supplies wood products to project contractors or subcontractors for on-site installation.'

Chlorofluorocarbons (CFCs) Chlorofluorocarbons are *haloalkanes* that are thought to aid in the depletion of the stratospheric ozone layer. They are carbon-based organic compounds (alkanes) which, in addition to carbon, include chlorine and fluorine atoms. Some are commonly referred to as *Freons*.

Cogeneration 'The simultaneous production of electrical or mechanical energy (power) and useful thermal energy from the same fuel/energy source such as oil, coal, gas, biomass, or solar.'

Commissioning (Cx) 'The process of ensuring that systems are designed, installed, functionally tested, and capable of being operated and maintained to perform in conformity with the owner's project requirements.'

Commissioning Plan 'A document defining the commissioning process, which is developed in increasing detail as the project progresses through its various phases.'

Commissioning Report 'The document that records the results of the commissioning process, including the as-built performance of the HVAC system and unresolved issues.'

Commissioning Specification 'The contract document that details the objective, scope, and implementation of the construction and acceptance phases of the commissioning process as developed in the design-phase commissioning plan.'

Commissioning Team 'Those people responsible for working together to carry out the commissioning process.'

Community 'An interacting population of individuals living in a specific area.'

Completed Design Area 'The total area of finished ceilings, finished floors, full-height walls and demountable partitions, interior doors, and built-in case goods in the space when the project is completed: exterior windows and exterior doors are not considered.'

Composite Wood 'A product consisting of wood or plant particles or fibers bonded together by a synthetic resin or binder (i.e., plywood, particleboard, OSB, MDF, composite door cores). For the purposes of LEED-NC requirements, products must comply with the following conditions:

1. The product is inside of the building's waterproofing system.
2. Composite wood components used in assemblies are included (e.g., door cores, panel substrates, plywood sections of I-beams).
3. The product is part of the base building system.'

Composting Toilet Systems 'Dry plumbing fixtures that contain and treat human waste via microbiological processes.'

Construction and Demolition (C&D) Debris 'Waste and recyclables generated from construction, renovation, and demolition or deconstruction of preexisting structures. Land-clearing debris including soil, vegetation, rocks, etc., is not to be included.'

Conventional Irrigation 'Most common irrigation system used in the region where the building is located. A common conventional irrigation system uses pressure to deliver water and distributes it through sprinkler heads above the ground.'

Curfew Hours 'Locally determined times when greater lighting restrictions are imposed. When no local or regional restrictions are in place, 10:00 p.m. is regarded as a default curfew time.'

Cutoff Angle Angle between the vertical axis of a luminaire and the first line of sight (of a luminaire) at which the light source is no longer visible (as per LEED-NC 2.1).

Daylighting 'The controlled admission of natural light into a space through glazing with the intent of reducing or eliminating electric lighting. By utilizing solar light, daylighting creates a stimulating and productive environment for building occupants.'

Development Footprint 'The area on the project site that has been impacted by any development activity. Hardscape, access roads, parking lots, nonbuilding facilities, and building structure are all included in the development footprint.'

Direct Line of Sight to Perimeter Vision Glazing 'The approach used to determine the calculated area of regularly occupied areas with direct line of sight to perimeter vision glazing. The area determination includes full-height partitions and other fixed construction prior to installation of furniture.'

Drip Irrigation 'A high-efficiency irrigation method in which water is delivered at low pressure through buried mains and submains. From the submains, water is distributed to the soil from a network of perforated tubes or emitters. Drip irrigation is a type of microirrigation.'

Ecosystem 'A basic unit of nature that includes a community of organisms and their nonliving environment linked by biological, chemical, and physical processes.'

Embodied Energy 'Energy that is used during the entire life cycle of the commodity for manufacturing, transporting and disposing of the commodity as well as the inherent energy captured within the product itself.'

Emissivity 'The ratio of the radiation emitted by a surface to the radiation emitted by a black body at the same temperature.'

Endangered Species 'An animal or plant species that is in danger of becoming extinct throughout all or significant portion of its range due to harmful human activities or environmental factors.'

Energy Conservation Measures (ECMs) 'Installations of equipment or systems, or modifications of equipment or systems, for the purpose of reducing energy use and/or costs.'

Energy Star 'The rating a building earns using the ENERGYSTAR Portfolio Manager to compare building energy performance to that of similar buildings in similar climates. A score of 50 represents average building performance.'

Environmental Attributes of Green Power 'Emission reduction benefits that result from green power being used instead of conventional power sources.'

Environmental Tobacco Smoke (ETS) 'Also known as secondhand smoke, consists of airborne particles emitted from the burning end of cigarettes, pipes and cigars, and exhaled by smokers. These particles contain about 4000 different compounds, up to 40 of which are known to cause cancer.'

Environmentally Preferable Products 'Products identified as having a lesser or reduced effect on health and the environment when compared with competing products that serve the same purpose.'

Environmentally Preferable Purchasing 'A U.S. federalwide program (Executive Order 13101) that encourages and assists executive agencies in the purchasing of environmentally preferable products and services.'

Erosion 'A combination of processes in which materials of the earth's surface are loosened, dissolved, or worn away and transported from one place to another by natural agents (such as water, wind, or gravity).'

Exhaust Air 'The air removed from a space and discharged to outside the building by means of mechanical or natural ventilation systems.'

Ex Situ Remediation 'Removal of contaminated soil and groundwater. Treatment of the contaminated media occurs in another location, typically a treatment facility. A traditional method of ex situ remediation is pump-and-treat technology that uses carbon filters and incineration. More advanced methods of ex situ remediation include chemical treatment or biological reactors.'

Flat Coatings 'Coatings that register a gloss of less than 15 on an 85-degree meter or less than 5 on a 60-degree meter.'

Fly Ash 'The solid residue derived from incineration processes. Fly ash can be used as substitute for portland cement in concrete.' Concrete is frequently made with around 25 percent by weight of its cementitious material as fly ash with the remainder portland cement.

Footcandle (fc) (1) A unit of illuminance equal to one lumen of light falling on a one-square-foot area from a one candela light source at a distance of one foot (as per LEED-NC 2.2) or (2) a measure of light falling on a given surface. One footcandle (1 fc) is equal to the quantity of light falling on a one-square-foot (1-ft^2) area from a one candela (1-cd) light source at a distance of one foot (1 ft). Footcandles can be measured both horizontally and vertically by a footcandle or lightmeter (as per LEED-NC 2.1). The second definition is more comprehensive.

Formaldehyde 'A naturally occurring VOC found in small amounts in animals and plants, but that is carcinogenic and an irritant to most people when present in high concentrations, causing headaches, dizziness, mental impairment, and other symptoms. When present in the air at levels above 0.1 ppm (parts per million), it can cause watery eyes; burning sensations in the eyes, nose, and throat; nausea; coughing; chest tightness; wheezing; skin rashes; and asthmatic and allergic reactions.'

Full Cutoff Luminaire A full cutoff luminaire has zero candela (0-cd) intensity at an angle of 90 degrees above the vertical axis (nadir) and at all angles greater than 90 degrees from nadir. Additionally, the candela per 1000 lamp lumens does not numerically exceed 100 (10 percent) at an angle of 80 degrees above nadir. This applies to all lateral angles around the luminaire (as per LEED-NC 2.1).

Functional Performance Testing (FPT) 'The process of determining the ability of the commissioned system to perform in accordance with the owner's project requirements, basis of design, and construction documents.'

GFA Gross floor area. See Square Footage of a Building.

Glare Sensation produced by luminance within the visual field that is significantly greater than the luminance to which the eyes are adapted, which causes annoyance, discomfort, or loss in visual performance and visibility (as per LEED-NC 2.1).

Glazing Transparent part of a wall and usually made of glass or plastic. Wikipedia, 10/10/06.

Glazing Factor 'The ratio of interior illuminance at a given point on a given plane (usually the work plane) to the exterior illuminance under known overcast sky conditions. LEED uses a simplified approach for its credit compliance calculations. The variables used to determine the daylight factor include the floor area, window area, window geometry, visible transmittance T_{vis}, and window height.'

Graywater or Greywater 'Term defined by the Uniform Plumbing Code (UPC) in its App. G, titled "Graywater Systems for Single-Family Dwellings," as "untreated household wastewater which has not come into contact with toilet waste. Graywater includes used water from bathtubs, showers, bathroom wash basins, and water from clothes-washer and laundry tubs. It shall not include wastewater from kitchen sinks or dishwashers." The International Plumbing Code (IPC) defines graywater in its App. C, titled "Graywater Recycling Systems," as "wastewater discharged from lavatories, bathtubs, showers, clothes washers, and laundry sinks." Some states and local authorities allow kitchen sink wastewater to be included in graywater. Other differences with the UPC and IPC definitions can probably be found in state and local codes. Project teams should comply with the graywater definitions as established by the authority having jurisdiction in their areas.'

Greenfield Sites 'Those that are not developed or graded and remain in a natural state.'

Greenfields 'Sites that have not been previously developed or graded and remain in a natural state.'

Greenhouse Gases 'Gases such as carbon dioxide, methane and CFCs that are relatively transparent to the higher-energy sunlight, but trap lower-energy infrared radiation.'

Halons 'Substances used in fire suppression systems and fire extinguishers in buildings. These substances deplete the stratospheric ozone layer.' Halons are haloalkanes as CFCs are, but with bromine as well as chlorine or fluorine groups. Two common ones are bromochlorodifluoromethane (halon 1211, CF_2BrCl) and bromotrifluoromethane (halon 1301, CF_3Br).

Heat Island Effect Effect that 'occurs when warmer temperatures are experienced in urban landscapes compared to adjacent rural areas as a result of solar energy retention on constructed surfaces. Principal surfaces that contribute to the heat island effect include streets, sidewalks, parking lots, and buildings.'

Horizontal View at 42 in 'The approach used to confirm that the direct line of sight to perimeter vision glazing remains available from seated position. It uses section drawings that include the installed furniture to make the determination.'

HVAC Systems 'Heating, ventilating, and air-conditioning systems used to provide thermal comfort and ventilation for building interiors.'

Hybrid Vehicles 'Vehicles that use a gasoline engine to drive an electric generator and use the electric generator and/or storage batteries to power electric motors that drive the vehicle's wheels.'

Hydrochlorofluorocarbons (HCFCs) 'Refrigerants used in building equipment that deplete the stratospheric ozone layer, but to a lesser extent than CFCs.' They are carbon-based organic compounds (haloalkanes) that in addition to carbon include chlorine and fluorine atoms and definitely hydrogen atoms. The hydrogen atoms are located in bonds that might have had chlorine atoms if they were CFCs and therefore have less of a chance of releasing chlorine atoms into the stratosphere. The chlorine atoms can cause chain reactions which destroy the ozone molecule.

Hydrofluorocarbons (HFCs) 'Refrigerants that do not deplete the stratospheric ozone layer. However, some HFCs have high global warming potential and thus are not environmentally benign.' HFCs are haloalkanes as CFCs are, but with only hydrogen and fluorine groups attached and no chlorine.

Illuminance Amount of light falling on a surface, measured in units of footcandles (fc) or lux (lx) (as per LEED-NC 2.1).

Impervious Surfaces 'Surfaces that promote runoff of precipitation volumes instead of infiltration into the subsurface. The imperviousness or degree of runoff potential can be estimated for different surface materials.'

Incident Light Light that strikes a surface. The angle between a ray of light and the perpendicular to the surface is the angle of incidence. Wikipedia, accessed 10/10/06, http://en.wikipedia.org/wiki/Main_Page.

Individual Occupant Spaces 'Typically private offices and open office plans with workstations.'

Indoor Adhesive, Sealant, and/or Sealant Primer Product 'An adhesive or sealant product applied on-site, inside of the building's weatherproofing system.'

Indoor Air Quality 'The nature of air inside the space that affects the health and well-being of building occupants.'

Indoor Carpet System 'Carpet, carpet adhesive, or carpet cushion product installed on-site, inside of the building's weatherproofing system.'

Indoor Composite Wood or Agrifiber Product 'Composite wood or agrifiber product installed on-site, inside of the building's weatherproofing system.'

Indoor Paint or Coating Product 'A paint or coating product applied on-site, inside of the building's weatherproofing system.'

Infrared Emittance (As defined in SSc7.1) 'A parameter between 0 and 1 that indicates the ability of a material to shed infrared radiation. The wavelength range for this radiant energy is roughly 5 to 40 μm. Most building materials (including glass) are opaque in this part of the spectru, and have an emittance of roughly 0.9. Materials such as clean, bare metals are the most important exceptions to the 0.9 rule. Thus clean, untarnished galvanized steel has low emittance, and aluminum roof coatings have intermediate emittance levels.'

Infrared or Thermal Emittance (As defined in SSc7.2) 'A parameter between 0 and 1 (or 0 and 100 percent) that indicates the ability of a material to shed infrared radiation (heat). The wavelength range for this radiant energy is roughly 3 to 40 μm. Most building materials (including glass) are opaque in this part of the spectrum and have an emittance of roughly 0.9. Materials such as clean, bare metals are the most important exceptions to the 0.9 rule. Thus clean, untarnished galvanized steel has low emittance, and aluminum roof coatings have intermediate emittance levels.'

In Situ Remediation 'Treatment of contaminants in place using technologies such as injection wells or reactive trenches. These methods utilize the natural hydraulic gradient of groundwater and usually require only minimal disturbance of the site.'

Installation Inspection 'The process of inspecting components of the commissioned systems to determine if they are installed properly and ready for systems performance testing.'

Interior Lighting Power Allowance 'The maximum light power in watts allowed for the interior of a building.'

Interior Nonstructural Components Reuse 'Factor determined by dividing the total area of retained interior, nonstructural components by the total area of the interior, nonstructural components in the completed design' (as per errata Spring 2007).

Invasive Plants 'Both indigenous and nonindigenous species or strains that are characteristically adaptable, aggressive, have a high reproductive capacity, and tend to overrun the ecosystems which they inhabit. Collectively, they are one of the great threats to biodiversity and ecosystem stability.'

Laminate Adhesive 'An adhesive used in wood and agrifiber products (veneered panels, composite wood products contained in engineered lumber, door assemblies, etc.).'

Landfill 'A waste disposal site for the deposit of solid waste from human activities.'

Landscape Area 'Area of the site that is equal to the total site area less the building footprint, paved surfaces, water bodies, patios, etc.'

Life-Cycle Analysis (LCA) 'An evaluation of the environmental effects of a product or activity holistically, by analyzing the entire life cycle of a particular material, process, product, technology, service, or activity.'

Life-Cycle Cost (LCC) Method 'A technique of economic evaluation that sums over a given study period the costs of initial investment (less resale value), replacement, operations (including energy use), and maintenance and repair of an investment decision (expressed in present or annual value terms).'

Life-Cycle Inventory (LCI) 'An accounting of the energy and waste associated with the creation of a new product through use and disposal.'

Light Pollution 'Waste light from building sites that produces glare is directed upward to the sky or is directed off the site.'

Lighting Power Density (LPD) 'The installed lighting power, per unit area.'

Local Zoning Requirements 'Local government regulations imposed to promote orderly development of private lands and to prevent land use conflicts.'

Low-Emitting and Fuel-Efficient Vehicles 'For the purpose of this credit (SS credit 4.3), vehicles that either are classified as zero emission vehicles (ZEVs) by the California Air Resources Board (CARB) or have achieved a minimum green score of 40 on the American Council for an Energy Efficient Economy (ACEEE) annual vehicle rating guide.'

Luminance Commonly called brightness or the light coming from a surface or light source. Luminance is composed of the intensity of light striking an object or surface and the amount of that light reflected back toward the eye. Luminance is measured in foot-lamberts (fL) or candelas per square meter (cd/m^2) (as per LEED-NC 2.1).

Mass Transit 'Transportation facilities designed to transport large groups of persons in a single vehicle such as buses or trains.'

Mass Transit Vehicles 'Vehicles typically capable of serving 10 or more occupants, such as buses, trolleys, light rail, etc.'

Metering Controls 'Generally manual on/automatic off controls which are used to limit the flow of water. These types of controls are most commonly installed on lavatory faucets and on showers.'

Microirrigation 'Irrigation systems with small sprinklers and microjets or drippers designed to apply small volumes of water. The sprinklers and microjets are installed within a few centimeters of the ground, while drippers are laid on or below grade.'

Mixed-Mode Ventilation 'A ventilation strategy that combines natural ventilation with mechanical ventilation, allowing the building to be ventilated either mechanically or naturally, and at times both mechanically and naturally simultaneously.'

Native (or Indigenous) Plants 'Plants that have adapted to a given area during a defined time period and are not invasive. In America, the term often refers to plants growing in a region prior to the time of settlement by people of European descent.'

Net Metering 'A metering and billing arrangement that allows on-site generators to send excess electricity flows to the regional power grid. These electricity flows offset a portion of the electricity flows drawn from the grid. For more information on net metering in individual states, visit the DOE's Green Power Network website at www.eere.energy.gov/greenpower/netmetering.shtml.'

Nonflat Coatings 'Coatings that register a gloss of 5 or greater on a 60-degree meter and a gloss of 15 or greater on an 85-degree meter.'

Nonoccupied Spaces 'All rooms used by maintenance personnel that are not open for use by occupants. Included in this category are janitorial, storage and equipment rooms, and closets.'

Nonporous Sealant 'A substance used as a sealant on nonporous materials. Nonporous materials do not have openings in which fluids may be absorbed or discharged. Such materials include, but are not limited to, plastic and metal.'

Nonpotable Water 'Water that is not suitable for human consumption without treatment that meets or exceeds EPA drinking water standards.'

Non-Regularly Occupied Spaces 'Corridors, hallways, lobbies, break rooms, copy rooms, storage rooms, kitchens, restrooms, stairwells, etc.'

Non-Roof Impervious Surfaces 'All surfaces on the site with a perviousness of less than 50 percent, not including the roof of the building. Examples of typically impervious surfaces include parking lots, roads, sidewalks, and plazas.' This definition is inconsistent with the typical definitions of impervious surfaces and the calculations in Chaps. 2 and 10. Typically a surface with a perviousness of less than about 10 percent is considered to be an impervious surface, and that is the assumption made in this text.

Non-Water-Using Urinal 'A urinal that uses no water, but instead replaces the water flush with a specially designed trap that contains a layer of buoyant liquid which floats above the urine layer, blocking sewer gas and urine odors from the room.' See Urinal, Non-Water-Using.

Off-Gassing 'The emission of volatile organic compounds from synthetic and natural products.'

On-Site Wastewater Treatment 'Use of localized treatment systems to transport, store, treat, and dispose of wastewater volumes generated on the project site.'

Open-Grid Pavement 'For LEED purposes, pavement that is less than 50 percent impervious and contains vegetation in the open cells.'

Open-Space Area 'Open Space Area is as defined by local zoning requirements. If local zoning requirements do not clearly define open space, it is defined for the purposes of LEED calculations as the property area minus the development footprint; and it must be vegetated and pervious, with exceptions only as noted in the credit requirements section. For projects located in urban areas that earn SS credit 2, open space also includes nonvehicular, pedestrian-oriented hardscape spaces.'

Open-Grid Pavement 'For LEED purposes, pavement that is less than 50 percent impervious and contains vegetation in the open cells.'

Outdoor Lighting Zone Definitions 'Definitions developed by IDA for the Model Lighting Ordinance that provide a general description of the site environment/context and basic lighting criteria.' The IDA is the International Dark-Sky Association and can be found at www.darksky.org.

Owner's Project Requirements (OPRs) 'Written document that details the functional requirements of a project and the expectations of how it will be used and operated; or an explanation of the ideas, concepts, and criteria that are determined by the owner to be important to the success of the project (previously called the design intent).'

Paints 'Liquid, liquefiable or mastic compositions that are converted to a solid protective, decorative, or functional adherent film after application as a thin layer. These coatings are intended for on-site application to interior or exterior surfaces of residential, commercial, institutional or industrial buildings.'

Pedestrian Access 'Implies that pedestrians can walk to the services without being blocked by walls, freeways or other barriers.'

Percentage Improvement 'Percentage improvement is the percent energy cost savings for the proposed building performance versus the baseline building performance.'

Perviousness 'The percent of the surface area of a paving material that is open and allows moisture to pass through the material and soak into the earth below the paving system.'

Phenol Formaldehyde Chemical that 'off-gases only at high temperature and is used for exterior products; although many of those products are suitable for interior applications.'

Porous Sealant 'A substance used as a sealant on porous materials. Porous materials have tiny openings, often microscopic, in which fluids may be absorbed or discharged. Such materials include, but are not limited to, wood, fabric, paper, corrugated paperboard and plastic foam.'

Postconsumer Descriptive of 'waste material generated by households or by commercial, industrial, and institutional facilities in their role as end-users of the product, which can no longer be used for its intended purpose. This includes returns of materials from distribution chains (source: ISO 14021). Examples of this category include construction and demolition debris, material collected through curbside and dropoff recycling programs, broken pallets (if from a pallet refurbishing company, not a pallet making company), discarded products (e.g., furniture, cabinetry, and decking), and urban maintenance waste (e.g., leaves, grass clippings, tree trimmings).'

Potable Water 'Water suitable for drinking and supplied from wells or municipal water systems.'

Preconsumer Content 'Material diverted from the waste stream during the manufacturing processes. Excluded is reutilization of materials such as rework, regrind, or scrap generated in a process and capable of being reclaimed within the same process that generated it (source: ISO 14021). Examples in this category include planer shavings, ply trim, sawdust, chips, bagasse, sunflower seed hulls, walnut shells, culls, trimmed materials, print overruns, overissue publications, and obsolete inventories. (Previously referred to as postindustrial content.)'

Preferred Parking 'Parking spots that are closest to the main entrance of the project exclusive of spaces designated for handicapped, or to parking passes provided at a discounted price.'

Previously Developed Sites 'Sites that previously contained buildings, roadways, or parking lots or were graded or altered by direct human activities.'

Primer 'A material applied to substrate to improve adhesion of a subsequently applied adhesive.'

Prior Condition Area 'The total area of finished ceilings, finished floors, full-height walls, and demountable partitions, interior doors, and built-in case goods that existed when the project area was selected; exterior windows and exterior doors are not considered.'

Private or Private Use 'LEED 2009 states that in these cases, private or private use refers to the plumbing fixtures in residences, apartments, and dormitories, private bathrooms in transient lodging facilities such as hotels and motels and also to private bathrooms in hospitals and nursing facilities. Public use restrooms in the transient lodging facilities, hospitals, and nursing facilities are not included.'

Process Water 'Water used for industrial processes and building systems such as towers, boilers and chillers.'

Property Area 'The total area within the legal property boundaries of a site that encompasses all areas of the site including constructed areas and nonconstructed areas.' However, for instances where the subject site area is part of a campus or other larger unit, the site area of the project can be defined by the project team, but must be consistent throughout the LEED certification process.

Proposed Building Performance 'The annual energy cost calculated for a proposed design, as defined in ASHRAE 90.1-2004 Informative App. G.'

Public Transportation 'Bus, rail or other transportation service for the general public, operating on a regular, continual basis that is publicly or privately owned.'

Rapidly Renewable Materials 'Materials considered to be an agricultural product, both fiber and animal, that takes 10 years or less to grow or raise, and to harvest in an ongoing and sustainable fashion.'

Rated Power 'The nameplate power on a piece of equipment. It represents the capacity of the unit and is the maximum a unit will draw.'

RCRA 'Resource Conservation and Recovery Act. RCRA focuses on active and future facilities. It was enacted in 1976 to give the EPA authority to control hazardous wastes from cradle to grave, including generation, transportation, treatment, storage, and disposal. Some nonhazardous wastes are also covered under RCRA.'

Receptacle Load 'All equipment that is plugged into the electrical system, from office equipment to refrigerators.'

Recycling 'The collection, reprocessing, marketing and use of materials that were diverted or recovered from the solid waste stream.'

Refrigerants 'The working fluids of refrigeration cycles. They absorb heat from a reservoir at low temperatures and reject heat at higher temperatures.'

Regionally Extracted Materials 'For LEED-NC purposes, materials whose source is within a 500-mi radius of the project site.'

Regionally Manufactured Materials 'For LEED-NC purposes, materials that must be assembled as a finished product within a 500-mi radius of the project site. Assembly, as used for this LEED definition, does not include on-site assembly, erection, or installation of finished components, as in structural steel, miscellaneous iron, or systems furniture.'

Regularly Occupied Spaces 'Areas where workers are seated or standing as they work inside a building. (They include office spaces, conference rooms, and cafeterias.) In residential applications it refers to all spaces except closets or storage areas, utility rooms, and bathrooms. (Bedrooms, living rooms, TV rooms, dining rooms, kitchens, media rooms, etc. would all be considered regularly occupied.).' Opposite of Nonregularly Occupied Spaces.

Remediation 'The process of cleaning up a contaminated site by physical, chemical or biological means. Remediation processes are typically applied to contaminated soil and groundwater.'

Renewable Energy Certificate (REC) 'Representation of the environmental attributes of green power. RECs are sold separately from the electrons that make up the electricity. RECs allow the purchase of green power even when the electrons are not purchased.'

Retained Components 'The portions of the finished ceilings, finished floors, full-length walls and demountable partitions, interior doors, and built-in case goods that existed in the prior condition and remained in the completed design.'

Reuse 'A strategy to return materials to active use in the same or related capacity.'

Risk Assessment 'A methodology used to analyze for potential health effects caused by contaminants in the environment. Information from the risk assessment is used to determine cleanup levels.'

Salvaged Materials 'Construction materials recovered from existing buildings or construction sites and reused in other buildings. Common salvaged materials include structural beams and posts, flooring, doors, cabinetry, brick and decorative items.'

Sealant 'Any material with adhesive properties that is formulated primarily to fill, seal or waterproof gaps or joints between two surfaces. Sealants include sealant primers and caulks.'

Sedimentation 'The addition of soils to water bodies by natural and human-related activities. Sedimentation decreases water quality and accelerates the aging process of lakes, rivers and streams.'

Shared (Group) Multioccupant Spaces 'Conference rooms, classrooms, and other indoor spaces used as a place of congregation for presentations, trainings, etc. Individuals using these spaces share the lighting and temperature controls, and they should have, at a minimum, a separate zone with accessible thermostat and an airflow control.'

Shielding A nontechnical term that describes devices or techniques that are used as part of a luminaire or lamp to limit glare, light trespass, and light pollution (as per LEED-NC 2.1).

Site Area Same as Property Area.

Site Assessment 'An evaluation of aboveground (including facilities) and subsurface characteristics, including the geology and hydrology of the site, to determine if a release has occurred, as well as the extent and concentration of the release. Information generated during a site assessment is used to support remedial action decisions.'

Solar Reflectance (Albedo) (As defined in SSc7.2) 'Ratio of the reflected solar energy to the incoming solar energy over wavelengths of approximately 0.3 to 2.5 μm. A reflectance of 100 percent means that all the energy striking a reflecting surface is reflected into the atmosphere and none of the energy is absorbed by the surface. The best standard technique for its determination uses spectrophotometric measurements with an integrating sphere to determine the reflectance at each different wavelength. An averaging process using a standard solar spectrum then determines the average reflectance (see ASTM Standard E903).'

Solar Reflectance Index (SRI) 'A measure of the constructed surface's ability to reflect solar heat, as shown by a small temperature rise. It is defined so that a standard black (reflectance 0.05, emittance 0.90) is 0 and a standard white (reflectance 0.80, emittance 0.90) is 100. To calculate the SRI for a given material, obtain the reflectance value and emittance value for the material. SRI is calculated according to ASTM E 1980-01. Reflectance is measured according to ASTM E 903, ASTM E 1918, or ASTM C 1549. Emittance is measured according to ASTM E 408 or ASTM C 1371. Default values for some materials will be available in the LEED-NC 2.2 Reference Guide.' Some example values are also available in the LEED 2009 Reference Guide.

For example, a standard black surface has a temperature rise of 90°F (50°C) in full sun, and a standard white surface has a temperature rise of 14.6°F (8.1°C). Once the maximum rise of a given material has been computed, the SRI can be computed by interpolating between the values for white and black. Materials with the highest SRI values are the coolest choices for paving. Due to the way SRI is defined, particularly hot materials can even take slightly negative values, and particularly cool materials can even exceed 100 (Lawrence Berkeley National Laboratory Cool Roofing Materials Database).

Square Footage of a Building 'Total area in square feet of all rooms including corridors, elevators, stairwells, and shaft spaces.' This is also referred to as the gross floor area (GFA). Typically GFAs are given from outer wall to outer wall. The inclusion or exclusion of parking garages has been defined in the LEED errata posted in the spring of 2007 as follows "Only 2 stories of a parking structure may be counted as part of building square footage. Surface parking (only 1 story of parking) cannot count as part of building square footage; this is to ensure efficient use of land adjacent to the building footprint. Both structured and stacked parking are allowable in square footage calculations."

Stormwater Runoff 'Water volumes that are created during precipitation events and flow over surfaces into sewer systems or receiving waters. All precipitation waters that leave project site boundaries on the surface are considered to be stormwater runoff volumes.' Note that this also means from the site through storm sewers.

Sustainable Forestry 'The practice of managing forest resources to meet the long-term forest product needs of humans while maintaining the biodiversity of forested landscapes. The primary goal is to restore, enhance, and sustain a full range of forest values—economic, social, and ecological.'

Systems Performance Testing 'The process of determining the ability of the commissioned systems to perform in accordance with the Owner's Project Requirements, Basis of Design, and construction documents.'

Tertiary Treatment 'The highest form of wastewater treatment that includes the removal of nutrients, organic and solid material, along with biological or chemical polishing (generally to effluent limits of 10 mg/L BOD_5 and 10 mg/L TSS).'

Thermal Comfort 'A condition of mind experienced by building occupants expressing satisfaction with the thermal environment.'

Threatened Species 'An animal or plant species that is likely to become endangered within the foreseeable future.'

Tipping Fees 'Fees charged by a landfill for disposal of waste volumes. The fee is typically quoted for 1 ton of waste.'

Total Suspended Solids (TSS) 'Particles or flocs that are too small or light to be removed from stormwater via gravity settling. Suspended solid concentrations are typically removed via filtration.'

Underground Parking 'A "tuck-under" or stacked parking structure that reduces the exposed parking surface area.'

Urea Formaldehyde 'A combination of urea and formaldehyde that is used in some glues and may emit formaldehyde at room temperature.'

Urinal, Non-Water-Using 'A urinal that does not use water, but instead replaces the water flush with a specially designed trap that contains a layer of buoyant liquid which floats above the urine layer, blocking sewer gas and urine odors from the room.'

Verification 'The full range of checks and tests carried out to determine if all components, subsystems, systems, and interfaces between systems operate in accordance with the contract documents. In this context, *operate* includes all modes and sequences of control operation, interlocks and conditional control responses, and specified responses to abnormal or emergency conditions.'

Visible Light Transmittance T_{vis} 'The ratio of total transmitted light to total incident light. In other words, it is the amount of visible spectrum (380 to 780 nm) light passing through a glazing surface divided by the amount of light striking the glazing surface. A higher T_{vis} value indicates that a greater amount of visible spectrum incident light is passing through the glazing.'

Vision Glazing 'The portion of exterior windows above 2 ft 6 in and below 7 ft 6 in that permits a view to the outside of the project space.'

VOCs (Volatile Organic Compounds) 'Carbon compounds that participate in atmospheric photochemical reactions (excluding carbon monoxide, carbon dioxide, carbonic acid, metallic carbides and carbonates, and ammonium carbonate). The compounds vaporize (become a gas) at normal room temperatures.'

Wetland Vegetation 'Plants that require saturated soils to survive as well as certain tree and other plant species that can tolerate prolonged wet soil conditions.'

APPENDIX C

Units

The following is a list of abbreviations or notations for units as found throughout this text.

Å	Angstrom (1×10^{-10} m)
acre	acre
ACH	air changes per hour (see App. B)
atm	atmosphere
Btu	British thermal units
cf	cubic feet
cfm	cubic feet per minute
cfs	cubic feet per second
cm	centimeter
cu. ft.	cubic feet
fc	footcandles
fpm	feet per minute
ft	feet
ft^2	square feet
ft^3	cubic feet
ft/min	feet per minute
ft^3/min	cubic feet per minute
ft^3/s	cubic feet per second
g	gram
gal	gallon
gal/min	gallons per minute
gpf	gallons per flush
gpm	gallons per minute
h	hour
in	inch
K	kelvin
kBtu	thousand British thermal units

kW	kilowatt
kWh	kilowatthour
L	liter
lb	pound
lf	lineal foot
m	meter
Met	metabolic rate
mg	milligram
mils	millimeters
mm	millimeter
MMBtu	million British thermal units
mol	mole
Pa	Pascal (1 Pa is approximately 0.004 in of water)
pm	picometer (1×10^{-12} m)
ppb	parts per billion (molar-based for gaseous phase; may be volumetric based for ideal gases—usually mass-based for aqueous phases, but should be specified)
ppm	parts per million (molar-based for gaseous phase; may be volumetric based for ideal gases—usually mass-based for aqueous phases, but should be specified)
psia	pounds per square inch absolute
s	second
sf	square feet
sq. ft.	square feet
square	100 ft^2 when referring to roofing
ton_{ac}	ton of gross ARI rated cooling (an energy unit: 1 ton_{ac} of cooling is 12,000 Btu/h)
W	watt
yd	cubic yard if referring to quantities of building materials or dumpster size
yr	year
μg	microgram (1×10^{-6} g)
μm	micrometer (1×10^{-6} m)
°C	degrees Celsius
°F	degrees Fahrenheit
°R	degrees Rankine
$	U.S. dollar
%	percent

Index

Note: Page numbers followed by *f* and *t* indicate figures and tables.

J

L

M

T